Popularizing Science

Popularizing Science

The Life and Work of JBS Haldane

KRISHNA DRONAMRAJU

OXFORD
UNIVERSITY PRESS

Oxford University Press is a department of the University of Oxford. It furthers the University's objective of excellence in research, scholarship, and education by publishing worldwide. Oxford is a registered trade mark of Oxford University Press in the UK and certain other countries.

Published in the United States of America by Oxford University Press
198 Madison Avenue, New York, NY 10016, United States of America.

© Oxford University Press 2017

All rights reserved. No part of this publication may be reproduced, stored in a retrieval system, or transmitted, in any form or by any means, without the prior permission in writing of Oxford University Press, or as expressly permitted by law, by license, or under terms agreed with the appropriate reproduction rights organization. Inquiries concerning reproduction outside the scope of the above should be sent to the Rights Department, Oxford University Press, at the address above.

You must not circulate this work in any other form
and you must impose this same condition on any acquirer.

Library of Congress Cataloging-in-Publication Data
Names: Dronamraju, Krishna R., author.
Title: Popularizing science : The life and work of JBS Haldane / Krishna Dronamraju.
Description: Oxford ; New York : Oxford University Press, 2016. | Includes bibliographical references and index.
Identifiers: LCCN 2016015631 | ISBN 9780199333929 (alk. paper)
Subjects: | MESH: Haldane, J. B. S. (John Burdon Sanderson), 1892–1964. | Genetics, Population | Biological Evolution | Biography
Classification: LCC QH455 | NLM WZ 100 | DDC 576.5/8—dc23
LC record available at https://lccn.loc.gov/2016015631

1 3 5 7 9 8 6 4 2
Printed by Sheridan Books, Inc., United States of America

I am pleased to dedicate this book to the late Naomi (Nou) Mitchison and her son, Professor Avrion Mitchison.

CONTENTS

Acknowledgments ix
Introduction xi

PART I   1920S

1. Family Background and Early Life   3

2. Charlotte and Sex Viri (Marriage and Scandal)   22

3. Eugenics and Predictions   54

4. Population Genetics   77

5. Evolutionary Biology   98

6. On Being a Guinea Pig   113

7. Chemical Genetics   128

8. Origin of Life   141

PART II   1930S

9. Human Genetics   157

10. The Marxist Years   166

PART III 1940S

11. Lysenko Controversy 187
12. Helen 207
13. Popularizing Science 216
14. Haldane and Huxley 235

PART IV 1950S

15. Relations with Other Scientists 245
16. Moving to Paradise (1957) 264

PART V 1960S

17. Life in Paradise (1957–1964) (Death) 279
18. An Indian Perspective of Darwin 320
19. Life with Haldane 328
20. Haldane and Religion 340
21. Impact of Haldane Today 349

Timeline 359
Index 365

ACKNOWLEDGMENTS

It is a pleasure to express my gratitude to the late Naomi (Nou) Mitchison, JBS Haldane's sister, for much family information and photographs as well as a Foreword to my previous book on Haldane in India. I am happy to dedicate this book to Nou and her son Avrion, who is a dear friend. I am also grateful to Av (Professor Avrion Mitchison) for numerous discussions and conversations over several decades that have benefited my thoughts and writings on Haldane.

Haldane's colleagues who were helpful in past discussions include Lionel Penrose, John Maynard Smith, Cedric A. B. Smith, Hans Kalmus, Harry Harris, N. W. Pirie, and M. S. Bartlett. It would be a serious omission not to mention the benefit and pleasure I have received in numerous discussions with my friend at the University of Wisconsin, James F. Crow, and his colleague Motoo Kimura (both now deceased)—undoubtedly the leading exponents of mathematical population genetics. I also acknowledge my debt to Ernst Mayr, who was most generous in his comments on Haldane's contributions to evolutionary biology in numerous personal conversations. Although occasionally disagreeing with Haldane's views, Mayr characterized Haldane as the "most brilliant person I ever met in my life" (Dronamraju, *Haldane, Mayr and Beanbag Genetics*, 2011).

It also gives me much pleasure to acknowledge my debt to Elof Carlson, formerly of SUNY and now at Indiana University, for many discussions and comparative interpretations of Haldane's scientific work with that of his mentor, Herman J. Muller, over several decades. Professor M. S. Swaminathan has been a valuable friend and colleague, whose commentary on Haldane's scientific work over the years has been most helpful.

The late Sir Arthur C. Clarke provided stimulating discussions and commentary on the futuristic writings of Haldane.

I thank Michele Wambaugh for many helpful discussions and suggestions that have enriched the book.

I also acknowledge the assistance of archivists and staff of following organizations: University College London; Wellcome Trust Library; Rockefeller Foundation, Sleepy Hollow, New York; Chemical Heritage Society in Philadelphia; and the Indian Statistical Institute Library in Kolkata.

For photographs, I am grateful to the following: Naomi Mitchison, Jim Watson, Klaus Patau, and the Indian Statistical Institute.

INTRODUCTION

Haldane: A Life of Science and Controversy

John Burdon Sanderson Haldane (JBS) was a brilliant scientist, a polymath who made important contributions to several sciences while possessing no academic qualifications in any branch of science. He was an unforgettable character who was noted for his gallantry and readiness to help others. Haldane lived a richly diverse and fulfilling life. It was filled with controversies and contradictions, including some self-made crises. Yet, because of his great intellect and knowledge, much was expected of him. His broad multidisciplinary knowledge and outlook helped to forge new scientific disciplines, such as population genetics, biochemical genetics, and immunogenetics as well as a new theory of the origin of life. Broad knowledge of several disciplines enabled Haldane to "cross-fertilize" on the intellectual plane, which has made the world a richer place today.

Haldane's ancestors excelled in politics, science, and literature. Haldane is an ancient Scottish surname dating back to around the twelfth century. His uncle was Viscount Haldane, who served as secretary of state for war and chancellor of the Exchequer. His father, John Scott Haldane, a respiratory physiologist at Oxford University, contributed much to improve working conditions for miners and divers. The younger Haldane was trained in science and physiological experimentation from an early age by his father.

Haldane's sister, Naomi Mitchison (1897–1999), was a distinguished and prolific author of numerous books on politics, travel, biography, and Scottish history.

Haldane identified himself with the working man. He possessed a prodigious memory, which enabled him to continue his popular writing while traveling without the need to consult any references. He often attracted the attention of the popular press, which followed him to record his quips and comments on various topics that interested him.

Starting in 1924 Haldane published a series of papers on the mathematical theory of evolution, which is generally recognized as his most important contribution to science. Simultaneously, Haldane was involved in teaching and research in biochemistry, especially enzyme kinetics, and wrote the first book on the subject, *Enzymes*.

Following in his father's footsteps, Haldane conducted a series of physiological experiments, testing the safety limits of breathing various poisonous gases. He was his own guinea pig in these dangerous experiments, which caused much pain and convulsions. Later, Haldane turned his attention to even more dangerous experiments when he was invited to investigate the loss of a submarine resulting in the death of many sailors. These experiments are described in the chapter "On Being a Guinea Pig."

Haldane's ideas impacted significantly on several other fields, including biometry, cosmology, animal behavior, cybernetics, origin of life on planet earth, epidemiology of infectious diseases, radiation effects, science in India, nonviolent biological research, popularization of science, and ethics.

Haldane suffered two major crises in his life; one was personal—when he wanted to marry Charlotte Franken, who was already married, and underwent a sensational trial with lurid details about his sex life—and the second was of a political nature, when he had to confront the suppression of genetics in the Soviet Union. Both crises are discussed in detail in this book.

Haldane and his second wife, Helen Spurway, moved to India in 1957. He studied Hindu classics and adopted Indian-style clothes as well as a vegetarian diet. He preached nonviolence as a way of life, suggesting several projects in nonviolent biological research for students. I was one of his last students.

The chapters in this book are primarily arranged chronologically, but they also follow various historical developments in Haldane's life and work in a logical fashion.

One of the last photos of the Haldanes in London before their move to India.

Popularizing Science

PART 1920S

1

Family Background and Early Life

Ancestors

John Burdon Sanderson (JBS) Haldane was born in Oxford to the physiologist John Scott Haldane and Louisa Kathleen Haldane (née Trotter), and descended from an aristocratic intellectual Scottish family. His younger sister, Naomi (fondly called 'Nou' by family and friends) became a celebrated writer. His uncle was Richard Haldane, 1st Viscount Haldane, a politician, and one time secretary of state for war; his aunt was the author Elizabeth Haldane. His father was a scientist, a philosopher, and a Liberal, and his mother was a Conservative. Haldane took interest in his father's work very early in his childhood. It was the result of this lifelong study of the natural world and his devotion to empirical evidence that he felt atheism was the only rational deduction available in light of all evidence, saying, "My practice as a scientist is atheistic. That is to say, when I set up an experiment I assume no god, angel or devil is going to interfere with its course. . . . I should therefore be intellectually dishonest if I were not also atheistic in the affairs of the world."[1]

Referring to his ancestors, JBS Haldane once wrote that, in 1620, his ancestor in the direct male line, John Haldane, and his brother James, went to fight for the United Netherlands. James was killed at the siege of 's Hertogenbosch in 1629, and it is recorded that the Prince of Orange put on mourning for him. John returned to Scotland and became the Scottish representative with Cromwell's army during the civil war in England.

The family Haldane, however, has been traced back to a much earlier date. Historians and genealogists have ascribed various origins to the family Haldane. Such experts on Scottish heraldry as Sir James Dalrymple and Alexander Nisbet have suggested its descent from an Anglo-Norman, Brien by name, whose son Bernard came to Scotland during the reign of King William the Lion (1165–1214), who gave him a manor on the border. This Bernard, son of Brien, was the undoubted founder of the Haldane family in Scotland. He was a frequent witness to royal and other charters, and these, through their being dated at different places, show that he belonged to the royal retinue and probably filled some post near the person of the king. That he was a man of birth and consequence is obvious from the fact that he appears as a witness in company with some of the most important persons in the kingdom.

Recent ancestors of JBS at first occupied a small fort lying at the foot of Glendevon, commanding a route from the bleak Highlands to the fair carse of Perth. JBS wrote, "Our main job was to stop the tribal people of the hills from raiding the cattle of the plainsmen; but perhaps once in a generation we went south to resist an English invasion, and at least two of my direct ancestors were killed while doing so."[2] The tradition persisted even after the Union.

More recently, the Haldanes pursued more peaceful lives. His great-grandfather James sold much of his family property to raise money for a mission in Bengal. His gentle grandfather Robert Haldane of Cloan married Mary Elizabeth Burdon Sanderson from a well-known Northumberland family. The second son of that union was the physiologist John Scott Haldane—father of JBS.

John Scott Haldane

John Scott Haldane[3] was an outstanding physiologist, who contributed much to the safety of miners and diving personnel. He graduated in medicine at Edinburgh University, where he was a member of the exclusive Eureka Club, an Edinburgh naturalist group which also included D'Arcy Thompson and W. A. Herdman.

Famous for intrepid self-experimenting that led to many important discoveries about the human body and the nature of gases, John Scott Haldane locked himself in sealed chambers breathing lethal cocktails of gases while recording their effect on his mind and body. He visited the scenes of many mining disasters and investigated their causes, often employing himself as his own "guinea pig." His early research was in Dundee, where he carried out investigations of air samples in houses, schools, and sewers.

Haldane was an international authority on ether and respiration, and the inventor of the gas mask during World War I. In 1907 he made a decompression apparatus to help make deep-sea divers safer and produced the first decompression tables after extensive experiments with animals. He was also an authority on the effects of pulmonary diseases, such as silicosis caused by inhaling silica dust.

He identified carbon monoxide as the lethal constituent of afterdamp—the gas created by combustion—after examining many bodies of miners killed in pit explosions, and was able to design respirators for rescue workers. He tested the effect of carbon monoxide on his own body in a closed chamber, describing the results of his slow poisoning. In the late 1890s, he introduced the use of small animals for miners to detect dangerous levels of carbon monoxide underground. The canary[4] in British pits was replaced in 1986 by electronic detectors.

Haldane pioneered study of the reaction of the body to low air pressures, such as that experienced at high altitudes. He led an expedition to Pike's Peak in 1913, which examined the effect of low atmospheric pressure on respiration. Later he joined his uncle Sir John Burdon Sanderson, Waynflete Professor of Physiology, at Oxford University.

John Scott Haldane has been variously described as kindly, courteous, and humanitarian but also eccentric and absorbed in his physiological experiments. He married Louisa Kathleen Trotter, a beautiful and strong-willed woman who was a passionate feminist. But she still believed in the white man's burden and founded the Imperialists'

Club in a left-leaning Oxford. They had two gifted children—Jack, who later became the famous JBS, and the prolific writer Naomi.

A frequent visitor to the household was Viscount Haldane,[5] brother of John Scott Haldane and a prominent Liberal imperialist who was a close associate of Asquith. He was educated at the University of Edinburgh and Gottingen University. He was a strong advocate of British commitments on the continent, and implemented the "Haldane Reforms," a wide-ranging set of reforms aimed at preparing the army for participation in a possible European war. He was for some time secretary of state for war and later lord chancellor. He was forced to resign in 1915, after being falsely accused of pro-German sympathies because of his German education.

Another visitor was John Scott Haldane's sister, Elizabeth Haldane, social reformer and philosopher. She was the author of several philosophical works and biographies, and she was the first female justice of the peace in Scotland.

JBS grew up in a household of science, politics, and social reform. From his mother he inherited his determination, commitment to a cause, and refusal to change his mind in the face of adversity, and from his father he inherited an open mind in scientific inquiries and the ability to reach decisions based on evidence and reason. Indeed, JBS was so open minded that he was one of the least dogmatic people I have ever met. It is certainly a rare quality among scientists. Even in situations where the solution was obvious he used to calculate at first the probabilities in favor of all the alternate hypotheses, no matter how unlikely they might appear to be.

Man of Violence

Years later, JBS wrote in an Indian newspaper:

> I am a man of violence by temperament and training. My family in the male line, can, I think, fairly be described as Kshattriyas.[6] . . . When I was a child my father read to me Scott's 'Tales of a Grandfather,' which are legends of the warlike exploits of the Scottish nobility, and trained me in the practice of courage. He did not do so by taking me into battles . . . but by taking me into mines. I think he first took me underground when I was 4 years old. By the time I was about 20, I was accompanying him in the exploration of a mine which had recently exploded, and where there was danger from poisonous gases, falls of roof, and explosions. So when in 1915, I was first under enemy shell fire, one of my first thoughts was "how my father would enjoy this."[7]

Childhood

"Jack was a beautiful child. In those days it was not done to cut small boys' hair short, so I kept his long lint-white locks until he went to school," Louisa Haldane[8] wrote in her autobiography, *Friends and Kindred*.

Figure 1.1 Haldane's mother wrote in her Autobiography that she "kept his long lint-white locks until he went to school."

Being the first child of the physiologist John Scott Haldane and Louisa Kathleen Trotter, JBS, or Jack, as he was known to the family, enjoyed the undivided attention of his parents until he was four years old, when his sister, Naomi, was born. That gave him a decided advantage, considering that his research career began at the age of two, as Haldane recalled years later in a brief autobiography. His beginnings are best described in his own words: "I was born in 1892. I owe my success very largely to my father, J.S. Haldane. He was perhaps best known as a physiologist. . . .I suppose my scientific career began at the age of about two, when I used to play on the floor of his laboratory and watch him playing a complicated game called 'experiments'—the rules I did not understand, but he clearly enjoyed it."[9]

Jack could read by his third birthday, and before the next, when injured from a fall, looking at the blood from a cut in his forehead, he asked the attending physician, "Is it oxyhaemoglobin or carboxyhaemoglobin?" By the age of five, he learned enough German from his nurse that he could leave small notes around the house, saying: "I hate you."

In her autobiography, *Friends and Kindred*, Mrs. Louisa Haldane recorded that her young son was encouraged to read the newspaper reports of the meeting of the British Association for the Advancement of Science before he was five years old. One visitor, the eminent biologist and friend of the senior Haldane, D'Arcy Thompson,[10] was

Figure 1.2 JBS as a school boy at the Dragon school in Oxford.

much impressed with the way the toddler handled and studied objects with a precocious and scientific expression. Jack displayed prodigious memory for anything that interested him, especially verse. In later life, during his scientific lectures, JBS often quoted entire passages from the classics in the original Latin; Dante and Virgil were among his favorites.

About the age of eight, young Jack was allowed to take down readings from the gas-analysis apparatus and later calculate the amounts of various gases in an air sample. Later he was promoted to making up simple mixtures of chemicals for experiments and, still later, to cleaning apparatus. Before he was fourteen, his father took him down a number of coal mines, and he spent some time under water both in a submarine and in a diving dress. As Jack got older, although still in his teens, his father discussed with him all his research and tried out his lecture course on him before delivering it to students. He also depended on his son's mathematical abilities to help in his data analyses.

Before he was ten, Jack learned the rudiments of stereoisomerism, acquiring the habit of thinking in terms of molecules. John Scott Haldane increasingly used his son's mathematical talents for his own research. On one occasion, his uncle, Sir John Burdon Sanderson, taught him the use of the slide-rule. Shortly afterward, John Scott Haldane, who was on an expedition, noticed that he had forgotten his log tables. But he was not concerned because, as he told his colleagues, Jack will calculate a set for us. And, apparently Jack sat down promptly and calculated a set of tables. Like many other stories about JBS, this story too is probably apocryphal, but Jack tried to live up to the legends.

From all accounts, especially JBS's autobiographical notes, it is clear that John Scott Haldane often used his son as an experimental subject to test his hypotheses and to impart science to young Jack during the process. In his later years, JBS recalled how he and his father were lowered into a pit in North Staffordshire where the air near the roof was full of methane, or firedamp, which is lighter than air. To demonstrate the effects of breathing firedamp, John Scott Haldane asked his son to stand up and recite Mark Anthony's speech from Shakespeare's *Julius Caesar*, beginning "Friends, Romans, countrymen." JBS wrote, "I soon began to pant, and somewhere about 'the noble Brutus' my legs gave away and I collapsed on to the floor, where, of course, the air was all right. In this way I learned that firedamp is lighter than air and not dangerous to breathe."

Education

Louisa Haldane explained in her autobiography, *Friends and Kindred*, "There was no 'Baby school' at the Oxford Preparatory School (later the Dragon School), however, Jack was able to recite more than enough of 'The Battle of Lake Regilus' that he was readily admitted to the lowest form where he was the youngest boy. Later, he passed the Latin exam also quite effortlessly to the great surprise of his teacher because he did not receive any formal instruction in Latin at school."[11]

Jack continued to perform brilliantly at school. Interest in science was encouraged by both his father and his uncle Richard (Viscount Haldane), who presented him with a chemical set. It contained enough ingredients to make a long series of explosions and "stinks," which kept him busy and, according to his mother, "out of mischief" long enough while she was recovering from an infection.

Young Jack's scholastic performance at school was a shadow of what was yet to come in his career. To say he was brilliant would be an understatement. He combined a prodigious memory and flashes of genius and originality in thinking that became evident at a young age, a rare feat in a boy of any age. He was in a class by himself, distancing himself by a wide margin from other children, and a puzzle to his teachers, who tried hard to keep pace with him. Rumor had it that he actually taught the math teacher at school!

Over the years, JBS displayed an unusual combination of shyness and brusque rudeness that kept others at bay, a technique he cultivated over the years to protect himself from unwelcome intrusions.

From his father, he learned or absorbed the scientific attitude almost by osmosis, including a capacity to describe whatever he felt and observed, regardless of the consequences; objective observation; disinterested and dispassionate description; and the logical deduction of one fact from another, and its consequences. But young Jack learned much more than science from his father. Through his long years of self-experimentation, Haldane senior taught himself how to ignore fear and pain. Curiously, although he did not seek pain in his work, when it occurred he greeted it with laughter.

Figure 1.3 JBS and father bonding from an early age.

John Scott Haldane rarely went to bed until late. He arrived at the lab at noon, fresh from breakfast, and then worked until the small hours. At times, he would rarely notice anyone or speak to them at dinner. Toward the end of dinner, a freshly baked cake would be brought in, and the professor would pick it up, completely oblivious of his guests, and disappear with it to his laboratory. His absent-mindedness was legendary. On one occasion, he arrived home late for an important dinner, and after apologizing, went to dress. When he did not return after a long time, his wife went to investigate and found him fast asleep in bed. He explained, "I suddenly found myself taking my clothes off, so I thought it must be time for bed."[12]

As he got older, Jack developed an uninhibited pugnacity, which, when combined with exceptional intelligence and prodigious memory, resulted in an arrogance that isolated him further from his classmates. An essential shyness and embarrassment resulting from his awareness of his own intellectual superiority, added to the complexity. Jack's response was obvious. He enlarged the gap and rejoiced in it.

His brilliance brought him to the top of the school, where he excelled in Latin, translation, arithmetic, and geometry, leading to armful of prizes including the First Scholarship to Eton. At Eton, he chose a broad selection of subjects for his studies, including Latin, Greek, German, French, history, chemistry, physics, and biology.

Years later, he was to comment that he knew no economics, psychology, or technology! In his broad selection of subjects he was supported by his father, but the headmaster, Canon Lyttleton, warned J. S. Haldane that his son was becoming a mere smatterer.

Haldane later wrote that he was not at all happy at Eton, although he was receiving an intellectually sound education. He was much tormented by the senior boys, one exception being Julian Huxley, as Haldane recalled later, whom he long remembered for giving him an apple, a mark of an exceptional favor from a senior to a junior.

Louisa Haldane, with her conservative leanings, approved of her son's choice of Eton, where many future prime ministers went. In *Friends and Kindred*, she wrote, "it was a terrible disappointment to me that Jack should never seem to be happy at Eton. In his second half . . . he seemed so miserable that I said if his father agreed he might perhaps be allowed to leave. He did not want that 'exactly'." Of course, he stayed on and completed his education successfully at Eton. Indeed, he was so successful that he became the captain, the most prestigious position at the school. His mother noted that he was happier when he changed his classical tutor for the science master, Mr. Hill.

In his *Memories*, Julian Huxley, who was Haldane's senior at Eton, wrote about the science master, M. D. Hill, generally known as "Piggy" because of his protruding nose. Julian wrote:

> I owe him immense gratitude. . . . I was lucky, for Piggy Hill was a genius as a teacher. He soon made me understand the excitement of Zoology, and I decided to specialize in the subject. Dear Piggy! He settled my career for me, and I have always been grateful to him. . . . Piggy Hill was a fussy little man, but inspiring. My brother Aldous began biology with him, and thus laid the basis for his constant fascination with science. In one of his early novels Aldous launches into a marvellous metaphysical disquisition, which actually derived from his hearing Piggy asking the lab boy for "the key of the Absolute"—the absolute alcohol being safely locked up in a cupboard, as the word Absolute is hidden in mystery.[13]

As captain of the school, Jack performed various social functions. One of the proudest moments for his mother was when he received the king on his first visit to Eton. JBS acquitted himself excellently, in spite of a gashed chin due to using a borrowed razor too vigorously.

Impact of Eton

In one way, Eton made JBS. As he recalled later, he received an excellent education. Yet, it left an indelible impression that is far from desirable. Many men have come through Eton unscathed and some have used its toughening process for their own benefit quite successfully. But not JBS Haldane. It left him even more irascible and unapproachable than he already was previously. It reinforced his independence and self-sufficiency, but also brought out a new sense of kinship with the persecuted minorities—perhaps to a degree that is surprising—and a belief that minorities are a generally persecuted lot. It reinforced his determination to fight authority, any authority, especially against the Government, as a matter of principle. He carried this attitude with him to India when he migrated there during his last years and was always inclined to take offense against the slightest comment, real or imaginary.

Yet, in a different way, the merciless torment at Eton shaped JBS's personality in such a way that caused enormous difficulties in his relations with other individuals and institutions, often involving him in needless controversies that wasted a substantial part of his life.

An equally difficult rift developed between JBS and religion, which he identified with authority. Several years later, in an article titled "When I Am Dead," he wrote, "My Catholic friends hope to survive death on the authority of the Church; some of my Protestant acquaintances rely on the testimony of the Bible. But they do not convince me, for the Church has taught doctrines which I know to be false, and the Bible contains statements—for example, concerning the earth's past—which I also know to be false."[14]

He was brought up in a household where religion was not an important part of daily life. Years later, in 1929, in a radio broadcast titled "My Philosophy of Life," JBS wrote, "As a child I was not brought up in the tenets of any religion, but in a household where science and philosophy took the place of faith. As a boy I had very free access to contemporary thought, so that I do not to-day find Einstein unintelligible, or Freud shocking."[15]

His sister, Naomi, once wrote, "We had, of course, been brought up without religious beliefs, although we went to New College chapel sometimes, and, indeed, both of us used to get prizes for scripture knowledge." However, absence of religious strictures had its own problems as well: "We had a set of strict ethical principles which were slightly harder to live up to because there was no supernatural sanction behind them." JBS later wrote, "I developed a mild liking for the Anglican ritual and a complete immunity to religion."[16]

With respect to their parents, Naomi added later: "We had no religious conflicts with our parents, but in time we began to wriggle in our ethical bonds, but lying, for instance, was apt to make us both rather uncomfortable."[17] At Cloan, the ancestral home in Scotland, on the occasions when he was the eldest male member of the family present, young Jack led family prayers that were quite inventive, including the cook and the butler and other household staff.[18]

While at Eton, JBS attended chapel services regularly, including the twenty-minute morning service, the ten-minute evening prayers, and the two-hour Sunday services, but was hardly touched by the proceedings.

In her autobiographical essay, "Beginnings,"[19] Naomi narrated an incident when she had to intervene quickly in an awkward family situation. Just after World War I, a love letter arrived for Jack but was mistakenly opened by the father as it was addressed to "J. Haldane," "who was a distinct puritan in these matters." Not knowing the circumstances, Naomi had to think quickly and invent a convincing lie about a highly suitable and marriageable friend, which, in her words, "pacified my father."

Eton facilitated selection of multiple subjects and JBS took advantage of that to his great satisfaction. He was able to switch from two years of classics to five terms of chemistry, and one term of physics followed by three terms of history followed by three terms of biology. He had no interest in economics or music. He was tone-deaf. His last classical master at Eton, H. MacNaughton, found him baffling. He wrote, "I shall be glad to be rid of him."[20]

Cherwell

Besides the Huxleys, Jack and Naomi's friends included Lewis and John Gielgud, who shared Naomi's early interest in acting. Their social prominence was more than matched by his academic success. In 1914, he achieved the rare distinction of attaining first class honors in both mathematics and classics at Oxford in 1914. Jack's background and academic record clearly made him an ideal candidate for a civil service position. Wisely, he chose science instead.

About the time Jack went to Eton, his parents moved from their modest house in North Oxford to their new home, "Cherwell" at the north end of the Banbury Road, ample and rambling, comfortable and ugly, on land sloping down to the river. Built to John Scott Haldane's specifications, Cherwell was to be their family home for sixty years. On the front, the eagle crest of the Haldanes was etched into the stone, with the single word of their family motto, "Suffer." This seems so natural and apt for JBS, for his life can be summed up in the terms "suffer" and "endure," which gained a new meaning upon his move to India during his last years.

Cherwell was a rambling mansion with 25 to 30 rooms depending on how they are counted, built for a bygone era when both servants and space cost little. It lacked one modern convenience, gas, which was forbidden by John Scott Haldane, who, as a gas referee, had seen a number of accidents caused by gas explosions. There was ample space on the ground floor for his study, which was always covered in plenty of papers on the carpet, desk, chairs, and elsewhere.

Also situated on the ground floor was the laboratory, complete with an airtight chamber with a sealable door and observation window for physiological experiments.

No one entered these premises except the physiologist and his young son. On one occasion, when he was asked if his study could be included in spring cleaning, J.S. replied, "Yes, as long as the papers are not disturbed." That was the end of any attempt to clean the study.

Diving Experiments

In 1908, John Scott Haldane recruited his son, while he was still at Eton, in diving experiments that made an enormous impact on the diving practice of the British Navy. Those experiments were the last stages of a program of "stage decompression" devised by the senior Haldane for bringing deep-divers to the surface. The experiments were carried out from H.M.S. *Spanker* in the Kyles of Bute off the Arran coast. It became a family enterprise of the Haldanes, and the whole family accompanied young Jack to Scotland.

Jack spent a considerable time on H.M.S. *Spanker* during the long series of dives the crew undertook under varying conditions.

Although he was not yet sixteen, Jack eagerly seized the opportunity to make a trial dive, however, because of his age, his name was carefully excluded from the official report. Although his frame was large, he was unable to provide a good fit for the full-scale diving suit available. Before long, water began to seep into his diving suit round his badly fitting ankle- and wristbands. However, he knew how to control the pressure by operating the valves, and remain submerged at the planned level, before being raised to the surface and dried out and warmed by blankets and a shot of whiskey.

A short time later, John Scott Haldane was invited to take part in the trials of a new submarine, which required a confidential assistant. While he was considering the possibilities, Mrs. Haldane wondered aloud, "What about Boy?" His dad turned to Jack and asked, "What is the formula for soda-lime?" Jack promptly answered and was on his way to his first submarine trials.

Many years later, before and during World War II, JBS, as subject-investigator, conducted a series of diving experiments to define the parameters for underwater physiology. These were mainly undertaken at the behest of the Admiralty, which invited JBS to investigate the loss at sea of a submarine, H.M.S. *Thetis*, in which many sailors lost their lives. An excellent summary of Haldane's diving experiments and their contribution to diving physiology were presented by two experts in the field, Albert Behnke and Ralph Brauer, in my book *Haldane and Modern Biology* (1968).[21] This subject is discussed in detail in chapter 6, "On Being a Guineapig."

Oxford University

When JBS went up to New College in the autumn of 1911, on a good mathematical scholarship, he was thus fully equipped; he knew enough chemistry and biology to do research, as well as a fair knowledge of history and contemporary politics. However, his reputation preceded him. Rumor had it that he had entered for a classical scholarship, had gone into the wrong room, and successfully answered the mathematical paper that was placed before him!

As JBS believed that no one can study mathematics intensively for more than five hours a day and remain sane, he attended, almost as a relaxation, Professor E. S. Goodrich's final honors course in zoology. Goodrich was not only an eminent scientist, having studied zoology under Ray Lankester, but also received earlier training in art at the Slade.[22] Many years later, JBS recalled that Goodrich's talent for drawing anatomical drawings in colors impressed him so much that they created a lasting interest in biology.

About her brother's early interest in science, Naomi wrote, "What led him to science? Mostly I suppose, our father. As children we were both in and out of the lab all the time. . . . My brother shared his father's scientific thought more and more, and when, about 1906, we moved to the house at the end of Linton Road, and my

Captain of the School, Eton, 1910. Jack and me

Figure 1.4 Jack and Naomi.

father had his own laboratory and his own colleagues and pupils, Jack was inevitably drawn in."[23]

Genetics

About the time JBS was attending Goodrich's lectures, his sister Naomi had a fall while riding. As she developed a slight allergy to ponies, it was suggested that she might take up guinea pigs as pets. Many years later, she wrote about their childhood in the book *Haldane and Modern Biology*.

In her contribution, "Beginnings," Naomi had recorded that they knew the pioneers of Genetics, William Bateson and Reginald Punnett by that time, and they met several other eminent scientists who came to see their father. There were Niels Bohr ("the great wild Boar") and Miss Christiansen from Denmark. Other colleagues came from Germany and Austria. Then there was Karl Pearson, of whom it was written (quoting Naomi):

> Karl Pearson is a biometrician
> And this, I think, is his position:
> Bateson and co,
> Hope they may go
> To monosyllabic perdition.

How did they get into genetics? She wrote:

> Certainly Jack was interested in genetics while he was still at Eton. Our first joint scientific experiments began when I was about twelve. I had by that time a number of guinea pigs and was making my own observations on their lives and loves. I could separate and mimic, so that they would answer back. Then came Mendelism, which at that time was easily understood, even by someone such as myself with no scientific knowledge except, of course, what I had picked up.[24]

Early genetics was, Naomi wrote, "relatively unmathematical. We talked in terms of dominants and recessives. Chromosomes had not come into their own; the cell mechanism was still obscure; but guinea pigs were a mine of information. . . . But there were

Figure 1.5 Author with Naomi.

more and more guinea pigs, and then we began to detect something that didn't quite fit in: linkage. . . . I remember very well the excitement of reading Morgan's book[25] when it first came out."[26]

Haldane read his first paper in genetics, reporting the first case of linkage in vertebrates, to a class organized by Goodrich in 1910–1911. As it was based on data collected by others, he was advised by R.C. Punnett to confirm the discovery by obtaining his own data. Jack and Naomi added rats and mice to their breeding experiments, and their paper on linkage was finally published in 1915[27] in collaboration with a colleague A. D. Sprunt, who was killed in World War I shortly afterward. Haldane long regretted the delay in its publication. Later, he wrote, "Had I been able to publish in 1910, it would have been simultaneous with Morgan's discovery of linkage in *Drosophila*," the first reported linkage outside the plant world.[28]

His late publication made it the first case of linkage in the vertebrates. He finished the last stages of the paper while fighting on the frontlines in France, which provided him a wonderful boast, "the only officer to complete a scientific paper from a forward position of the Black Watch."

An interesting footnote to this episode was his concern that he might be killed on the front before finishing the paper. He wrote to William Bateson, a leading authority on Mendelism and director of the John Innes Horticultural Institution, where JBS himself was to work later.

Dear Prof. Bateson:

I have been doing some work on reduplication (linkage)* in mice, along with A.D. Sprunt and my sister, who is now carrying on, as we are both at the front. The factors concerned are C colour and E dark eye. Our latest from Ce.cE × ccee was 3 dark eyed, 24 pink eyed (albino and Cee) suggesting 1:3:3:1 repulsion between C and E. If I am killed could you kindly give my sister help if she wants it.

Yours sincerely,
J. Haldane, 2/Lt.

First Paper in Physiology

JBS was simultaneously engaged in research in genetics and physiology while he was still a student at Eton and Oxford, studying classics among other subjects. His first scientific publication in 1912 was a paper in the *Journal of Physiology*.[29] On one occasion, John Scott Haldane had written to the Eton master (Whitworth) asking him to allow his son to visit London to help in his physiological research. He wrote, "He has been giving me very great help in the mathematical part of a rather important physiological investigation of which I am giving an account at the meeting of the Physiological Society. The paper is in our joint names and Jack's part is a very important one as he has evolved an equation which has thrown light into what was a very dark region." Finally, he stated the reason: "I should like him to be present if possible, partly to fortify me against possible attacks from people who know the higher 'mathematics'." The paper was later published in the *Journal of Physiology*.

His first research paper in the *Journal of Physiology*, and another investigation in progress in genetics, and attending scientific meetings with his famous scientist-father—everything seemed to indicate that JBS was all set for a scientific career. However, no one could have guessed what came next. Nothing is simple and uncomplicated in JBS's life. At the end of his first year at Oxford, he surprised everyone by switching from mathematics and biology to "Greats," or *Literae Humaniores*, which includes studies in the first part, of Latin and/or Greek language; and in the second part, a choice of varied disciplines of literature, Greek and Roman history, philosophy, archeology, and linguistics.

The switch was supported by John Scott Haldane because he knew his son better than anyone else. JBS had not left science but found a way to improve his literary skills and writing with an economy and precision. It formed a solid foundation for his later career in science popularization. JBS later wrote that the subjects studied had little relation to modern life, and that the successful "Greats Man," with his high capacity for abstraction, makes an excellent civil servant, prepared to report as unemotionally on the massacre of millions of African natives as on the constitution of the Channel Islands. It was, in fact, ideal training for the physiologist recording his own reactions to abnormal or painful experiments. And a successful Greats Man was considered ideal to carry on certain jobs, because the isolation of his studies from the world in which he lived encouraged a dispassionate attitude to affairs.

Oxford Life

JBS made lasting friendships at Oxford, including the Gielguds (both John and Lewis), Huxleys (both Julian and Aldous, and Trev and Gervas Huxley), and Dick Mitchison, who later married his sister Naomi. JBS was not good at skilled games, but he rowed frequently. His sister recalled, "Not fancying other games, he rowed in Torpids . . .; he had to have a special oar built as he cracked an ordinary one. At the end of the training period there were bonfires and much drink; when with liquor he could and did speak entirely in blank verse—or sometimes in rhyme."

Naomi also drew attention to the fact that her brother was expert at climbing in and out of College, breaking a few of the heavy Stonesfield slates with which New College cloisters were roofed, and once got a spike into himself. Later, when he became a don himself, after World War I, he had some of the easier routes closed, as they did not "create sufficient problems to be of educational value."[30] She wrote further about her close relationship with her brother Jack. Among other things, "Jack and I, oddly enough, did a lot of dancing together, but almost always Viennese waltzes, in which we felt it was cheating to reverse; the big interest was which of us first would get too giddy to stand. In spite of no musical ear and all that, we did feel the beat of dancing, as we did of poetry, perhaps of drumming."

JBS was much happier at Oxford than he had been at Eton. He loved to quote poetry, first classics, especially Dante and Virgil, later Milton, and much later Housman and Yeats.

At Oxford, he was first drawn into politics, joining the University Liberal Club and taking part in debates, earning a note of commendation from his uncle, Viscount Haldane. He became a member of the University Co-operative Society, volunteering behind its counter several times.

One incident at Oxford that brought him on to the political stage was a portent of his activities to come later. In 1913 horse-drawn trams were still being used for public transport. The staff went on strike, demanding higher wages. The strikers' positions were taken by blacklegs. Agitation supporting the strikers continued for three days. The police made baton charges to disperse the crowds. As he was busy elsewhere, Haldane was unable to participate in the protests, but on the fourth day he was able to do so. On the fourth evening when the streets were quiet, he walked up and down Cornmarket St. chanting the Athanasian creed and the hymeneal psalm *Eructavit cor meum* in a loud but unmelodious voice. A large crowd collected. According to Haldane, the police at that point "ineffectively pushed pious old ladies into the gutter. The trams failed to penetrate the crowd, and their horses were detached and ran away. The strike was successful, and the trams were replaced by motor buses which proved to be a profitable enterprise paying higher wages."[31]

For his modest role in the civil disobedience Haldane was fined two guineas—"the first case for over three centuries when a man was punished in Oxford for publicly professing the principles of the Church of England."[32]

By 1914, John Scott Haldane was already making plans for his son's future. As soon as he successfully completed the "Greats," he was to return to science and sit for an

Figure 1.6 Author with Jim Watson, artist Dorothy Hood, and Mrs. Elizabeth Watson at Cold Spring Harbor.

Figure 1.7 Jim Watson with Haldane's sister Naomi (Nou) Mitchison. *Double Helix* was dedicated to her.

Figure 1.8 Author with Haldane's nephew Avrion Mitchison, walking near University College London in 2011. Photo by Michele Wambaugh.

exam in physiology after two years. Then he would join Harrison or Abel in the Oxford laboratory. The first part was easier for JBS to achieve, but the second part was, as he put it, "somewhat overshadowed by other events," namely World War I.

Notes

1. J. B. S. Haldane, *Fact and Faith* (London: Watts, 1934).
2. J. B. S. Haldane, personal communication to the author.
3. John Scott Haldane (1860–1936) was the father of J. B. S. Haldane and his sister Naomi. He was a famous Scottish physiologist who experimented on himself in sealed chambers, breathing lethal gases, such as carbon monoxide, while recording their effect on his mind and body. Haldane investigated the scenes of many mining disasters, studying lethal effects of poisonous gases. He investigated gas warfare during World War I, which led to his invention of the first gas mask. Among his numerous other contributions, he identified carbon monoxide as the lethal constituent of afterdamp, designed respirators for rescue workers, helped determine the regulation of breathing, and founded the *Journal of Hygiene*. In 1907 Haldane made a decompression chamber to help make deep-sea divers safer and produced the first decompression tables after extensive experiments with animals. In the late 1890s, he introduced the use of small animals for miners to detect dangerous levels of carbon monoxide underground, either white mice or canaries.
4. The canary (*Serinus canaria domestica*) is a domesticated form of the wild canary, a songbird of the finch family that originated from the Canary Islands. While wild canaries are a yellowish-green color, domestic canaries have been selectively bred for a wide variety of colors. Canaries were once regularly used in coal mining as an early warning system. Canaries have been extensively used in brain research and in basic research to understand how songbirds produce song. They have also served in research for understanding the brain's learning and memory process.
5. Richard Burdon Haldane, 1st Viscount Haldane of Cloan (1856–1928), was a Scottish lawyer, philosopher, and statesman who instituted important military reforms while serving as British secretary of state for war (1905–1912). He was a member of the House of Commons from 1885 until his elevation to the peerage in 1911. The speedy mobilization of the British Expeditionary Force in August 1914 was largely the result of his planning. As Anglo-German relations were deteriorating, Haldane went to Berlin in February 1912 on a well-publicized but ineffectual mission concerning British neutrality and the relative naval strength of the two countries. Haldane became lord chancellor in H. H. Asquith's Liberal government. In Ramsay MacDonald's first Labour Party government (January–November 1924), he once more served as lord chancellor. With the Fabian Socialists Sidney and Beatrice Webb, Haldane was one of the founders of the London School of Economics in 1895. In *The Reign of Relativity* (1921) he dealt with the philosophical consequences of Albert Einstein's theories of physics.
6. *Kshattriyas* occupy an important position, the warrior caste, that is second only to *Brahmins* in the traditional Hindu caste system. Traditionally, they have been members of fighting forces and occupied forts and other strategic locations concerned with repelling foreign invaders.
7. "Some Reflections on Non-violence," *The Hindu* of Madras (India), 1959, reproduced in *What I Require from Life*, ed. K. R. Dronamraju (Oxford: Oxford University Press, 2009), 132–39.
8. Louisa Kathleen Haldane (1863–1959), wife of the physiologist John Scott Haldane and mother of J. B. S. Haldane and Naomi Mitchison, was politically conservative and an Empire loyalist in an Oxford that was steadily moving to the left. She tried to inculcate conservative philosophy and Empire loyalty in her children, but both Jack and Naomi moved to the left as they got older. When Jack became a Marxist in the 1930s, it was a terrible disappointment to his mother. See L. K. Haldane, *Friends and Kindred* (London: Faber & Faber, 1961), 248.
9. "Autobiography in Brief" was first published by Haldane in the *Illustrated Weekly of India*, Bombay, in 1961, but was later reprinted in Krishna R. Dronamraju, ed., *Selected Genetic Papers of JBS Haldane* (New York: Garland, 1990), 19–24.

Family Background and Early Life

10. Sir D'Arcy Wentworth Thompson (1860–1948) was a Scottish biologist, mathematician, and classics scholar. A pioneering mathematical biologist, he is mainly remembered as the author of the 1917 book *On Growth and Form*, which Peter Medawar called "the finest work of literature in all the annals of science that have been recorded in the English tongue." Peter Medawar, *Pluto's Republic* (Oxford: Oxford University Press, 1982), 240. Thompson wrote *On Growth and Form* in Dundee, mostly in 1915, though wartime shortages and his many last-minute alterations delayed publication until 1917. The central theme of the book is that biologists of its author's day overemphasized evolution as the fundamental determinant of the form and structure of living organisms, and underemphasized the roles of physical laws and mechanics.
11. From his mother's autobiography, L. K. Haldane, *Friends and Kindred* (London: Faber & Faber, 1961), 216.
12. Haldane, *Friends and Kindred*, 210.
13. J. S. Huxley, *Memories* (London: Harper & Row, 1970), 50–51.
14. J. B. S. Haldane, *Possible Worlds and Other Essays* (London: Chatto & Windus, 1927), reprinted in J. M. Smith, *On Being the Right Size and Other Essays* (Oxford: Oxford University Press, 1985), 26.
15. A BBC broadcast in November 1929. The other speakers in the same series were G. Lowes Dickinson, Dean Inge, Bernard Shaw, H. G. Wells, and Sir Oliver Lodge. Haldane's essay is included in the collection titled *The Inequality of Man and Other Essays*, J. B. S. Haldane (London: Chatto & Windus, 1932), 211–24.
16. Naomi Mitchison, "Beginnings," in *Haldane and Modern Biology*, ed. K. R. Dronamraju (Baltimore: Johns Hopkins University Press, 1968), 304.
17. Mitchison, "Beginnings."
18. Naomi Mitchison, personal communication, 1984.
19. Mitchison, "Beginnings."
20. J. B. S. Haldane, personal communication.
21. A. R. Behnke and R. W. Brauer, "Physiologic Investigations in Diving and Inhalation of Gases," in Dronamraju, *Haldane and Modern Biology*, 267–75.
22. The Slade School of Fine Art (informally "The Slade") is a world-renowned art school in London as a department of University College. The school was founded in 1868 when the lawyer and philanthropist Felix Slade (1788–1868) bequeathed funds to establish three chairs in fine art. The Slade art collection was started when the yearly prizes awarded to top students were combined with a collection scheme in 1897 and the works that won the Summer Composition Prize and the Figure and Head Painting Prizes began to be kept by the school. Works by students and staff of the Slade School of Fine Art form the basis of the University College, London Art museum today.
23. Mitchison, "Beginnings," 302.
24. Mitchison, "Beginnings," 303.
25. T. H. Morgan, *The Mechanism of Mendelian Heredity* (New York: Henry Holt, 1915).
26. Mitchison, "Beginnings."
27. J. B. S. Haldane, A. D. Sprunt, and N. M. Haldane, "Reduplication in Mice," *Journal of Genetics* 5 (1915): 133–35.
28. J. B. S. Haldane, "An Autobiography in Brief." Reproduced in *Selected Genetic Papers of J.B.S. Haldane*, ed. K. R. Dronamraju (New York: Garland Publishing Inc., 1990), 19–24.
29. J. B. S. Haldane, "The Dissociation of Oxyhaemoglobin in Human Blood during Partial CO Poisoning," *Journal of Physiology* 45 (1912): 22–24.
30. Mitchison, "Beginnings," 301.
31. J. B. S. Haldane, personal communication, 1959.
32. J. B. S. Haldane, personal communication, 1960.

2

Charlotte and Sex Viri (Marriage and Scandal)

Haldane's Cambridge period, 1923–1933, was both the time of his worst personal crisis—which led to his dismissal from his biochemistry readership at Cambridge University—and also his most rewarding period scientifically, when he produced first-rate research in population genetics, biochemistry, and physiology.

It all began one fine afternoon in the summer of 1924, when an attractive young journalist, Charlotte Franken,[1] came to Haldane's rooms at Trinity College,[2] Cambridge, to interview him for London's *Daily Express*. Not finding Haldane in his rooms, Charlotte wandered around the grounds, enjoying the scenery and architecture. She was duly impressed by the College's distinguished history and traditions.

The Great Gate of the College is the main entrance to the college, leading to the Great Court. A statue of the college founder, Henry VIII, stands in a niche above the doorway. In his hand he holds a table leg instead of the original sword, and myths abound as to how the switch was carried out and by whom. In 1704 the University's first astronomical observatory was built on top of the gatehouse. Beneath the founder's statue are the coats of arms of Edward III, the founder of King's Hall, and those of his five sons who survived to maturity.

Professor Frederick Gowland Hopkins

Haldane was invited, in 1922, by Professor Hopkins (later president of the Royal Society and Nobel laureate) to Cambridge as reader in biochemistry. In his "Autobiography in Brief," Haldane wrote, "I was his second-in-command for 10 years and supervised the work of about twenty graduate students—much of which was first rate. ... Perhaps my own most important discovery was ... a substance for which carbon monoxide competes with oxygen, now called cytochrome oxidase, ... However, my enunciation of some of the general laws of enzyme chemistry may have been more important."[3]

Haldane's appointment as the first Sir William Dunn Reader in Biochemistry at Cambridge University came as a surprise to some of his colleagues. According to Weatherall and Kamminga,[4] who wrote a history of biochemistry at Cambridge

University, Haldane was not the first choice. The professor of biochemistry, Frederick Gowland Hopkins, first approached Edward Mellanby[5] for the post, but Mellanby was unwilling to abandon his research program into the etiology of rickets at King's College, London. Haldane's appointment was an imaginative one, especially because Haldane lacked formal qualifications in biochemistry or in any of the natural sciences. In fact, he did not possess a degree in any branch of science, having qualified in classics at Oxford. His publications until then were in blood chemistry, especially in relation to the physiology of respiration, and genetics. As reader, Haldane was Hopkins's second-in-command in the Dunn Institute at a salary of 600 pounds per annum.

The Dunn Institute was formally opened by Lord Balfour on May 9, 1924, the funds for constructing the building having been bequeathed by the estate of Sir William Dunn in 1920. Hopkins expressed his idea of an ideal institute:

> The Institute I have in mind would steal something from the activities (usually, however, minor activities) of various existing biological Institutes, but would justify the theft by a highly profitable combination and co-ordination of the stolen materials. I am in fact only pleading for one of those reconstellations in intellectual pursuits which the progress of knowledge often makes desirable or necessary. I am sure that the one in question is now desirable.[6]

However, Hopkins further added that he had already created an Institute such as this in Cambridge for the purpose of teaching the students an unusual scientific outlook and creating a research program on broad, biological lines.

By the summer of 1923 JBS was settling down at Cambridge. He carried a heavy load of teaching and research in biochemistry. But, being Haldane, he was not satisfied with these tasks. In his spare time, he started writing papers on the mathematical theory of natural selection, which formed the foundations of what came to be known as "population genetics" later on. Haldane's research in that field was independent of the contributions of R. A. Fisher and S. Wright. Haldane was an outstanding teacher. Some of his students of that period, for instance, Joseph Needham, J. H. Quastel, V. B. Wigglesworth,[7] and N. W. Pirie, became eminent scientists in their own right.

Enzyme Kinetics

Haldane's research at the Biochemical Institute at Cambridge is notable for its contribution to enzyme kinetics. What has been called the "Michaelis theory" (also called the "Victor Henri theory") assumes that the combination of enzyme and substrate corresponds always to equilibrium. Haldane considered this an improbable assumption. In collaboration with G. E. Briggs, he derived, in 1925, the basic law of steady-state kinetics, which is still used for treating enzymatic catalysis. In his seminal work, *Enzymes*, Haldane (1930)[8] wrote that his treatment of the subject was limited to enzymes concerned in the complicated processes of alcoholic fermentation and respiration. He died on December 1, 1964, but lived long enough to write a preface to the posthumously published paperback edition by the MIT Press in

August 1965. Professor Rene Wurmser of the University of Paris praised it as a very influential book that played a significant role in the formation of two generations of biochemists.

Recollections of Haldane were narrated several years later by his former students and associates from his Cambridge years. Some of these are quoted below. The following account is from "Confessions of a Biochemist" (23–24), the reminiscences of J. Murray Luck,[9] Professor of Chemistry Emeritus, Stanford University, and founder of *Annual Reviews*.

> I spent the summer of 1922 ... I learned to my surprise that I had won an 1851 Exhibition Scholarship—nominated by the University of Toronto—for several years of postgraduate study in Britain. I chose Cambridge University, to which I was duly admitted as a research student after acceptance by Gonville and Caius College. Biochemistry was to be my field of research. What an interesting collection of biochemists occupied the laboratory! Each was different from the next, and each, in his own way, contributed to the exciting, entertaining, and brainy atmosphere of the place. I can now recall only Sir Gowland Hopkins (the lovable Hoppy), Rudolph Peters, J. B. S. Haldane, Marjorie Stephenson, The Hon. Mrs. Onslow, R. A. McCance, Dorothy Moyle (later Mrs. Needham), Joseph Needham, Tim Hele, Robin Hill, Malcolm Dixon, H. F. Holden, Bill Pirie, J. H. Quastel, Margaret Whetham, (later Mrs. Bruce Anderson), Vincent Wigglesworth, and my research partner for a time, Trilok Nath Seth. There were others, many others, each pursuing his or her research in glorious independence.
>
> Hoppy, with much on his mind, seldom knew what each of us was up to but was always interested in hearing about our biochemical doings, and was ever affectionately concerned about our welfare. Haldane was a walking encyclopedia. He came to Cambridge in 1922 as Reader in Biochemistry and soon knew what everyone was doing. In his constant roamings about the laboratory, and in frequent chance encounters, he would discuss with remarkable insight the intricacies of one's research activities.
>
> Hoppy suggested to Robin Hill and me that it might be interesting, as a starter, to collaborate with Haldane. It was indeed interesting! He strongly believed in "being one's own rabbit." As such, he swallowed in three days a 3.5-liter aqueous solution of 85 grams of calcium chloride to induce a good acidosis. Robin and I were responsible for analyzing the great man's urine. He developed an acidosis that was noteworthy. I recall swimming in the river Cam on a Sunday during the height of the 13-day experiment. Haldane was also there. Soon, a punt, bent on ascending the river, made its approach. Seated therein were Hoppy, who had been knighted but recently, Lady Hopkins, and two distinguished-looking guests. Haldane at once swam under and around the punt, describing in his booming voice his experiment on acidosis: "I am now excreting the most acid urine that has ever been excreted." "Yes, yes," replied Hoppy, rubbing his brow in characteristic fashion. Later it emerged that someone, somewhere, had reached a slightly lower pH, but Haldane took his "defeat" in stride.

In my post-Cambridge years, I have felt greatly indebted to Haldane for the introduction in physiology he gave to four or five of us from the lab. We met in his rooms in Trinity on frequent occasions. Questions and more questions always preceded a very informal but informative "lecture" by Haldane. At the first session, I remember he started off with the query "How big do you think my liver is?" He weighed 100 kg. We answered with widely differing percentages of his body weight. "How much blood do you suppose I have?" Answers: A few pints up to a few gallons. "How may one determine the blood volume?" And so on. At the end of the evening, there was always an unorthodox and entertaining summary by Haldane.

As much of Haldane's scientific work, certainly his most important work on the mathematical theory of natural selection, is of theoretical nature, some people came to believe that he was not much good at experimentation. Although Haldane's work in his later years (1933–64) was mostly theoretical, his early research during the years 1910–32 involved much experimentation in physiology and biochemistry as well as directing research in plant breeding for studying the biochemical genetics of plant pigments. In the following communication, the distinguished biochemist Joseph Needham, who was a disciple and later colleague of Haldane at Cambridge, commented on this aspect of Haldane's research:

> I don't think I would myself want to subscribe to the view that J. B. S. Haldane was never much good as experimentalist. I always understood he did a great deal of good work on submarine physiology with his father in World War I; and then in the 'twenties and early 'thirties he did a lot of work on salts, acids and bases in the human body with students such as V. B. Wigglesworth and C. E. Woodrow. I remember once meeting him on the stairs of the Biochemical Institute which he was descending in an apparently drunk condition, but when I asked if I could help him, he said "Oh, it's nothing. It will pass, I'm only about 80% sodium Haldanate at the present moment." It remains true, I think, that in his biochemical work he always remained very much the respiratory physiologist, as one could tell from the Douglas bags and such equipment in his room.
>
> Then of course there was all his work in theoretical evolution studies and genetics, and mathematical statistics, done with (or was it against?) R. A. Fisher. Certainly J. B. S. Haldane had a great gift for choosing young people who would do brilliant work, notably Rose Scott-Moncrieff on the genetics of the anthocyanin pigments in flowers.
>
> But it is also very true that throughout his life he did help others, not only as a colleague or as an editor, but also as a research supervisor. I can say this from personal experience, because he was my research supervisor when I was starting work in the Biochemical Institute in 1922 or 1923. I am sure I owe him a great deal for his help. Nor do I think it would be fair to omit the great talent which J. B. S. Haldane had for comic verse; this was published year after year in *Brighter Biochemistry*, the humorous journal that came out between 1925 and 1933 in the Biochemical Institute.[10]

Another disciple of Haldane, Professor R. B. Fisher wrote:

> He (Haldane) was one of my examiners for my doctorate. In the viva his colleague objected very strongly to my use of regression analysis to work out the relation between nitrogen ingestion and uric acid excretion in pigeon because I had not made duplicate analyses. I couldn't get over to him that attempts to sample the excreta introduced more error and that the analysis of the results of single estimations produced an acceptable measure of random error. Jack (Haldane) took over and spent half an hour giving his colleague an excellent tutorial on the subject of regression analysis which saved me.
>
> The other story concerns his concept of his duty as a Reader in Biochemistry. Biochemical Society meetings used to be delightful when he attended because he would get up at least once and congratulate Dr. X on his paper and say he wondered if Dr. X was aware of the findings of Dr. Y—working on some apparently quite different topic—because there were some striking similarities in the two pieces of work, which he would proceed to explain. These interventions were usually sufficiently cogent to persuade me that I really ought to read the *Biochemical Journal* the whole way through.
>
> I suppose that you know the story about Jack and Kinematic Relativity. Milne in Oxford published a model in which space was supposed to expand steadily as time passed. Jack wrote a letter to *Nature* in which he wondered whether it had occurred to Professor Milne that if this were true the solar system could have arisen at a sufficiently remote epoch as a result of the collision of a single photon with sun. *Nature* carried a series of congratulatory letters for some weeks.[11]

Public Figure

Haldane had become a public figure by the mid-1920s. Initially, he enjoyed fame and recognition as the nephew of Viscount Haldane, former lord chancellor, and the son of the physiologist John Scott Haldane as well as the scion of a score of other famous relatives. However, by the mid-1920s, he had come into his own, as the brilliant writer of the controversial *Daedalus*, with its revolutionary ideas in eugenics and test-tube babies, ideas later copied by Aldous Huxley in his fictional work *Brave New World* and other articles in the popular press. Furthermore, he quickly acquired the reputation as a witty and entertaining public speaker who could be relied on to produce quotable comments on any subject or situation at a moment's notice. Physically, he was as impressive as intellectually formidable. Tall and heavily built, Haldane, with his massive balding forehead and a thick sandy moustache, reminded one of a charging bull, an impression that he still carried in his 60s, when I first met him in India. To his colleagues and reporters he presented this formidable exterior. For close family and friends, especially children and youngsters, however, he was warm and kind-hearted.

He was especially good at entertaining small children throughout his life. Yet, there was something about Haldane that one felt was totally alien to the human species. This is best illustrated in the description of Mr. Codling in Ronald Draper's novel *The Flying Draper*. Codling was not only a man of genius but one who seemed to be set apart from the rest of humanity, and there was clearly a limit to his friendship. Fraser wrote of Codling, "It was like being friends with a fish, or a bird, or a half-human god."[12]

Charlotte Franken

Charlotte Franken (1894–1969) was the daughter of Jewish immigrants. She was born in London on April 27, 1894. Her father, Joseph Franken, was a fur dealer from Germany. Charlotte had a strong desire to become a writer, and in October 1916 her story "Retaliation: A Revenge by Hypnotic Suggestion," was published in the *Bystander* magazine. On July 30, 1918, Charlotte married Jack Burghes, and soon afterward gave birth to a son, Ronnie Burghes. She was planning to attend Bedford College for Women to study languages when her father's business suffered a financial collapse. Instead she enrolled in a shorthand and typing course at a business school in London. This led to work as a secretary at a concert agency.

In 1920 Charlotte began working as a freelance journalist for the *Daily Express* until being given a full-time post by Lord Beaverbrook. Charlotte contributed several articles on the role of women. A feminist, Charlotte was particularly critical of women politicians such as Nancy Astor and Margaret Wintringham, who she believed had a disappointing record in the House of Commons. Her Jewish immigrant parents first lived in Sydenham but later moved to a comfortable house in South Hampstead. Her father regularly attended the Sunday Wagner concerts in the Crystal Palace.

Charlotte was a journalist, war correspondent, and prolific writer. By the time of her death in 1969, Charlotte had written twenty books, including a draft of *Madame de Maintenon: Uncrowned Queen of France*, nine translations, and numerous radio broadcasts and plays. She was Britain's first woman war correspondent in 1941. In spite of her numerous publications and distinction as a writer, she remained an unrecognized and unknown figure in the literary and cultural history of her time. But a biography by Judith Adamson, *Charlotte Haldane: Woman Writer in a Man's World* (1998), brought together a much-needed biographical account of her life and work. This excellent work provides us a rare opportunity to evaluate Charlotte's view of her life, especially her marriage years with Haldane. Adamson interviewed Charlotte's son, Ronnie Burghes; his wife, Betty; and their daughter, Louie, in England. Ronnie died in 1997, a few years after these interviews were conducted. It is important to caution the reader that any information that Charlotte imparted to Ronnie about Haldane and their marriage is likely to be colored by her disappointment with Haldane during their marriage and subsequent divorce. Haldane left Charlotte as soon as his young student, Helen Spurway, entered his life. She was twenty years younger than Charlotte and twenty-four years younger than Haldane!

First Meeting

It was while pursuing her career as a journalist that Charlotte first came into contact with Haldane. In 1924, while considering an article on the social implications of the ability to predetermine the sex of newborn, Charlotte came across an article in the American magazine *Century*. It was an abridged version of Haldane's book *Daedalus, or Science and the Future*, which became an instant bestseller on both sides of the Atlantic. Charlotte's own thinking and her admiration for the writings of H. G. Wells prepared her well. The predictions and speculations contained in *Daedalus* excited her imagination, to say the least. She was particularly interested in Haldane's statement that "the biologist is the most romantic figure on earth. . . . With the fundamentals of ectogenesis in his brain, the biologist is the possessor of knowledge that is going to revolutionize human life."[13] According to her biographer Adamson, "She knew immediately 'This is my man!' and took *Daedalus* to her editor, J.B. Wilson, to convince him of Haldane's newsworthiness."[14] She was interested in gathering scientific information for the book she was writing, but she was also clearly intrigued and attracted by the man behind *Daedalus*.

Charlotte's first meeting with Haldane, which she later described as "high romance," took place on a sunny Saturday in Cambridge. Haldane never responded to her request for an interview, but she was determined to see him anyway. She formed a mental picture of an aging bearded professor. But what she found was quiet different—a man who appeared to be larger than life, with a huge domed balding head, fiercely bushy eyebrows, bold and roman nose and a thick moustache. He was only a year and half older than she was. In a BBC program, she said, "There was, in fact, very little conversation, for when he talked it was like listening to a living encyclopedia."[15] She was thoroughly overwhelmed by Haldane's charm, brilliance, and encyclopedic knowledge. Upon hearing of his physiological experiments in which he swallowed acid sodium phosphate, she declared that he was positively heroic! Haldane, for his part, noted later, "To my astonishment, the resulting paragraph in the *Daily Express* not only kept to the facts but, as had been stipulated, did not mention my name. For this, and other reasons, I fell in love."[16]

On that summer afternoon, Charlotte came to see Haldane with two aims; she wanted to write an article about Haldane for the *Daily Express* and she was keen on seeking Haldane's advice about a new book she was planning to write. She was planning her first novel, *Man's World*, in which a couple would be able to choose the sex of their children. She wanted to make sure that she understood the scientific part correctly. She had read a shorter version of Haldane's book *Daedalus*, which contained predictions of test-tube babies, in the American *Century Magazine* and decided that Haldane was her man. In her autobiography, *Truth Will Out*, Charlotte wrote, "I took the article by J.B.S. Haldane to my news editor, J.B. Wilson, and easily convinced him that it was—to use the Fleet Street cliché—a great story. I received the assignment I aimed at—to interview the author."[17] But it took her considerable effort to locate the whereabouts of Haldane.

She wrote, "At the offices of the Oxford University Press and the Cambridge University Press, I found in reference books the name of a Professor J.S. Haldane, but on looking up his distinguished scientific publications, I concluded that he was not the one I was after." Furthermore, when she did find the right man she discovered to her great surprise that "J.B.S. appeared to have no academic degrees."[18] She discovered that J. B. S. Haldane was a Fellow of New College, Oxford, and that he was Reader in Biochemistry at Cambridge.

Charlotte wrote a "diffident and polite request" seeking an interview with Haldane but received no reply. After a couple of weeks, she decided to approach him personally. On a hot summer afternoon in 1924, she took a train to Cambridge. Inquiries at the University eventually led her to the narrow staircase of Nevile Court at Trinity and to the white door with Haldane's name painted next to it. As Haldane was out for several hours, she wandered around Wren's Great Court, King's College Chapel, and the Backs. By the time Haldane arrived, she was "hot, footsore, physically depressed and weary," as she put it, but was fully rewarded upon meeting Haldane. Her first meeting with Haldane was certainly memorable. In her own words, the door was opened "by an enormous figure that loomed over me and invited me to come in. The room was large and high and exquisitely proportioned. Bookshelves occupied one entire side of it. I was completely over-awed by the largeness of the man and, nearly scorched by the blaze of his intellect, I felt my inconsiderable culture shrink to Lilliputian insignificance."[19]

Upon learning the purpose of her visit, Haldane loaded her with advice, references, and books. Fortunately, he learned only as he was escorting her to the bus that she

Figure 2.1 JBS and Charlotte.

was a newspaper reporter as well as a novelist. He made a few pungent comments about newspaper reporters in general, but added that he had taken 10 cc of his own blood that morning. Charlotte wrote, "But it soon became clear that my don and mentor was not interested merely in the cultivation of my mind. I was wholly in love with his mind, and when, with a charming affectation of eighteenth-century gallantry he 'implored my favours' I did not withhold them. In any case, coyness was not natural to me."[20]

At that point, Haldane was 33 and Charlotte was 28. However, before they could marry several obstacles had to be surmounted.

Charlotte foresaw that marrying Haldane would be a step up for her and her four-year-old son both socially and financially. She was keenly aware that through Haldane she would be able to climb the social ladder, which she could not have achieved otherwise. She wrote, "I had been introduced to this exalted circle as the result of J.B.S.'s plan to marry me. When he first mentioned it, I was overwhelmed by conflicting emotions."[21] In the deeply class-ridden English society of the 1920s, Charlotte was deeply conscious of her Cockney[22] accent. Haldane frankly admitted his reservations, the most important being her Cockney accent and her profession—journalism. He grew up with a deep mistrust of journalists and journalism. But their mutual love and a strong desire for marriage overcame all objections. There was one problem, however. She was already married to another man, Jack Burghes, who refused to agree to a divorce. On her part, she wrote that her marriage had already ceased to exist for some time, and was shocked when her husband was not readily agreeable to a divorce. He threatened further that she would permanently lose their only child, Ronnie, if she went through with it. Haldane was a fighter and in the fight that preceded his marriage to Charlotte he was not intimidated by the prospect of becoming a co-respondent and the ensuing notoriety. On the contrary, he welcomed any opportunity that would shock the establishment, especially by expressing his disapproval, in his own unique way, of the stuffed shirts of Cambridge society of the 1920s.

Their troubles were just beginning. As Charlotte saw them, in 1925 the English divorce laws were both legally antiquated and morally indefensible. Desertion was not considered an adequate reason for a divorce. Adultery had to be committed, and proven in the court with all the sordid details. Divorce still carried a heavy social stigma. It was something to be avoided at all costs, especially by professional men and women. The eminent Cambridge biologist William Bateson, now running the John Innes Institution, commented on hearing of the case involving Haldane, "I am not a prude, but I don't approve of a man running about the streets like a dog."[23]

Both Charlotte and JBS braced themselves for such adverse reactions. It was especially damaging for an academic like Haldane, whose fortunes lay at the mercy of the petty gossip that goes on in a small university town. Haldane later wrote, "I informed the Vice-Chancellor of Cambridge University that I was about to commit this act, to which he replied: 'Oh!'"[24]

The great event was scheduled for an auspicious day in February 1925. Haldane and Mrs. Charlotte Burghes duly arrived at the hotel. But she did not like its appearance and wanted to move to another hotel. However, it was important not to lose the private detective who was playing the third wheel in this drama. He was spotted in the hotel lobby by Charlotte with her keen journalistic eye and was redirected to another hotel. He obliged, carrying one of their suitcases. Next morning he showed up with the newspapers outside their hotel room. Her husband Jack Burghes found out that she and Haldane spent a night together in the same bedroom at a local hotel and filed the expected petition against his wife. Charlotte and Haldane responded to the charge of adultery by denying it. In fact, in view of Charlotte's later claim that Haldane was impotent, what actually happened in that bedroom remains a mystery. When the case was heard later, in October 1925, Lord Merrivale granted a decree to Mr. Burghes, awarding 1,000 pounds in damages against Haldane.

Sex Viri

Although now they were free to marry, Haldane had to resolve another crisis. On the grounds of immoral behavior, a committee of Cambridge University, acting under the University Statutes, informed Haldane that unless his resignation was received by the vice-chancellor by a certain date, he would be deprived of his position as reader in biochemistry. The committee was called *Sex Viri*, or, as Haldane has explained, six men, not "Sex Weary."[25] (To the relief of everyone, that committee was abolished many years ago.) The letter was signed by the vice-chancellor. The last sentence of the letter, which induced Haldane to fight, advised him to resign quietly to "avoid publication of the judgement."

Haldane appeared before the *Sex Viri*, arguing that he had not broken up a home, since there had been no home to break up. However, the committee went ahead and deprived him of the readership. Under the University Statutes, Haldane appealed to a tribunal of five, appointed by the Council of the Cambridge Senate. By now, Haldane's fight had become a major center of attention at the university, many people taking sides on the issue of academic freedom and the private life of dons. He was supported by the head of his department, the much respected and recently knighted Frederick Gowland Hopkins, who refused to appoint another individual to fill the readership. Furthermore, Haldane was backed by the National Union of Scientific Workers. Haldane was represented by Stuart Bevan, later Conservative MP for Holborn, and his witnesses included his father, John Scott Haldane, and sister, Naomi. The tribunal was presided by Justice Avory. Not surprisingly, the court's majority verdict was in favor of Haldane.

The judges added, however, "This decision is not to be taken as any expression of opinion that adultery may not be gross immorality within the meaning of the statute." Haldane later commented that the tribunal disapproved of the action taken by the *Sex Viri* because their verdict was delivered before hearing all the evidence. For whatever reasons, Haldane was reinstated. In fact, his success was also a success for all others in similar situations. University faculty in Britain have not been threatened since because of their behavior in private lives.

According to her biographer, Charlotte decided immediately to consider it her duty to shape and promote Haldane's future career in popular writing. She soon found out that in addition to "a phenomenal intellect, a colossal memory, and admirable gifts of verbal fluency," Haldane had an "unusual enthusiasm for the popularizing of science, and a firm conviction that this was both desirable and necessary."[26] In *Truth Will Out*,

Charlotte wrote that Haldane accepted her suggestion that he should practice scientific popularization as a hobby in his spare time. She decided to become his secretary and agent and help him to earn extra income through popularization. She wrote, "As I had anticipated, there was a large and keen market, particularly in the United States. American magazines gladly paid fifty guineas for an essay of a few thousand words, which he would scribble at his leisure . . . and I would type out and sell."[27]

Comments on Their Marriage

Commenting on their marriage, JBS later said they were "happy for some years." They differed mainly in their scientific interests. Haldane was devoted to science all his life. Charlotte's main interests were in writing and journalism. Charlotte wrote in her autobiography, *Truth Will Out*,

> During the course of many discussions with him [Haldane] in the early years of our marriage, it became clear to me that science could not give me the answers I was seeking, because I was asking the wrong questions, in the wrong quarter. I was basically preoccupied with the problem of the "Why" of the Universe. But science could only answer, and to a limited extent still, the questions directed towards the various ways in which certain universal laws worked. Its business was to deal with the "How"? rather than the "Why"? . . . Its chief tools were not verbal but mathematical concepts. And for mathematics I had, unfortunately, no natural aptitude. In consequence, I realized, I would never be able to learn science, but only about science.[28]

But they shared political interests. She wrote, "Our interest in politics had always been a strong one. We both were and had been, long before we had met, socialists and, to use the horrible cliché that, at about that period, began to pass into the language, 'left-wing intellectuals.' Temperamentally also, we were strongly inclined to radicalism."[29]

Entertaining Children

Except for the disappointment of being unable to have children, the years of Haldane's marriage with Charlotte were among the happiest of his life. As he watched his stepson, Ronnie, playing in the grounds and growing up, Haldane was constantly reminded of his own inability to be a father. On the other hand, those were also the years of important scientific productivity for Haldane. Not only was he continuing to publish his important series of papers on the mathematical theory of natural selection, which established his preeminence as a first-rate scientist, but also he was simultaneously making important contributions to enzyme chemistry and the physiology of sensory thresholds. In addition, Haldane took on the additional part-time appointment at the John Innes Horticultural Institution, where he initiated pioneering research on the biochemical genetics of anthocyanin pigments.

Those were the halcyon years in Haldane's life. He was in his thirties, establishing his first home away from his parents, but not yet entangled in Marxist politics. It appears that Haldane had only two major goals during that period: pursue his scientific career and enjoy his family life with Charlotte. It was also during those years that Haldane matured from a young scientist to a full-fledged professional, culminating in his election to the Royal Society of London and the publication of his seminal work *The Causes of Evolution* in 1932.

The Roebuck House

For their first residence in Cambridge and their first home as a married couple, Charlotte found Roebuck House in Old Chestertown. It still preserved some of its previous charm as an old Fen village. Charlotte aptly put it, "There was a Lover's Lane. There was a country inn of great character, next to an old village tree with a wooden bench running around it, on which the local elders gathered as they had done for centuries past, to browse and gossip."[30] Roebuck House used to be an inn of substance, "The Roebuck." Charlotte wrote that parts of it, housing the furnaces and potting shed, were still authentic Elizabethan, much of the rest was Stuart and Georgian of not very good style, and had a Victorian exterior "in the shape of a wooden porch and other offensive items," which she removed. However, there were other aspects that made it attractive to Charlotte—a flower garden that ran down to the banks of the Cam, beautiful well-proportioned cellars and kitchen, and a kitchen garden enclosed by brick walls. It contained a vinery, where grapes ripened from October to December, and two long glasshouses. Furthermore, there was an enormous old-fashioned Victorian conservatory attached to the house, and in the rose garden, there was another smaller, modern double greenhouse.

Their drawing room became Charlotte's salon or, in Haldane's words, "Chatty's addled salon," which became the meeting place for an odd assortment of characters, poets, writers, and scientists. One of them was a young student of biochemistry, Martin Case, who was invited by JBS to live in his house instead of in university digs. He was an excellent jazz pianist who often entertained the gathering. Case soon became an "uncle" to Ronnie, and eventually Charlotte's lover. And another member of the group was young Malcolm Lowry, a chronic alcoholic who achieved fame later as the author of several books including *Under the Volcano*, which was made into a motion picture directed by John Houston. Charlotte and Lowry became drinking buddies, if not lovers. Haldane never joined their parties, which were too noisy for his taste. He was tone deaf and was not able to enjoy the music.

JBS soon accepted an appointment as part-time head of the genetical department at the John Innes Horticultural Institution at Merton, near Wimbledon, which was founded by his mentor William Bateson in 1910. Haldane continued breeding work on primulas, especially *Primula sinensis*, which was first initiated by Bateson several years earlier. There were about ten thousand primulas at Merton. Charlotte readily agreed to take a thousand plants into the greenhouse, and to observe the results of crosses involving them, and the segregation of various factors that was taking place among them, such as leaf, stalk, and flower shape and color, and other characters. She made the observations under the supervision of JBS, and wrote in her autobiography, "I found this hobby engrossing and enjoyable."[31] Although Haldane encouraged her to take up the scientific work seriously, Charlotte wrote that she did not feel qualified, partly because of her lack of mathematical ability, and also because of her keen interest to continue her career in writing. Perhaps because of that disappointment, Haldane later chose a scientist, Helen Spurway, as his second wife.

Haldane's Cambridge period was marked by his support of liberal groups, especially anticlerical movements. One of them was the Cambridge University Benskin Society, which seemed to be devoted to the uninhibited enjoyment of beer and bawdy and ribald verse and song. It served the purpose of a channel for blowing off steam periodically for rather more serious and academically minded undergraduates and younger graduates. The meetings always culminated in uproarious and noisy endings with the inevitable ill-feeling among the neighbors and police involvement. The meetings were usually held in various members' rooms in one of the colleges. However, the society was eventually forbidden to hold meetings anywhere on the university premises. Upon hearing this, Haldane immediately offered Roebuck House as the future meeting place, and both he and Charlotte were elected to the society. Haldane was accorded the honor of being the "Sanitary Inspector," but he insisted on being called the "Insanitary Spectre." Such activities, at first glance, may seem out of place in the life of a man of Haldane's intellect, however, his junior colleague Martin Case explained that because of the vast intellectual gulf between Haldane and the common herd, he took delight, almost pathetic delight, in being accepted and assimilated at this rather unintellectual level.

On one occasion, Martin Case, who was staying at Roebuck House, was charged with dangerous driving, Haldane stepped in. He told the concerned student, "If you care to leave this to me, I think I see a possible course of action that *might* just...." Nothing further was said.

Haldane ascertained that the only witness for the prosecution, a security guard, spent his daytime hours in a certain pub. On the day of the Court hearing when his case was to be heard, Martin arrived before time and waited for the proceedings to start. In due course the prosecution called the witness in question. However, what appeared before the court was not the security guard known to the student but an apparition of sorts, some likened it to a scarecrow, who stumbled into the box. His speech was so slurred that no one could understand what he was saying. When asked to narrate his version of the incident, this is what he was reported to have said with great difficulty: "Ecumuptheillikefuckinellanwentarscovertip." An official said sternly, "You cannot say that sort of thing here. Kindly tell us what happened in a way that the Court can accept as evidence." This seemed to have irritated the witness. He repeated the same answer in a much louder voice, except that he belched violently, retched, and sank to the floor of the witness box from which he had to be extricated "like a winkle."

Everyone in the courtroom was too incredulous and flabbergasted to laugh or react in any way. Next day Martin ran into Haldane in the lab and Haldane inquired casually about the Court proceedings. He was a little disappointed to hear that the security guard made it to the Court.

Living with Haldane

Haldane's married years with Charlotte at Cambridge were blissful at first. However, after a few years it became clear that Haldane could not father a child. Both of his marriages were childless. Haldane's stepson, Ronnie Burghes, was interviewed by Charlotte's biographer, Judith Adamson. Ronnie told Adamson that he was fond of his stepfather but he was afraid of his bad and unpredictable temper, which was a barrier to "real affection." He said, "You never knew when he would lash out. Nini [the maid] once made the mistake of going into his study to dust. All hell broke loose and no one ever went in again uninvited. The room was an enormous mess, the floor buried under dusty papers. He would even get angry at dinner and take his plate out on the steps to eat."[32] Haldane showed no physical affection toward Ronnie, but he used to take Ronnie on long hikes on which he used to identify plants. They used to play games, rowed on the Cam, and swam together. Strangely, even when Ronnie won an athletic prize at school, Haldane showed no particular interest. However, Ronnie's badminton skills impressed Haldane, as he was never able to beat him.

Charlotte's biographer, Adamson, commented that, for Haldane, it was as if his sister Naomi was replaced by Charlotte's child. Years later, Naomi recollected that Ronnie "was having a rather bad time as a stepson."[33] Bertrand Russell's wife, Dora Russell, recalled that she could not help feeling some anxiety for Ronnie, with such a dominant personality for a stepfather.[34] Charlotte said that Haldane competed with everyone in everything, children included. When Ronnie wrote a fairy tale at the age of twelve, Haldane felt compelled to write stories for children himself, and these were eventually published under the title *My Friend, Mr. Leakey* in 1937.

Charlotte wrote that Haldane could never realize what a terrifying effect his size, his dome, and his bushy eyebrows had on smaller and lesser mortals.[35] I should reiterate that Charlotte's recollections of their married life may have been influenced by her disappointments and subsequent divorce. We should keep this in mind while assessing her comments on Haldane. Haldane's former colleague at Cambridge, Norman Pirie, also commented on this aspect of the Haldane-Charlotte relationship.

One of Charlotte's last comments was that a recently published biography (by Ronald W. Clark) of Haldane had failed to point out that "it was I who made him the greatest scientific journalist of the day." In his biographical account of Haldane, a former colleague and friend of Haldane, Norman Pirie (1969) wrote that he found "the statement that childlessness warped his [JBS's] character ... surprising. Still more surprising is the suggestion that Haldane was a sexual braggart, not very proficient in actual performance.... It is possible ... that ... suggestion is based on gossip from a disappointed woman."[36] Charlotte responded by saying that Dr. Pirie was never a member of the Haldane's intimate circle. "Only myself and Dr. Helen Spurway Haldane know the facts about JBS's sex life."[37] She died six weeks later.

Life in Cambridge had its ups and downs. Old Cambridge found Haldane rather a bore from misunderstanding his manners. When he dined at the High Table,

Haldane was in the habit of discussing the most intimate details of his life in a loud booming voice. He was totally oblivious of any social convention. He once rushed to join the dinner table while carrying on an experiment, carrying a flask of urine, which he placed on the table. While these acts did not endear Haldane to his colleagues, he was at least tolerated because he was a "Haldane," and was allowed a certain degree of latitude in behavior. However, Charlotte's case was different. There were three strikes against her: as a woman, as a Jew, and as a wife. And she tried harder to be recognized as a professional in her own right. Years later, her son, Ronnie, told her biographer Adamson, "In those days, you were either somebody or nobody if you were a woman. Charlotte wanted to be somebody. She had to be aggressive." As can be imagined, there had been a lot of "inbreeding" among the dons, and as Adamson put it, "In the end acceptance into Oxbridge was a matter of birth as much as of manner."[38] Charlotte got on reasonably well with Haldane's father, but there was a definite chill between her and Mrs. Haldane. Although she was known to call her husband "SP" or Senior Partner, Mrs. Haldane was known to "rule the roost." She deplored the marriage of Julian Huxley because he married a Swiss girl, Juliette Bailott. Juliette wrote, "The English have never approved of foreigners. . . . to marry one was the 'original sin'."[39] Indeed, both the Huxley brothers—Julian and Aldous—married foreign girls.

In *Truth Will Out*, Charlotte wrote of Haldane's bad temper, especially in the mornings. If Haldane read something in *The Times* that he found disagreeable, he used to find a pretext to pick a quarrel with Charlotte. But she found a way to dodge him, by hiding in the bathroom until he cooled off. She wrote, "It was an interesting fact that Haldane, whose life was passed in scientific experiment and control could never learn to control himself.[40]

Haldane's research in genetics stimulated Charlotte's interest to translate into English a brilliant book on genetics and environment by the German authority Professor J. Lange, titled *Verbrechen als Schicksal*. It was a pioneering study of twins, both dizygotic and monozygotic, some of whom became criminals. It was written in a colloquial style and was based on observations of thirteen cases of "identical" criminal twins. In his foreword to the English edition, *Crime and Destiny*, Haldane wrote:

> What would be the effect on human conduct if the view that crime is destiny were generally adopted? Supporters of indeterminism state that a belief in fatalism should logically yield to a blind acceptance of events and a refusal to struggle either against external circumstances or defects of character. I cannot myself see the cogency of this view. My will may not be free in certain senses of that word, but it is at least my own. I regard my character and my environment as equally predestined, and get quite a lot of quiet fun out of the attempt to prove that the former is the more important. As a matter of historical fact, fatalism does not conduce to weakness of will. The opposite is true. Among the ranks of the fatalists must be reckoned Mohammed and his successors, who conquered from the Atlantic to the Indus in a century, the leaders of the reformation, the founders of the New England States, Napoleon, Lenin, . . .

We all have our weak spots, and it is well worth while sowing a few wild oats if we can find out what they are and act upon the knowledge. The consistent believer in free-will repents, and hopes that his will may keep him out of the same sin in the future. The intelligent fatalist regards his lapses with a certain tolerance, but acts on the knowledge of his own character which he gains through them....

So much for the individual. But what would be our attitude to the errors of our fellows if we adopted a strict determinism as a general view of life? To answer that question we must first remember an elementary fact. Praise and blame, which are very powerful social motives, are largely reserved for those sides of contact which they can in fact influence. We blame people for being lazy or vicious, and this does on the whole have an effect in making them more industrious and sober. We do not blame them for being stupid or physically weak, and it would be useless to do so. But the fact that certain sides of conduct can be influenced in this way need not lead us into dubious metaphysics. The determinist will go on blaming his erring brother, but the blame will be more than half pity. And he will avoid moral indignation. I find certain kinds of conduct in others disgusting, but there is no reason why I should lose my temper about them. And I know that an attitude of moral indignation is peculiarly ineffective in bringing about a change of heart in others. On the contrary, it is an ideal excuse for cruelty.[41]

Travels

JBS and Charlotte traveled widely, especially in Russia and Spain, but they had their political differences as well. Her disappointment with Stalin's Russia, which later, in the postwar years, turned into an anticommunist tirade, and her "psychoanalysis" of communist mentality did not endear her to Haldane and his left-wing admirers. However, according to Charlotte's biographer, Adamson, the most serious damage to their relationship was caused early in their marriage by their inability to have a child. In 1926, Charlotte accompanied Haldane to the Twelfth International Physiological Congress in Stockholm. The French biochemist Rene Wurmser attended that Congress and described his first impression of Haldane, which I have described on page.[42]

In the early years of their marriage they traveled widely. Charlotte wrote,

During the first years of my marriage to J.B.S. which took place in 1926, we travelled a great deal.... We made frequent trips to various European countries, as tourists, lecturers, or to attend scientific congresses. Our honeymoon was spent in Switzerland, where we both gave lectures for the Sir Henry Lunn tourist organization.... The following year we attended the Genetical Congress in Berlin. In 1928, JBS was invited to the Soviet Union, at the instigation of Prof. N.I. Vavilov,[43] the leading Russian plant geneticist, and at that time an influential figure in political circles.[44]

Visiting Russia

Two years after their marriage, in 1928, Haldane was invited to the Soviet Union by the leading Russian plant geneticist Professor N. I. Vavilov, who was not only a respected geneticist but also an influential figure in political circles. Neither Haldane nor Charlotte was a Communist at that time. Vavilov had worked under Bateson at Cambridge University. He was at the height of his fame at the time of Haldane's visit. The British geneticist S. C. Harland described Vavilov in glowing terms:

> He had worked on genetics with Bateson in England; he could speak at least ten languages ... gifted with immense curiosity and a truly gigantic intellect. ... Ideas sprang from his mind ... a man who was entirely selfless in his personal life ... Vavilov occupied the posts of President of the Academy of Agricultural Sciences and Director of the Institute of Applied Botany for some twenty years ... founded 400 large research institutes, each staffed by up to 300 research workers. ... By 1930 Vavilov had created a world collection of every known economic plant. I myself saw the great wheat collection of 26,000 varieties growing near Leningrad.[45]

Vavilov was their host in Moscow and Leningrad in the summer of 1928. He arranged lectures by Haldane in both cities. His hospitality was lavish. Charlotte wrote: "Owing to his influence, we were allowed to make a private visit to Lenin's tomb on Red Square, as well as to the churches and State Museum in the Kremlin, where we were shown the State coaches of the Czars, and the little sable crown, with its huge diamond cross, of Ivan the Terrible."[46] They visited Vavilov's famous institute and plant collections in Leningrad and met with the eminent cytologist G. D. Karpetchenko. Others included the eminent biochemist Professor Bach and his pupil Lina Stern, who was then the director of the Institute of Physiology. Charlotte wrote that almost all the scientists spoke fluent German, French, or English.

In *Truth Will Out* Charlotte wrote that Haldane did not share her disdain for the Soviet world.[47] He regarded their attitude toward science and scientists as of the most outstanding quality. The scientists and the factory workers were the most favored classes. Vavilov, Levit, and other eminent scientists were enjoying excellent opportunities for research and exalted positions in society. But Charlotte's first impression of Soviet life was not favorable. She was very conscious of police surveillance and felt an instant relief upon departing. However, she admired certain aspects, such as its "racial policy of social and political equality for all Soviet citizens."[48]

Political Transformation of Haldane

By the end of her Cambridge years (1932), Charlotte began taking interest in the activities of the small core of Communists at the University and enjoyed holding long discussions with Marxist students. Charlotte had recorded that she was "indoctrinated" with Marxism by a young Jewish male. She described him as her "protégé" (most likely her lover) during the last years of their Cambridge period (1932–1933). In the meantime, Haldane was appointed to the Chair of Genetics at University College London (UCL). Charlotte, too, thought it was a good idea, because she was getting tired of "stuffy" Cambridge life and would be happier in London, to which she had been accustomed before marrying Haldane. In *Truth Will Out*, Charlotte wrote, "I myself had become a rebel against Cambridge intellectual and social snobbery. After a few years I was profoundly bored with both. My spiritual and emotional quest was driving me on as ruthlessly as ever; I had found no inner peace beside the rather turgid and shallow waters of the Cam. My personal life contained another frustration; for as the years passed it seemed less and less likely that I would have the children for whom I had longed."[49]

Another reason for Charlotte's interest to move to London was her desire to send her 13-year-old son to St. Paul's School in London.

In 1932, Haldane was on a long tour of America, attending the International Congress of Genetics at Ithaca, New York, and visiting the California Institute of Technology in Pasadena, California, and Columbia University in New York afterward. Charlotte, in the meantime, moved to 16 Park Village East, a Crown-owned house in London with "terraced flower beds running down to the canal on the north-east corner of Regents Park."

In the spring of 1933, Haldane took up his position as professor of genetics at University College. He claimed at once that University College is "as full of bloody Communists as Cambridge," and "I tried to keep out of politics, but the support given by the British Government to Hitler and Mussolini forced me to enter the political world."[50] With the rise of Nazism, both Haldane and Charlotte felt that the days of carefree, cultured, and secure life were over. About 1935, soon after Helen Spurway entered JBS's life, first as a student and later his wife, Charlotte contributed a series of articles to the *Sunday Express*, under the heading, "Is It Possible to Be Happily Married?" Quite simply put, she concluded that marital success was contingent on three factors: having children, wife's financial independence, and surviving the "danger" period. In the meantime, the new "danger" on the horizon came in the form of Helen Spurway, who announced her intention to take a degree and to marry Professor Haldane.

Meanwhile, events in the larger world dictated their activities and social life. Hitler's rise to power led to a number of refugee scientists who found haven in the Haldanes' home. Among them were Boris Chain, who later shared a Nobel Prize with Alexander Fleming and Howard Florey for their work on penicillin, and others who joined Haldane's department at UCL, including Hans Gruneberg, Ursula Philip, and Hans Kalmus.

Haldane and Charlotte continued their political activities. They attended the League of Nations scientific and literary conference in Madrid and became deeply entangled in the Spanish civil war. But it was in their attitudes toward the new Russia that Haldane and Charlotte differed sharply. Commenting on their earlier visit to Russia, Adamson wrote, "Russia interested her but she had no regrets at leaving. With hindsight, she realized that her need for an exit visa, was symptomatic of what was to come. . . . JBS was more enthusiastic at his first glimpse of the Soviet world, but it was the commitment of the State to science and scientists not the communist system which had turned his head."[51]

Childless Couple

Haldane had been known to express his deep regret, on more than one occasion, of not being able father a child. Charlotte, too, was eager to have more children, especially with Haldane. She wrote, "The deepest and strongest bond J.B.S. and I had in common was this philoprogenitiveness."[52] As he watched Charlotte's son, Ronnie, playing in the grounds, Haldane said to an old friend, "How I wish I had a son like that." It has been said that Haldane's childlessness tended to warp his character out of shape. His sister, Naomi (Nou), suggested that perhaps a bad attack of mumps in childhood may have been responsible for his infertility. Charlotte told her daughter-in-law, Betty Burghes, that Haldane was impotent. She believed that if Haldane had been able to father the children they both wanted their marriage would have lasted longer and remained happier. It was also clear that her relationship with her mother-in-law, Louisa Haldane, would have decidedly improved had she been able to give birth to a young Haldane. When Julian's Swiss wife, Juliette Huxley, gave birth to her first child, Anthony, Mrs. Haldane's previous hostility suddenly vanished. She became welcoming, understanding, and even helpful. Her attitude toward Charlotte, on the other hand, remained unchanged.

Sexual Prowess

Ironically, Haldane bragged about his sexual prowess on more than one occasion. As a young teacher at Cambridge he used to cover his clumsiness in the laboratory by saying to his female students that he was a skilled exponent at using his "paternal apparatus." In his final essay, "On Being Finite," Haldane wrote, "Most of my joyful experiences have been by-products. Thus I have enjoyed the embraces of two notoriously beautiful women. In neither case was there any wooing. After knowing one another for some time we felt like that, and no words or gifts were needed. I later married a colleague."[53] Haldane's bragging about his sexual prowess was in total contrast with his real life. It may have been a consequence of the need he felt to compensate for not being able to produce an offspring! It was tragic and pathetic and was totally out of sync with the rest of his life, which we all came to know and admire!

Their marriage was tragic. Soon it began to unravel. Charlotte began a life of extramarital affairs, which earned her the antipathy and envy of many local women. It was widely believed that Aldous Huxley's[54] Shearwater, the biologist in *Antic Hay*, who was constantly immersed in his experiments while his friends took his wife to bed, was based on Haldane's life at Cambridge.

Haldane's fame and importance as a scientist overshadowed Charlotte's writing career. Adamson identified four reasons that offered a possible explanation for the nonrecognition of Charlotte as a writer. First, although her novels were distinctly modernist, she did not follow the new narrative techniques developed by Virginia

Woolf and other writers that were generally accepted as the hallmarks of social change. Second, her writings in the popular press were not regarded as high journalism. They were largely directed at popularizing social, scientific, and philosophical issues. Third, her public defection from the British Communist Party in 1941, when German tanks were in the Moscow suburbs, stirred up such anger and hostility that her name is unmentionable among the leftist intellectuals even today. Fourth, her marriage to a great scientist tended to eclipse her own achievements, and, as Mrs. Haldane, her own writings were mentioned (if at all) as a footnote to Haldane's wider accomplishments.

Nevertheless, an evaluation of the nature of Charlotte's activities and her life with Haldane are helpful to an understanding of Haldane's life during the critical years of his Cambridge period and his subsequent move to London. This was especially true of his politics. Although by 1933 Haldane had already began taking a serious interest in the works of Marx, Engels, and Lenin, his conversion to communism was slow and gradual. It was not until 1938 that he took Marxism seriously enough to discuss its relationship with scientific progress. In the meantime, Charlotte joined the Communist Party of Great Britain (CPGB), embracing its many activities, while maintaining deep reservations about the state of Soviet society under the repressive reign of Stalin. There is evidence that indicates that, as a Jew, she admired Karl Marx, with whose Jewish background she could identify.

It was clear that while Haldane, together with Lancelot Hogben, J. D. Bernal, and Joseph Needham, was in the forefront of the younger generations' socialist aspirations during the 1930s, he was far from becoming a full-fledged Marxist and a Communist. Charlotte's influence on Haldane's political evolution during those years is obvious. Events during the Spanish civil war and her son Ronnie's decision to fight with the anti-Franco forces brought Haldane into closer contact with the revolutionary left. Haldane himself did not allow his name to go forward for the membership of the CPGB until 1942. As he resigned from the Party in 1950, his association with the CPGB lasted just eight turbulent years, including his chairmanship of the editorial board of *Daily Worker*.

Divorce

Charlotte was granted the divorce decree on November 26, 1945. Both she and Haldane had filed cross-petitions on grounds of desertion, but Haldane later withdrew his, whereupon she was granted a divorce with costs. Charlotte had stated that their marriage was happy until Haldane returned from Spain in 1936, when she found him to be increasingly irritable and difficult to please. According to Charlotte, during the air raids Haldane had gone to Julian Huxley's shelter at the London Zoo, telling her that there was no room there for her. In November 1940, he moved to Harpenden, where the Rothamsted Experimental Station is located. Apparently, he told her that she could visit him occasionally. But it is already clear that Haldane was very much preoccupied by his young disciple and future wife, Helen Spurway, by then.

Alimony settlement was complicated by the casual manner in which Haldane treated all personal financial affairs. Soon after the divorce was granted, Haldane married Helen Spurway and gave her the Roebuck House as a wedding present. Records showed that even before the divorce was granted, Haldane began transferring his assets to Spurway, and made a new will leaving everything to her in the spring of 1944. The divorce proceedings and the subsequent alimony arrangements were being discussed at a time of great scientific and political crisis in Haldane's life. Haldane's position on the ticklish matter of the Lysenko controversy and his slow break with the Communist Party were yet to come, but already by 1945 the strain could be felt because of the Lysenko controversy (see p. 187).

A footnote to their marriage and divorce—when Haldane was making plans to move to India in 1957, it had been mentioned that his move was motivated by a desire on his part to stop making alimony payments to Charlotte. I have listed the possible reasons for Haldane's move elsewhere (see Chapter 16), but a desire to stop paying Charlotte's alimony was not among them. The timing of the move was due to the confluence of several factors, especially his disappointment with Soviet genetics and a gradual withdrawal from Marxism and communism, an offer of suitable appointments for Helen Spurway and himself from P. C. Mahalanobis at the Statistical Institute in Calcutta, opportunities for biological research in India, forthcoming superannuation from UCL, and finally, the trigger—Helen's arrest in connection with the police dog incident coming on the heels of the Suez crisis (see Chapter 16).

Brighter Biochemistry

Haldane had always shown great talent for writing comic verse, and the Cambridge period was no exception. The following comic verses were contributed by Haldane year after year to *Brighter Biochemistry*, the humorous journal that came out between 1924 and 1933 at the Biochemical Institute at Cambridge.

1924, no. 2

I cannot synthesize a bun
By simply sitting in the sun;
I do not answer "yes, yes, yes,"
If I am offered meals of S;
I must admit I always flee
When offered drinks of NH_3;
I fear that $NaNO_2$
Would turn my haemoglobin blue;
And you are really quite mistaken
To give me nitrate, save in bacon;
The synthesis of tryptophane
My tissues find too great a strain;
And I metabolize no more
On breathing things like CH_4;
While even Barcroft, as you know,
Could never oxidize CO;
No men, no guinea pigs, and few crows,

Can make a simple thing like sucrose;
But readers, rhyzostomes and rats
Are fairly good at making fats.
So I shall concentrate on this.
My most effective synthesis.

The nutritive and therapeutic value of the simpler glucides
Though Pope and Dixon have denied
The virtues of formaldehyde,
Yet it would not become the bard ill
To sing the polymers of Yadil.
    2. Before I climb a peak or high col
       I drink the aldehyde of glycol;
    3. I should require a pen like Cicero's
       To tell how fond I am of glycerose.
    4. While my immunity to CO's
       Due to my eating lots of threose.
    5. My inability to tie bows
       Is due to insufficient ribose.
    6. A certain bug of Mrs Callow's
       (And that alone) can live on allose.
    7. Full many a mystic and adept owes
       His psychic powers to sedoheptose.
    8. The use of alpha-gluco-octose
       Has made my nails grow through my sock toes;
    9. But Emil Fischer's ghost alone knows
       What comes of taking mannononose.
    10. The whole laboratory echoes
       With my demand for glucodecose.
    11. Oh that I had a hundred stylos
       To write the praise of glucoxylose!
    12. If in my skull a hole or crack arose
       I'd ask for fifty grams of saccharose.
    18. My room resounds with fervid rightos
       When people give me melezitos.
    24. Employés of the London Gas Cos.
       Write testimonials to verbascose.
       Though, thanks to editorial vetoes
       I cannot sing of any ketose,
    6n. Yet many a higher polysaccharide
       Has made a beauteous damsel blacker eyed.

1925, no. 3

Report to the Secretary of the Sir William Dunn Trustees for the year 1924–1925
Sir, on the upper floor the classes
Included genii and asses.

The former got out tryptophane,
The latter poured it down the drain.
The well-known author, Mr. Cole,
(We hope again to see him whole)
Besides the classes that he took
Re-wrote his admirable book,
Next door the firm of Seth and Luck
Found that ammonia comes unstuck
From carbamide in pigs and pugs
As easily as in beans and bugs.

Professor Hopkins down below, too,
Has measured the uptake of O_2
By proteins from every organ
In Barcrofts, watched by Ruse and Morgan,
The contents of our cranial domes
Were studied by the spouses Holmes,
Who find each intellectual gent. owes
His mental powers to a pentose.
Miss Robinson and R. McCance
Have made a notable advance
In dealing with tyrosinase,
And the queer laws which it obeys.
Aided by Anderson and others
Our saccharologist Carruthers
Attacked the problem of rotation
Of glucose during activation.
The indefatigable Harrison
Found out how glutathione carries on
Quite free from iron, in all sorts
Of lovely gadgets made of quartz.

And on my arm I'll keep a spot shorn
For venepuncture by Miss Watchorn.
Zoophilists need never feel
Much pity for the dogs of Hele
(Even for those whom Nature dooms
To drinking compounds made by Coombs).
Although they sometimes look dejected,
Their barking powers are unaffected.

I next pass on to Mrs. Onslow,
Whose knowledge lays all mere male dons low.
I know I'd rather meet a lion* in
My path than talk on anthocyanin.

I am afraid the Lab. has lost a
First-rate researcher in Miss Foster.

A pancreatic anti-ferment
Came quite near causing her interment;
Her ether (she was not to blame)
Exploding far from any flame.

Attached hereto there is a pastel
Portraying Dr. J. H. Quastel
Surrounded by his bugs protesting
Against the work they're given when resting.

Wooldridge and Woolf (who will not rhyme)
Assisted in this sordid crime.
Still harder were the problems set 'em
By Misses Stephenson and Whetham;
For data on how their bugs feel
Ask Timothy (not Dr. Hele).
Assisted by Miss Sylva Thurlow
(Now, I regret to say, on furlough)
Our demonstrator, Dr. Dixon,
Tried several entertaining tricks on
Catalysis of linked reactions,
A subject which the vulgar hack shuns.
Thanks to some work of Mr. Hill's
Iron is banished from our pills.
Pale people wishing to be pink
Can make fresh pigment out of zinc,
Or, if their doctors think it proper,
Iron may be replaced by copper.
   *Or leap from some queer crag or funny cliff,
   Or smell mercaptans made by Tunnicliffe.
Our Mr. F. J. W. Roughton
Has watched his haemoglobin spout on
Its colour changing as it goes
From purple to a lovely rose.
I should be worse than a barbarian,
If I omitted the Librarian;
The ways impartially she probes
Of publishers and anaerobes.

He would be an extremely bold 'un
Who'd steal a tube from Mr. Holden;
Armed with a bludgeon and a scimitar
He sits and guards the polarimeter.
What Winter got from all those goats
Is no fit subject for my notes;
Nor shall I soil these pages with
The horrid things found out by Smith,

Because Part I must never know
How glucose sends blood-sugar low,
While insulin may make it rise
To everybody's great surprise.
The work of Messrs. Kay and Irving
To me, at least, is most unnerving;
Since oxalate makes glucose leak in
To corpuscles, despite my shrieking
They want to centrifuge my veins,
And see how much the cell contains.
The victims of Havard and Reay
Ran up and down the stairs all day,
And each became and expert bleeder
Under the influence of the Reader.

I now descend below the floor,
Passing the prep. room and the Store,
To where the ever-patient Perkins
Studies the very curious workings
Of *Sacculina* which destroys
The theory that boys will be boys;
And Thomson gets out various dyes
From wings of moths and bufferflies.
I often wonder into which hell
S. Coleridge would condemn our Mitchell
For playing interesting tricks
Suggested by the fiendish Hicks
Upon the nitrogen partition
Of rats with copious imbibition.
Next I proceed to sing the Murrays;
One dealt with the Professor's worries;
The other roasted unhatched chicks
Next door to where the rats of Hicks
Ran round and round in little wheels
Between their tryptophaneless meals.
Then N.J.T. and D. M. Needham
Transfix amoebae, yes, and bleed 'em,
Besides determining their buffering,
And thus preventing human suffering.
Although this year there's little to be
Remarked about the rats or Ruby,
In next year's Lab. notes, when I write 'em in,
I hope to tell of a new vitamin (perhaps).
So though no doubt you often hate us
For asking for new apparatus,

I am quite sure you will relent
On learning how the money's spent.

1930, no. 7

Advice

He will tell you at once how long a gnat lives, how far down into the sea the sunlight penetrates, and what the soul of an oyster is like (Lucian).
If I were Donald D. Van Slyke
I should employ a motor bike
To shake my apparatus when
I estimate amino N.

If I were Phoebus A. Levene
My phosphatides would be quite clean.
With my ten fingers and my ten toes
I'd separate desoxypentose.

If I were Stanley Benedict
My methods would be very strict,
But I should often find a hole in
Those of my colleague Otto Folin.

If I were Mr. H. D. Dakin
My protocols would need no faking,
I'd demonstrate to every nation
The truth of β-oxidation.

And if I were Herr Waldschmidt-Leitz
I'd make some peptone from egg whites
And then determine all the steps in
Its degradation by erepsin.

And if I were Professor Parnas
I'd say to all my colleagues "Garn, ass
The source of tissue NH_3
Was not found out by you, but me."

And if my name were Monsieur Bourquelot
You may be sure that I should work a lot
At getting glucosides from fruits
And leaves and flowers and stems and roots.

And if I were Professor Euler
I'd boil some yeast in a large boiler
And crystals, I would then produce
Of pure cozymase from its juice.

But since, alas, I'm merely me
I only publish in B.B.

1930, no 7

Tetrapeptides

"What," said Grangousier," my little rogue, hast thou been at the pot that thou dost rime already?" (F Rabelais, md).

The reader's father
Would very much rather
Believe in the Whole
Than in chemistry or the soul.

Sir Walter Fletcher
Was not a lecher
After a vamp
Of Imogen's stamp.

When wicked men
Drained Coe Fen
The supply of frogs
Went to the dogs.

An excised muscle
Does not hustle
Though retaining quite a
Lot of *vis insita*.

Embryonic bone
Can now be grown
Without any loss for days
Of its store of phosphatase.

1931, no 8

Cautionary tale

Marcus, who criticised his Betters
Few students find a doom as dark as
That which befell the wretched Marcus,
He met with such a horrid fate
Because he criticised the great.
For though his station was so lowly
He read de Sitter and de Broglie,
And had acquired the curious knack
Needed to understand Dirac.
The READER (with extreme sagacity)
Makes no correction for fugacity
In cases where experiment
Is only right to ten per cent,
And often, in a course of lectures,
Does not refer to the conjectures
Of those who prove by calculation
That enzymes work by radiation.

When, to his horror, Marcus missed 'em,
It made his autonomic system
Work overtime. He wept, he blushed,
Adrenaline in torrents rushed
Into his blood, his glucose rose,
Great drops of sweat ran down his nose,
His hair stood up as if in fright,
His spleen contracted, and turned white,
And rising in his place he spoke:
"Sir, things have gone beyond a joke.
I cannot stand a person who is
Deaf to the dulcet tones of Lewis:
I also feel it is a scandal
When he does not refer to Randall,
And I should like to ask him why
The names of Hückel and Debye
Are rarely heard, nor yet are those
Of Fermi, Schrodinger, and Bose,
And when he says enzymes are unstable
He makes no reference to Constable.

His formula about activity
Contains no term of relativity."
Staggered by so much erudition
The READER gave up his position.
And went to edit "The Spectator".
Marcus next tried the DEMONSTRATOR.
"Sir," said he, "I must really beg
For fuller facts about the egg.
Please tell me whether phosphatase
Appears at six or seven days.
Were Iljin's linnets wild or tame,
And what was Goblev's Christian name:
Can papists eat (and not repent)
The eggs of monotremes in Lent?"
The Demonstrator groaned and shook,
He burned his sketchy little book,
And with a most profuse apology
Resigned the study of oölogy.
Soon after this the Cabinet,
Feeling the world was in his debt,
Thought the PROFESSOR should be sent
To the Upper House of Parliament,
And that without undue delay.
So on a pleasant autumn day
(It was the vigil of St Luke)
He was transformed into a duke.

The KING, the Minister of Health,
Lord Rothschild (famous for his wealth),
Jack Jones, the Bishop of Hong Kong,
Miss Amy Johnson, and the Bong
Of Bhoojumgunga all came here
To make Sir F.G.H. a peer.
Among the guests was Marcus, who
Now gave all lectures in Part II;
He did not know 'twas not the THING
To go and criticise the KING.
So when His Majesty came round,
While others bowed and made no sound
Marcus (I shudder to relate)
Said, "Sir, your tie is not quite straight."
Two Beefeaters arrested him
And placed him in a dungeon dim,
Where he was bound in chains and fetters
For disrespect unto his betters.
Soon afterwards (not without reason)
He was found guilty of high treason.
His entrails, at the execution,

Were placed in Ringer-Locke solution,
Where they continued to display
Motility throughout the day.
After evisceration, he
Was given to the M.R.C.,
And, till his heart began to fail,
Perfused by Gaddum and by Dale,
Who found that (as they always thought)
His R.Q. stayed at 1.0.

Oh readers, learn from Marcus' fate
And do not criticise the great.

Notes

1. Charlotte Haldane Franken (1894–1969) was a British feminist author. Her second husband was J.B.S. Haldane. Charlotte was born in London. Her parents were Jewish immigrants, her father, Joseph, a German fur trader. She married Jack Burghes in 1918 and they had a son, Ronnie. In 1924 she interviewed the biologist J.B.S. Haldane for the *Daily Express* and they soon became friends. She then had a scandalous divorce from her husband, before marrying Haldane in 1926. During this time, she also worked as a journalist and editor of the antifascist magazine *Woman Today*. After a wartime trip to the Soviet Union, she became disillusioned with socialism, which J.B.S. still believed in, writing about it in *Russian Newsreel*. The Haldanes separated in 1942, divorcing in 1945. J.B.S. later married Helen Spurway. Charlotte died in 1969 of pneumonia.
2. Trinity College, Cambridge, has many notable alumni, including six British prime ministers and several heads of other nations, as well as Isaac Newton, Bertrand Russell, George VI, and Ludwig Wittgenstein. The college was founded by Henry VIII in 1546. The king duly passed an Act of

Parliament that allowed him to suppress (and confiscate the property of) any college he wished. Rumor had it that the universities used their contacts to plead with his sixth wife, Catherine Parr. The queen persuaded her husband not to close them down, but to create a new college. The king did not want to use royal funds, so he instead combined several existing colleges to form Trinity.
3. J. B. S. Haldane, "An Autobiography in Brief," *The Illustrated Weekly of India* (Bombay), 1961. Reprinted in *Selected Genetic Papers of J.B.S. Haldane*, ed. K. R. Dronamraju (New York: Garland, 1990), 19–24.
4. M. Weatherall and H. Kamminga, *Dynamic Science: Biochemistry in Cambridge 1898–1949* (Cambridge, UK: Wellcome Unit for the History of Medicine, 1992).
5. Sir Edward Mellanby (8 April 1884–30 January 1955) was a British nutritionist who discovered vitamin D and the role of the vitamin in preventing rickets in 1919. He was professor of pharmacology at the University of Sheffield and consultant physician at the Royal Infirmary in that city. He served as the secretary of the Medical Research Council from 1933 to 1949. He was a fellow of the Royal Society. His publications include *Nutrition and Disease: The Interaction of Clinical and Experimental Work* (Edinburgh and London: Oliver and Boyd, 1934).
6. Weatherall and Kamminga, *Dynamic Science*, p. 37.
7. Vincent Wigglesworth, (1899–1994) was a British entomologist who made significant contributions to the field of insect physiology. His early training in biochemistry was provided by Haldane at Cambridge University. He studied metamorphosis. His most significant contribution was the discovery that neurosecretory cells in the brain of the South American kissing bug (*Rhodnius prolexus*) secrete a growth hormone that regulates the process of metamorphosis. This was the first experimental confirmation of the function of neurosecretory cells. Wigglesworth developed a coherent theory of how an insect's genome can selectively activate hormones that determine its development and morphology.
8. J. B. S. Haldane, *Enzymes* (London: Longmans, Green, 1930).
9. J. Murray Luck, "Confessions of a Biochemist," *Annual Reviews of Biochemistry* (1981), 50, 1–22. J. Murray Luck (1899–1989) was professor emeritus of chemistry at Stanford University and founder of *Annual Reviews*. He published more than two hundred scientific papers. He studied biochemistry at Cambridge University in England in the laboratories of Sir Gowland Hopkins and J. B. S. Haldane. Luck earned his PhD in 1925 with a thesis titled "The Origin of Blood Ammonia." He was a member of the Chemistry Department of Stanford University from 1926 until his mandatory retirement in 1965. However, he continued to do research at Stanford as professor emeritus of chemistry until his death in 1993. Luck's specialty was in the role of proteins in carcinogenesis. He was also a world authority on histones. Luck is probably best known for his work with the Annual Reviews. He launched the Annual Reviews of Biochemistry in 1932 and founded Annual Reviews Inc., a nonprofit enterprise devoted to the advancement of science through the publication of critical reviews and analyses of the rapidly expanding volume of scientific research literature. Luck served as editor-in-chief and secretary-treasurer of Annual Reviews Inc. from 1939 to 1969 and then was on the board of directors. In 1976, at the age of seventy-six, he founded the Society for the Promotion of Science and Scholarship, a nonprofit corporation for scholarly publishing, with special interests in British and European studies.
10. Joseph Needham, personal communication.
11. R. B. Fisher, letter, November 9, 1983, Communicated to the author by J. M. Mitchison (Haldane's nephew).
12. Ronald Draper, *The Flying Draper* (London, Fisher, 1924).
13. J. B. S. Haldane, *Daedalus, or Science and the Future* (London, Kegan Paul, 1923), 77.
14. Judith Adamson, *Charlotte Haldane: Woman Writer in a Man's World* (New York: Macmillan, 1998), 39.
15. Charlotte Haldane, BBC interview, London, September 11, 1965.
16. Haldane had a long record of friction with the press. He generally regarded the press reports as sloppy, superficial, and irresponsible. Quite often, he found that his speeches and scientific statements were misrepresented in the popular press. On the other hand, he was especially talented at attracting the attention of the daily press by getting into controversies or by making witty, pungent, and controversial pronouncements. Although complaining frequently about the inaccuracies in the press reports, Haldane enjoyed the publicity generated by his

conflicts with the press. Of course, his stature as a major scientist and his articles in the popular press also contributed to his fame in the popular press.
17. Charlotte Haldane, *Truth Will Out* (New York: Vanguard Press, 1950), 20.
18. C. Haldane, *Truth Will Out*.
19. C. Haldane, *Truth Will Out*.
20. C. Haldane, *Truth Will Out*, 20.
21. C. Haldane, *Truth Will Out*.
22. The term "Cockney" has both geographical and linguistic associations. Geographically and culturally, it often refers to working-class Londoners, particularly those in the East End. Linguistically, it refers to the form of English spoken by this group. Cockney is characterized by its own special vocabulary and usage, and traditionally by its own development of "rhyming slang." It is generally agreed that to be a true Cockney, a person has to be born within hearing distance of the bells of St. Mary le Bow, Cheapside, in the City of London. This traditional working-class accent of the region is also associated with other suburbs in the eastern section of the city, such as the East End, Stepney, Hackney, Shoreditch Poplar, and Bow. The Cockney accent is generally considered one of the broadest of the British accents and is heavily stigmatized. It is considered to epitomize the working-class accents of Londoners and, in its more diluted form, of other areas. The area and its colorful characters and accents have often become the foundation for British "soap operas." One of the most popular soaps set in this region is the BBC's *East Enders*; the characters' accents and lives within this television program provide wonderful opportunities for observers of language and culture. It is surprising that Charlotte's Cockney accent bothered Haldane. Later on, when he embraced Marxism, he would have welcomed it, or even bragged about it, as a sign of his close association with the proletariat.
23. William Bateson (1861–1926) was an English biologist who coined the term "genetics" in 1906 for the new science of inheritance. Bateson introduced several other terms in genetics that are commonly used today, including "zygote," for the individual that develops from the fertilized egg; "homozygote" and "heterozygote"; and "F1" and "F2" for first and second filial generations following hybrid crosses, as well as the term "allelomorph" (later changed to "allele"). In his book *Materials for the Study of Variation* (1894), Bateson showed that biological variation exists both continuously, for some characters, and discontinuously for others.
24. J. B. S. Haldane, Personal communication, 1959.
25. *Sex viri*, was the term for a panel of minor judges constituted in ancient Rome. It was recreated in recent times and was used by Cambridge University to evaluate whether Haldane had violated the statutes of the university by indulging in immoral behavior such as adultery, which was serious enough to warrant dismissal from his position at the university.
26. C. Haldane, *Truth Will Out*, 21.
27. C. Haldane, *Truth Will Out*, 21.
28. C. Haldane, *Truth Will Out*, 31–32.
29. C. Haldane, *Truth Will Out*, 37.
30. Adamson, *Charlotte Haldane*, 25. Charlotte looked forward to the spacious grounds to start her own floral and vegetable gardens. She wrote: "I had a 'green thumb' but the heavy work of mowing the long, beautiful lawn, of digging and trenching and planting bush fruit and vegetables was too much for me. When we first went to Roebuck House, the old gardener, who was nearly eighty, and his wife, lived in the cottage attached to it. Apart from mowing the lawn in Summer, and pottering around for the rest of the time, he had been, for several years past, merely acting as caretaker. My horticultural ambitions terrified the poor old man, but he had an excellent come-back. The first time I asked him to dig the vegetable garden, he literally threw a fit. It was my turn to be terrified! I discovered, however, that he suffered from petit mal, a mild form of epilepsy. So I pensioned him off." C. Haldane, *Truth Will Out*, 26.
31. C. Haldane, *Truth Will Out*, 27.
32. Adamson, *Charlotte Haldane*.
33. Adamson, *Charlotte Haldane*, 69.
34. Dora Russell, *The Tamarisk Tree* (London, Virago, 1985), 182.
35. Adamson, *Charlotte Haldane*, 69.
36. N. W. Pirie, Letter to the Editor, *The Listener* (BBC, London, 1969).

37. Adamson, *Charlotte Haldane*, 197; Charlotte Haldane, letter to BBC's *The Listener*, January 30, 1969.
38. Adamson, *Charlotte Haldane*.
39. Juliette Huxley, *Leaves of the Tulip Tree: Autobiography* (London: Murray, 1986).
40. C. Haldane, *Truth Will Out*, 68.
41. J. B. S. Haldane, Foreword, *Crime and Destiny* (Allen & Unwin, 1931) (by the German authority Professor J. Lange, titled *Verbrechen als Schicksal*.)
42. K. R. Dronamraju (ed.), *Haldane and Modern Biology* (Baltimore: Johns Hopkins University Press), 313.
43. Professor Nikolai Ivanovitch Vavilov (1887–1943) was a botanist, plant breeder, geographer, and geneticist. He received his early training in genetics under William Bateson at Cambridge University. During World War I, he began the second line of his life's work: the exploration of cultivated plants. In 1916 he visited Persia and surrounding countries collecting cereals, the systematic relationships of which he had already studied experimentally. In 1917 he became a professor of agriculture, botany, and genetics at Saratov University, southeast of Moscow, and built up an outstanding research department. Here in the 1920s Vavilov developed his law of homologous series—the idea that related species develop similar variations. In 1921 he was appointed as head the Branch of Applied Botany at St. Petersburg, where Vavilov built his outstanding All-Union Institute of Plant Breeding, an international center in plant selection and genetics. Vavilov set up more than four hundred research institutes, the combined staff totaling twenty thousand by 1934. His journal, the *Bulletin of Applied Botany, Genetics, and Plant Breeding*, became a leading international publication in its field. He was greatly interested in the geographic distribution and variability of plants, and his immense collections were to be the raw materials for making new crops for various regions. Vavilov's collections formed the basis for new theories on the origin of cultivated plants, and laid the foundation for the future improvement of crop plants and of one of the largest and oldest seed banks in the world today. His potato collection led to the establishment of the British Empire Potato Collection, on which potato breeding in Britain was later based. Vavilov was founder of the Academy of Agricultural Sciences, 1929, and its first president, 1929–1935; head, 1930–1940, of the Genetics Laboratory, later Institute of Genetics, of the Academy of Sciences of USSR; and an active member of the Geographical Society of USSR, serving as president, 1931–1940. However, with the rise of the pseudoscientist Trofim Lysenko, who was supported by Stalin, Vavilov lost all his positions. He was officially denounced, and Mendelian genetics itself was banned in the Soviet Union. Vavilov disappeared by 1940 and was reported to have died of starvation and ill health in a concentration camp in Siberia in 1943.
44. Charlotte, *Truth Will Out*, on their travels.
45. Radio broadcast, later published in *The Listener*, December 9, 1948.
46. C. Haldane, *Truth Will Out*, 41.
47. C. Haldane, *Truth Will Out*, 51.
48. C. Haldane, *Truth Will Out*.
49. C. Haldane, *Truth Will Out*, 54.
50. Adamson, *Charlotte Haldane*, 91.
51. Adamson, *Charlotte Haldane*, 71.
52. C. Haldane, *Truth Will Out*.
53. J. B. S. Haldane, "On Being Finite," *Rationalist Annual* (1965): 203.
54. Aldous Huxley, *Antic Hay* (New York: Carroll & Graf, 1990).

3
Eugenics and Predictions

> Eugenic organizations rarely include a demand for peace in their programmes, in spite of the fact that modern war leads to the destruction of the fittest members of both sides engaged in it.
>
> J. B. S. Haldane

Haldane enjoyed speculating various possibilities for human evolution and methods of human intervention to improve the genetic quality of human life. He was well known for his bold and daring speculations regarding such possibilities. He was optimistic in 1932, when he discussed the subject of eugenic applications in his book *The Inequality of Man and Other Essays*:

> Eugenics is at present the only possible way of improving the innate characters of man ... if innate human diversity is an ineradicable fact, the ideal society is one in which as many types as possible can develop in accordance with their possibilities. So far every society has tended to idealize one particular type. We cannot expect nature to start improving our innate abilities once more. The usual fate of a species in the past has not been progress, but extermination, very often after deteriorating slowly through long periods.... There is no reason to suppose that man will escape it unless he makes an effort to do so.[1]

Daedalus, or Science and the Future

Haldane's first book stands out for its originality, daring, and scientific predictions. *Daedalus, or Science and the Future*[2] was a slim octavo volume that was an instant success. It was a slightly extended version of his address to the Heretics Society at Cambridge on February 4, 1923. It was first published by Kegan Paul in London in 1923, and a year later by E.P. Dutton in New York. A year later fifteen thousand copies had been sold.

Haldane's eugenic ideas began to take shape while he was still an Oxford undergraduate in 1912. He wrote an essay on the future applications of science, but did not pursue it further until he was invited nine years later by the New College Essay Society in Oxford. He updated it, incorporating later developments in science as well as his own wartime experiences including the new and deadly chemical warfare. He revised it once again in 1923, when he was invited to address the Heretics Society at Cambridge University. As good luck would have it, his audience included

Charles Kay Ogden (1889–1957), the well-known polymath and English linguist, philosopher, and writer, and the inventor and propagator of *Basic English*. Ogden was a founder of the Heretics Society in Cambridge in 1909, which questioned traditional authorities in general and religious dogmas in particular. The Society was nonconformist and open to women. Previous speakers included George Bernard Shaw and G. K. Chesterton. On February 4, 1923, Haldane's lecture to the Society was a speculative and generally optimistic vision of future scientific developments. It spurred in 1924 a less optimistic response from Bertrand Russell, *Icarus, or the Future of Science*. Other speakers who followed Haldane included Virginia Woolf and Ludwig Wittgenstein. Ogden built up a position as editor for Kegan Paul, publishers in London. *Daedalus, or Science and the Future*, was the first of the Today and Tomorrow series.

Haldane's address to the Heretics was a thrilling and daring presentation of what might be forthcoming in science and how it might affect the future human society in health, social, and cultural traditions such as marriage and reproduction, religion, and political institutions. Haldane wrote speculative essays on science and eugenics repeatedly throughout his life, but the passion and bold imagination that he displayed in *Daedalus* was never to be seen again. Haldane wrote in his introduction to *Daedalus*, "It is the whole business of a university teacher to induce people to think." Haldane achieved that goal admirably through his writings, lectures, personal contacts, and, above all, the example of his own life and work, which many others have emulated.

Haldane's criticism of the war, indeed any war, was implicit in his opening lines of *Daedalus*:

> As I sit down to write these pages I can see before me two scenes from my experience of the late war. The first is a glimpse of a forgotten battle of 1915. It has a curious suggestion of a rather bad cinema film. Through a blur of dust and fumes there appear, quite suddenly, great black and yellow masses of smoke which seem to be tearing up the surface of the earth and distintegrating the works of man with an almost visible hatred. These form the chief parts of the picture, but somewhere in the middle distance one can see a few irrelevant looking human figures, and soon there are fewer. It is hard to believe that these are the protagonists in the battle. One would rather choose those huge substantive oily black masses which are so much more conspicuous, and suppose that the men are in reality their servants, and playing an inglorious, subordinate, and fatal part in the combat. It is possible, after all, that this view is correct. (pp. 23–24)

Of great interest were Haldane's biological predictions, especially revolutionary advances in molecular biology and reproductive biology that came true in the 1970s and beyond. His prediction of genetic manipulation and in vitro fertilization (ectogenesis),[3] resulting in the mass production of "test-tube" babies, deeply influenced several scientists and writers. One of them was his friend Aldous Huxley, who copied these ideas in his *Brave New World*.

Daedalus also contained Haldane's predictions of a number of inventions, such as new sources of energy, novel methods of food production, and widespread use of psychotropic drugs, among others. Most importantly, *Daedalus* raised ethical and moral concerns resulting from the application of an everchanging science in a society that was becoming increasingly science oriented in early twentieth century. At least one commentator, the physicist Freeman Dyson, observed many years later, "My copy of Daedalus once belonged to Einstein. ... He evidently grasped at once the main message of Haldane's book, the message that the progress of science is destined to bring enormous confusion and misery to mankind unless it is accompanied by progress in ethics."[4]

Controversy

Haldane first tasted fame with *Daedalus*. Ironically, a few years earlier, when his friend Julian Huxley's axolotl experiments created sensation in the popular press, Haldane warned him that he would lose credibility as a serious scientist! Now, with *Daedalus*, it was Haldane's turn. He loved controversy, and *Daedalus* was perfect for that purpose. *Daedalus* attracted controversy right from the beginning. It dealt with revolutionary biological intervention and the production of "test-tube" babies at a time when even the mention of "birth control" in the public media caused an uproar. As late as 1926, Julian Huxley was rebuked by the head of the British Broadcasting Corporation (BBC), Lord Reith,[5] for mentioning birth control on BBC radio.[6] When the first birth control clinic in North America was opened in Brooklyn by Margaret Sanger[7] in 1916, it became a center of much controversy and conflict, and a similar fate awaited another birth control clinic that was opened by Marie Stopes[8] in London shortly afterward. Haldane's *Daedalus* appeared in the midst of these sweeping social changes.

Daedalus quickly became the staple subject for gossip in the college halls and the tea rooms around Oxford. It soon gained notoriety as the butt of donnish jokes and ridicule in an Oxford for which it seemed to have been tailor-made.

The notoriety stirred up by Haldane's biological predictions caused much anguish to his parents. In his *Memories*, Julian Huxley (1970) quoted the following letter from Mrs. Louisa Kathleen Haldane[9] (Haldane's mother):

Cherwell, Oxford
Dear Julian:

I find the SP [she always called her husband the Senior Partner] is frightfully upset about *Daedalus*. Will you abstain altogether from poking fun at him on account of it? And if you can do so, keep people off the subject altogether when he is about?

I knew he'd object, but had no idea till to-day how really unhappy he is—
odd people these Liberals and no accounting for them!

But an' you love me, keep people off him, or he'll hate you all! (which is not only sad for him but extremely inconvenient for me.)

Yours aff:
LKH.

Unlike his father, JBS immensely enjoyed the sensation it caused and took great pleasure in expounding his views loudly at every opportunity, quickly clearing tea shops of the more susceptible churchgoers. His great skill at scientific popularization quickly became evident, establishing him as one of the most eminent (and most sought after) scientist-writers for the rest of his life. *Daedalus* was soon followed by a number of popular essays, which were later published in the book *Possible Worlds and Other Essays* (1927). This collection contains one of Haldane's most famous essays, "On Being the Right Size," which Arthur C. Clarke described as a "perfect example of his lucidity and the breadth of his interests." It was also in 1927 that a text book, *Animal Biology*, was published with Haldane and Julian Huxley as coauthors. Both Haldane and Huxley continued to popularize science throughout their lives, contributing numerous articles to the popular press in several countries.

In response to Haldane's *Daedalus*, Bertrand Russell wrote *Icarus, or The Future of Science* in the following year. Russell disagreed with the benevolent view of science presented by Haldane, arguing that science is not a substitute for virtue and reason. According to Russell, technological applications make the world less safe, not more secure.

Lord Birkenhead

It was typical of Haldane that he attracted (or sought after[10]) controversy for much of his life. A few years after the publication of *Daedalus*, he was deeply involved in a major controversy with the aging F. E. Smith, first Lord Birkenhead[11] and former lord chancellor and secretary of state for India. In 1930, Birkenhead's book *The World in 2030* appeared, and might have passed almost unnoticed except for the fact that it happened to reach Haldane, who reviewed the book in the *Weekend Review* under the heading "Lord Birkenhead Improves His Mind." Haldane commented, "Certain of the phrases seemed unduly familiar. Where had I seen them before? Finally I solved the mystery. They were my own." Haldane stated that he had noted a total of forty similarities between Birkenhead's book and *Daedalus*. Haldane concluded "I will not say ... Plagiarism, but ... a certain lack of originality ... because it carries with it corollaries which I find unthinkable." Haldane even suggested a tongue-in-cheek explanation, that both he and Birkenhead must have seen the same original—including one written 48 hours hence!

A wiser man would have let the matter drop at that stage because Haldane contemplated no legal recourse, but not Lord Birkenhead. Instead, he responded by attacking Haldane in the columns of one of the most widely circulated newspapers, the *Daily Express*, which naturally led to the sudden popularity of *Daedalus* among the general public, who would otherwise have remained blissfully ignorant of its existence. Birkenhead claimed that Haldane lacked historical knowledge and that he (Birkenhead) was following in other men's footsteps. Haldane retorted (in the *Weekend Review*) that his knowledge of history had earned him a first in *literae humaniores* at Oxford. He noted further that he had no objection to Birkenhead following in other men's footsteps but that "I object to ... stealing my boots to do so, and I am amused when they do not know how to put the boots on."

That was the end of the controversy, one of many in Haldane's life. He enjoyed them all.

Years Preceding *Daedalus*

In 1901, when Haldane was only eight years old, his father took him to a Royal Society conversazione, where Haldane heard A. D. Darbishire lecture on the recently rediscovered work of the Austrian monk Gregor Mendel (Haldane 1960, personal communication). That particular encounter with the beginnings of genetics struck an intellectual chord in the young Haldane's mind. It provided him an alternative intellectual interest and excitement.

At first, he made important contributions to the physiological experiments of his father. Many years later, in 1963, writing about the scientific method, Haldane stated, "A piece of research directed by a good scientist should leave one with high standards of accuracy and integrity which one can transfer to other fields of science.... I think that for most of us an occasional change is desirable because we are apt to think that the topics which, very rightly, excited us in our twenties, are still the most important."

Beginning in 1912, Haldane published fifteen papers before *Daedalus*. They were mostly in human respiratory physiology, especially the physiological effects of raising or lowering the blood pH. His physiological papers continued the kind of research that his father followed for many years. But the two genetic studies, published in 1914 and 1919, were clearly much more original and important: first, a discovery of linkage in the vertebrates; second, the development of the first mapping function. So it is obvious that his heart (and talent) lay in genetics, not in physiology.

Shortly after the publication of *Daedalus* in 1923, the first paper in his great series The Mathematical Theory of Natural Selection was published. It was one of the founding papers of population genetics. Although continuing to tread the path laid down by his father in physiology, Haldane was keenly interested in making a mark of his own independently in a different branch of science, and the young science of genetics provided that opportunity. This was perhaps his way of establishing his own identity as a scientist, separate from his father, which was to be expected in a young scientist, especially one with a famous scientist-father.

Daedalus was Haldane's first publication of his eugenic views. It is of interest that he chose to do so as soon as he moved to Cambridge, away from New College, Oxford, where he was a Fellow in physiology, still deeply under his father's influence. While his early encounters with Mendelian genetics and appreciation of the fundamental nature of the biochemical basis of gene action may have led to eugenic speculations, the social and political milieux of those years may also have had a decided impact on Haldane.

Sociopolitical Climate

The timing of Haldane's *Daedalus* is of interest. What prompted Haldane to write *Daedalus* at that particular time? Until then his publications were what one would normally expect from a university teacher, papers in physiology and genetics published in traditional scientific journals. The answer may be found in the sociopolitical climate of that period.

Haldane's *Daedalus* appeared amid a series of rebellious cultural, scientific, and literary events of the postwar period. The suffragette movement was in full swing both in London and New York. The years following the end of World War I saw great social changes. Henry Ford ushered in the mass production of the Model T. War and major disasters have always speeded up change dramatically and created acceptance of cultural changes once unthinkable. There was a new generation of young people who had seen the horrors of war firsthand. Like Haldane, young men had seen the reality of trench warfare. Women who had acted as nurses had seen terrible injuries and mental suffering. Ironically, out of the war came medical progress in the form of cosmetic and plastic surgery. There grew a general feeling that life was short and should be enjoyed. The war also started the process of breaking down longstanding class barriers. It was subtle at first, but the erosion had begun.

The position of women in Britain and America was changing. In 1918, after the war ended, women over thirty were given the vote if they were householders. By 1928 all women over twenty-one were given the vote. Slowly women were breaking down old attitudes. The war had given ordinary working women an alternative to domestic employment. They found they liked working on the land, in factories, and on buses.

In 1921 Marie Stopes, a distinguished paleobotanist, opened Britain's first birth control clinic and, with the aid of her second husband, Humphrey Roe, she went on to found an entire chain of clinics with chapters in Australia, New Zealand, and South Africa. She was a passionate promoter of women's rights and women's sexual pleasure and a staunch supporter of eugenics. Ironically, her marriage to the botanist Reginald Ruggles Gates was annulled because of her husband's impotence.

In the United States, Margaret Sanger founded the American Birth Control League (ABCL) in 1921. Her first birth control clinic in Brooklyn, which she founded in 1916, was closed by the police.

The tenor of those times can be readily seen in the events and the cast of characters surrounding Haldane during the years preceding and following World War I. Haldane and his sister Naomi (Lady Mitchison) were inevitably drawn into wider intellectual circles, starting with their close friends the Huxleys—Julian and Aldous. Naomi wrote and produced plays. In her contribution to my book *Haldane and Modern Biology*,[12] she wrote, "I dragged them all into acting, for at that time I was forever writing plays, first *Saunes Bairos* about an imaginary country in the Andes where eugenics was highly organized by the priesthood." The topic of eugenics received special attention in Haldane's *Daedalus*:

> The eugenic official, a compound, it would appear, of the policeman, the priest, and the procurer, is to hale us off at suitable intervals to the local temple of Venus Genetrix with a partner chosen, one gathers, by something of the nature of a glorified medical board. To this prophecy I should reply that it proceeds from a type of mind as lacking in originality as in knowledge of human nature. . . . It is moreover likely, as we shall see, that the ends proposed by the eugenist will be attained in a very different manner. (p. 35)

It is clear that the eugenic ideas that Haldane expressed in *Daedalus* were very much a part of their social and intellectual culture during the preceding several years. In her essay, Naomi recorded that her brother was interested in genetics while still at Eton (1905–1911).

Lady Ottoline Morrell and Garsington

In their social circles, intellectual boundaries were blurred—pursuit of science, literature, and drama as well as numerous other activities brought them all together. Naomi was close to the Gielguds as well—both Lewis and John. Through Aldous and Julian, the Haldanes were drawn closer to the intellectual centers of Bloomsbury, especially the left-leaning group that found a generous hostess in Lady Ottoline Morrell[13]—married to the politician Philip Morrell, but well known to have many lovers, among them Bertrand Russell, D. H. Lawrence, and Julian Huxley.

Ottoline's biographer Miranda Seymour commented that no English country house of the war years is more famous than the manor house in Garsington near Oxford as it blossomed under the occupancy of Ottoline and her husband Philip from 1915 onward. Neither of them inherited a great deal of money, but they were generous to a fault in entertaining some of the best minds of their time. Ottoline entertained at the manor house, complete with Italianate gardens that she designed herself. She was a superb hostess to numerous intellectuals, many of whom subsequently enjoyed success and fame: the writer Virginia Woolf (who took pleasure in mocking Ottoline behind her back); the philosopher Bertrand Russell; the poet T. S. Eliot; the journalist and literary critic Desmond MacCarthy; the writer Lytton Strachey; the artist Duncan Grant; the economist John Maynard Keynes; the biologist Julian Huxley and his brother, the writer Aldous Huxley, who was a close friend of Haldane's sister Naomi; Vanessa and Clive Bell; Margot Asquith; Katherine Mansfield; the poet W. B. Yeats; David Cecil; Eddy Sackville-West; Maurice Bowra; the artists Wyndham Lewis and Augustus John; and a young au pair from Belgium, Juliette Bailott (who later married Julian Huxley), who looked after Ottoline's daughter, Julian.

Although Haldane was not a member of that group, the literary and artistic mores of *Daedalus* were no doubt influenced by the intellectual activities and climate of that period. Social emancipation, especially the suffragette movement and sexual liberation, were spearheaded by that group. This change was, in part, due to a rejection of the hypocrisy and guilt associated with sex that was characteristic of the preceding era. Writing of his own experience, Julian Huxley commented, "This battle between sexual attraction and a puritanical sense of guilt. . . . The whole climate of the Edwardian age, with its hypocritical suppression of everything 'nasty', fostered this conflict between instinct and reason. My schooldays, with their smutty stories and my concern over masturbation, had induced an underlying sense of moral guilt about sex which took me years to outgrow."[14]

Lady Ottoline's close friendship with D. H. Lawrence was a source of much irritation to his German wife, Frieda. Adding insult to injury, Ottoline told everyone that the marriage of Lawrence and Frieda was a mistake. Controversy was very much a hallmark of that group. Lawrence's book *The Rainbow* was published in 1915. One of the characters ridicules a soldier's uniform, while the author glorifies lesbian behavior. And it was dedicated to Lawrence's German sister-in-law. These facts were not congenial to the wartime climate. The book was widely attacked in the press as a menace to the war effort against Germany. The director of public prosecutions ordered all remaining copies of the book destroyed. The presiding magistrate, who lost his own son on the war front, had little sympathy for Lawrence. The publisher of *The Rainbow*, Algernon Methuen, was fined ten guineas and was ordered to withdraw the novel.

Haldane's sister, Naomi Mitchison,[15] stated that there was a widespread feeling of rebellion among the young, who were mostly supportive of Lawrence and the right of freedom of expression. She was among those present at the trial of Lawrence and his publisher.

Some years later, a similar fate awaited Lawrence's *Lady Chatterley's Lover*,[16] which was based on his affair with Ottoline. Another of Ottoline's lovers, Bertrand Russell, was a pacifist who went to prison for refusing to serve in the war. On the scientific side, Julian Huxley, too, stirred up notoriety and controversy when he discovered that the metamorphosis of the axolotls, an amphibian that retains the gills and the broad swimming fin of its tadpole state lifelong, could be speeded up by feeding them on thyroid gland. Huxley's brief note, which appeared in *Nature* in 1920, immediately became the subject for headlines in the popular press. One headline read: "Young British Scientist Finds Elixir of Life."

Daedalus was another landmark in a series of sensational events and publications that emanated from the rebellious left-wing intellectual elite during the pre– and post–World War I years. From the establishment's point of view, they were no doubt outrageously irritating incidents. Clearly there was a connection between the wartime experiences and several publications of that period. *Daedalus* opens with a description of the battle scene from World War I.

Haldane's *Daedalus* served several purposes: His predictions of genetic engineering and mass production of "test-tube" babies have had a significant impact on scientists and writers alike;[17] Haldane's discussion served as a warning of what was yet to come in the future and of the need to be prepared for the ethical dilemmas and challenges that will be confronting humanity; *Daedalus* was Haldane's attempt to shock the establishment, which undoubtedly delighted him; and through his predictions, Haldane was suggesting lines of research that would be desirable.

Haldane emphasized that a eugenically minded government would consider proper feeding as an important aspect of its administration. No eugenic measure could have much effect if an inadequate diet caused physical or mental disabilities. Another fact of social inequality is an educational system that does not recognize the large innate inequalities of intelligence. An educational system designed by a biologist would recognize that inequality. An ideal society based on biological knowledge would recognize that large innate differences exist and that we must take both genetics and environment into consideration in matters of eugenic planning.

Figure 3.1 Aldous Huxley was a close friend of JBS and his sister Naomi (Nou) at Oxford.

Sterilization

Haldane wrote on eugenics and related issues throughout his life, adding new ideas and ethical consequences in successive papers.[18] Indeed, with the growth of knowledge of human genetics, he became more and more skeptical of any immediate benefits of eugenic applications. In fact, he was averse to using the term "eugenics" because of its association with the racist policies of Nazi Germany before and during the World War II.

As a mathematical biologist, Haldane was well aware of the slow rate of change that would be brought on by a eugenic program. In this respect, Haldane's position was quite different from that of Julian Huxley and Herman J. Muller, who were more enthusiastic about the prospects of eugenics. Furthermore, Haldane noted, "Eugenic organizations rarely include a demand for peace in their programmes, in spite of the fact that modern war leads to the destruction of the fittest members of both sides engaged in it."[19] From about 1930 onward, a small group of pioneers, including Haldane, developed the methodology of formal human genetics, establishing its foundations on a firm scientific footing.[20] Haldane profoundly disagreed with the views of those who were recommending wholesale sterilization of mentally defective individuals, petty criminals, and the chronic unemployed, the so-called social problem group.[21]

Even when he seemed to have a common interest initially with a correspondent regarding eugenic methods, Haldane was quick to withdraw support when sterilization was mentioned as one of the means of achieving a eugenically improved society. Such was the case with his exchange with a certain Ethna Donnelly in Los Angeles. At first, he was supportive of what he thought was a project on ectogenesis, but when Donnelly mentioned "sterilization of all below a certain stage of mental and physical fitness," he quickly backed off, saying, "I am in complete disagreement with your views on sterilization, which, in my opinion, are to quote your own words, 'actuated by cruel impulses or goaded by fear' which are almost certainly based on ignorance of the facts. If, therefore, you are going to hook-up your project with that kind of propaganda, you can count me out."[22]

During the heyday of his Marxist association, Haldane viewed earlier eugenic programs in terms of a class struggle, as sterilization and other eugenic measures proposed would impact disproportionately on the economically disadvantaged. An unexpected consequence of the rigor of methodology of human genetics, which was established by the mathematical analyses of Haldane, Hogben, and others, was that it led to the destruction of class-bound eugenics.[23]

Address to the Eleventh International Congress of Human Genetics, 1963

In 1963, Haldane was invited to address the Eleventh International Congress of Human Genetics, held in The Hague in the Netherlands.[24] He began by explaining that his connection with the Netherlands goes back to the seventeenth century. In 1620 his ancestors in the direct male line, John Haldane and his brother James, went to the Netherlands to fight for the United Netherlands. James was killed at the siege of 's-Hertogenbosch in 1629, and John returned to Scotland and later became the Scottish representative with Cromwell's army during the civil war in England. However, Haldane claimed, with his tongue firmly in cheek, that he may have some unknown cousins in the Netherlands! Haldane claimed that he possessed about 1/256th of John's autosomal genes and may have a fairly good copy of his Y chromosome as well. It was typical of Haldane to start a lecture on human genetics with a comment about his own genes.

Haldane pointed out that, for a variety of reasons, men and women find it harder to accept the implications of genetics than those of other scientific disciplines. First, genetic findings may conflict with or reinforce two powerful emotions, the sexual and parental. Second, they may conflict with or reinforce theories that support the class structure of human societies and the divisions between societies, or alternatively with theories that regard class and national divisions as evil. The findings of genetics may conflict with religious doctrines as well, but it is not as troublesome as the conflicts brought on by the studies of astronomy and evolution. Haldane emphasized that it is impossible to speak on the applications of human genetics without offending someone but he would hope to give it impartially.

Haldane pointed out that the earliest attempts to apply genetic ideas to human societies are found in ancient Indian writings, some of which attempted to justify the hereditary caste system. Plato, in his Republic, formalized the ideal of a hereditary ruling class, with occasional rejection of unfit members and the inclusion of superior ones. His idea was based on the successful breeding of dogs to improve their stock. However, historical examples tell us that hereditary aristocracies do not last very long.

The important point that Haldane emphasized is that generalizations based on race are usually misleading and untrue. For instance, when we say that the ancient Greeks were great mathematicians we are thinking of about twenty men, but we know nothing about the mathematical ability of the average Greek. Genetic bases of racial comparisons are not meaningful, because each race may be superior to others in its own environment with respect to a certain trait. Socially important innate characteristics can be reversed by changed circumstances. Haldane cited the example of Norwegians and Danes, whose ancestors (Vikings) thought it disgraceful to die otherwise than in battle, but today they are most peaceful people, and have, in fact, low murder rates.

Race Crossing, Letter to Tom Driberg, MP

On the subject of the desirability of race crossing, especially unions between whites and negroes, Haldane maintained an open mind, saying that the available data (in 1942) were inadequate to justify any conclusions. He was responding to an inquiry from Tom Driberg,[25] labor Member of the Parliament.

Prospects for Eugenics

Haldane had little patience with many eugenic programs. One of his objections is that they did not take into account the extreme slowness of the evolutionary process. He wrote, "The development of population genetics has begun to show us why evolution is so vastly slower a process than any of Darwin's contemporaries believed. So far from thinking that human or other animal populations are evolving at an observable speed, it is often justifiable to think of them as being in equilibrium. Natural selection, based on genetical differences, is certainly occurring in human populations, but it is very nearly balanced by mutation, segregation, and migration." He pointed out further that no eugenic program (short of the prevention of mutation) would reduce the frequency of rare dominant and sex-linked recessive abnormalities in human populations that are kept in being by mutation. On the other hand, developing tests for heterozygosity in the female relatives of hemophiliacs would reduce their frequency but it would not abolish the disease.

Haldane stated that Muller and some others who agree with him believed that there is a similar equilibrium for most human recessive genes that, in heterozygous condition, lower longevity and fertility. However, it is certain that at some loci several alleles remain in being because heterozygotes are biologically fitter than either homozygote. Haldane's calculations showed that gene frequencies can be stabilized in more situations than is generally realized.

Haldane believed that it was desirable to prevent the production of the most socially inefficient 5 percent or so of most populations. There are two quite distinct methods of negative eugenics: (1) First, persons with well-marked dominant and sex-linked recessive defects should not have children; nor should the daughters of hemophiliacs and others with similar sublethal sex-linked conditions. Eugenists should encourage such people to intermarry, after thorough training in the use of contraceptives, or voluntary sterilization of one of the partners. Haldane wrote, "Eugenic organizations could do good work by introducing pairs of people for both of whom fertile marriage was contra-indicated."[26] (2) Second, whole populations should be screened for rare and harmful autosomal recessive genes. Although there is no hope of eliminating the "load" we carry for many generations, it should be possible to discourage marriages between persons heterozygous at the same locus. If, for example, 2 percent of all babies are homozygous for an unwanted recessive, this would only mean that 8 percent of intended marriages would be contraindicated. Some people have argued that such a policy would lead to increased frequency of undesirable recessive genes, which are now being eliminated because the homozygous recessives are inviable or sterile. However, the increase, which would be due to mutation, would be very slow.

Haldane agreed with Muller that, as long as artificial insemination is practiced, genetically superior donors should be chosen. However, he was rather skeptical concerning whether such methods would greatly improve the genetic condition of our species. One problem is that we do not know the genetic determination of several desirable characters, nor how rapidly genetic homeostasis would prevent their realization.

At a symposium in Princeton in 1948, Haldane commented on the method suggested by Muller for eugenic improvement: "Dr. H. J. Muller has suggested a method for the radical improvement of the human race, involving the widespread use of artificial insemination. I guess that if I were made eugenic world dictator I should have one chance in a hundred of choosing the right path. Dr. Muller is ten times as good a geneticist as I, so he might have one chance in ten, but not, I think, much more."[27]

Haldane further wrote, "We do not at present know how to recognize any but the crudest forms of innate human inequality. Our school examinations, and to a large extent our university examinations, are largely tests for precocity and memorization rather than for more important qualities."

Among other predictions, Haldane suggested that organ grafting holds great promise. He suggested two other possible applications of human genetics. First, *clonal reproduction*: This involves the culture of one or more cells from an adult so as to produce another individual of the same genotype. Second, *control of development*: Since genes act by initiating biochemical processes, in the absence of a gene it should be possible to produce its effects. Haldane wrote that it is conceivable that in future our descendants may discover nongenetic methods of controlling human ontogeny other than the very crude methods of intellectual and physical training. "If a way can be found, by prenatal or postnatal treatment, to convert what would otherwise have developed into an ordinary person into a moral or intellectual genius, there may be no need for further human biological evolution."[28]

In his Norman Lockyer lecture, "Human Biology and Politics," Haldane[29] stated that although he would like to see a State medical service that would benefit the poor, the middle-class patient would be better off with a capitalist type of medical organization. In the same address, Haldane suggested preventive eugenic measures that could be implemented on the basis of the then newly acquired knowledge of human genetics, considering both dominant and recessive conditions. Haldane emphasized the multiplicity of causes for various types of congenital defects, including mental defects, and warned of the dangers of irresponsible sterilization of whole categories of people as it would open the door for government intervention with serious consequences.

Haldane then tackled the often-posed question of whether the mean innate intelligence of the population is declining: "Men and women born into one economic class are constantly passing into a richer one if they possess more innate intelligence than the average of their class, into a poorer one if they possess less. But the poor breed faster than the rich. Hence the innately stupid breed faster than the innately clever, and the mean innate ability of the population is falling."[30] Haldane considered this doubtful for the following reasons: First, he compared the populations of Moslems, Christians, and Jews in Islamic countries. During the last twelve centuries, followers of the prophet Mohammed who acquired great wealth have practiced polygamy. On the other hand, Christians and Jews in those countries have practiced monogamy. Hence, Haldane argued that Moslems would have acquired greater commercial ability than members of other religions and that a Turk would generally beat a Jew or an Armenian in a commercial deal. Since that is not the case, Haldane concluded that if the rich in England bred faster than the poor the resulting population would not necessarily possess greater innate ability. Second, the phrase "innate ability" is meaningless. A may prove abler than B in a particular environment but not in all, except in a few exceptional cases such as when B is a microcephalic idiot. In the same essay, Haldane[31] predicted that application of the data of human biology to politics and ethics will probably be more complex than application of the data of physics to industry.

Haldane's views on eugenic applications became less speculative and more realistic with time. The bold speculations of *Daedalus* were replaced by a cautious and sober attitude in later years. In his address to the Princeton University Symposium, Haldane[32] evaluated prospects for both negative and positive eugenics.

Eugenic Views of Haldane and Muller

It is interesting to compare the views of Haldane and Muller on the possible application of eugenics for the betterment of humanity.[33] Their lives and careers were remarkably similar. Both made important contributions to genetics and evolution. Both were socialists who expressed admiration for Marxism and both were disappointed in its outcome, although Haldane took longer than Muller. Both were seriously concerned about the genetic future of humanity and the need to apply scientific knowledge for its betterment. Both were atheists and materialists in their outlook. Both had a wide range of interests outside science. Both were fearless in their outspokenness in condemning social injustice. Both left their countries in disgust and disappointment.

Haldane's *Daedalus* is a man's world, reflecting the men's club mentality of the British society in the 1920s. Muller, on the other hand, emphasized the role of women in society and the need to acknowledge their contribution. Muller believed in opening new opportunities for women to follow their professions and express their genetic talents. Muller's commitment to feminism stems from his first wife, Jessie Muller, a

PhD in mathematics, who was dismissed from the faculty of the University of Texas on the birth of their son. Muller was deeply affected by that incident and the irrational prejudice against women who desired to pursue both a family and a career. Muller's views on a eugenic program, which were summed up in his book *Out of the Night*, were dated at least a decade later than Haldane's *Daedalus*, and that difference may account for some of the differences in their views.

There were other significant differences between Muller and Haldane. While Haldane favored ectogenesis, Muller supported the use of cultured sperm, which was preferably kept in storage until twenty years after the death of the donor to allow a careful evaluation of the donor's worth, among other reasons. He foresaw some 250,000 inseminations from each voluntarily chosen male donor and repeating that process with each generation's best-endowed individuals for intelligence and personality traits. Muller was interested in selecting not only for intelligence but also for other desirable qualities, such as compassion, cooperation, and other qualities that promote social equality and justice. Muller noted further that sexual selection could be used to prevent X-linked disorders in a relatively few generations. Muller's commitment to eugenics was much more intense than Haldane's. Muller maintained a serious interest in eugenics from his undergraduate years that lasted throughout his life. Haldane and Muller had similar eugenic goals but differed in their methods and possibilities.

Haldane[34] returned to the subject of hypothetical biological intervention, but not systematic eugenic programs as advocated by Muller and Julian Huxley. In a paper contributed to the CIBA Foundation Symposium on the future of man in 1963, Haldane speculated on the possibilities for human evolution in the next 100,000 years. He considered the incorporation of synthesized new genes into human chromosomes, duplication of existing genes to perpetuate the advantages of heterozygosity (hybrid vigor), and intranuclear grafting to enable our descendants to incorporate many valuable capacities of other species without losing their human capacities. For instance, the disease-resisting quality of many species could be incorporated without losing human consciousness and intelligence. Haldane cited gene grafting as a means to induce various desired phenotypes suited for special tasks. One of these tasks was special adaptation for long-distance space travel. The following comment was typical of Haldane's approach: "A regressive mutation to the condition of our ancestors in the mid-pliocene, with prehensile feet, no appreciable heels, and an ape-like pelvis, would be still better." With reference to the unlikely prospect of encountering high gravitational fields, Haldane wrote, "Presumably they should be short-legged or quadrupedal. I would back an achondroplasic against a normal man on Jupiter."[35]

Among all the major biologists who discussed future eugenic possibilities, Haldane was unique in emphasizing the inadequacy of our technical knowledge of human genetics.[36] With respect to the possibilities of human evolution over the next thousand years, he wrote,

> It may take a thousand years or so before we have a knowledge of human genetics even as full as our present very incomplete knowledge of organic chemistry. Till then we can hardly hope to do much for evolution. . . . If the

capacity for consciousness and control of physiological processes is prized by posterity, steps will probably be taken to make it commoner, and it may be that ten thousand years hence our descendants will differ from us not only in achievements but in capacities and aspirations, to so great an extent that it is useless to attempt to follow them further.[37]

Science and Ethics

The distinguished physicist Freeman Dyson noted that Einstein and Haldane were almost alone in the early twentieth century to discuss the ethical issues that are brought on by scientific progress. First, Haldane's discussion in *Daedalus* dealt with a whole range of ethical issues relating to the application of science to the conduct of wars in the twentieth century.[38] Second, he raised a whole range of questions concerning the application of biological knowledge for the purpose of reproductive intervention. In what was clearly his first foray into the controversial subject of eugenic applications, Haldane appears to be mocking the kind of eugenic selection that has been (and still is) advocated by many followers of the eugenics movement, an attitude that is consistent with his later writings.

In 1924, Haldane's approach to the genetic betterment of the human species was based on hypothetical technical advances in what was later called molecular genetics.[39] He argued that the separation of sexual intercourse and the reproductive process will facilitate greater flexibility in manipulating the human genome. Haldane was well ahead of his time in discussing these biological and ethical issues. He predicted that the term "parent" acquires a new meaning in such a society. Directed mutation and control of in vitro fertilization would lead to a mass production of humans with specialized skills and talents. The genetic revolution that Haldane anticipated has begun. It is occurring a few decades later than he predicted.

A major underlying theme in *Daedalus* is the fact that new scientific discoveries are constantly changing our ethical outlook. Ethics are bound by time, knowledge, and culture. New ethical duties arise from the application of new discoveries. Haldane was especially concerned with the ethical impact of biological discoveries. There is a clear warning in *Daedalus* that progress in science must go hand-in-hand with progress in our ethical outlook. Haldane returned to the subject of the impact of science on ethics repeatedly in his writings.

In his book *The Inequality of Man and Other Essays*, Haldane listed five different ways in which science can impact on ethical situations:

> In the first place, by its application it creates new ethical situations. Two hundred years ago, the news of a famine in China created no duty for Englishmen.... Today the telegraph and the steam-engine have made action possible, and it becomes an ethical problem what action, if any, is right.
>
> Secondly, it may create new duties by pointing out previously unexpected consequences of our actions.... We may not all be of one mind as to whether

a person likely to transmit club-foot or cataract to half his or her children should be compelled to abstain from parenthood.

Thirdly, science affects our whole ethical outlook by influencing our views as to the nature of the world—in fact, by supplanting mythology.

Fourthly, ... anthropology ... is bound to have a profound effect ... by showing that any given ethical code is only one of a number practiced with equal conviction and almost equal success.

Finally, ethics may be profoundly affected by an adoption of the scientific point of view; that is to say, the attitude which men of science, in their professional capacity, adopt towards the world. This attitude includes a high (perhaps an unduly high) regard for truth, and a refusal to come to unjustifiable conclusions ... agnosticism."[40]

In his foreword to my book *Haldane's Daedalus Revisited*, the Nobel laureate Joshua Lederberg[41] drew attention to a kind of imperialism of the present over the future; the complex problem of intergenerational responsibility, a legacy of technology about whose merits they (the future generations) had no voice in deciding. Our children surely deserve adequate ethical and moral preparation to deal with new technologies.

In his commentary on *Daedalus*, physicist Freeman Dyson[42] drew attention to the fact that both Haldane and Einstein shared common concern in this regard. Einstein wrote,

> A positive aspiration and effort for an ethical-moral configuration of our common life is of overriding importance. Here no science can save us. I believe, indeed, that over-emphasis on the purely intellectual attitude, often directed solely to the practical and factual, in our education, has led directly to the impairment of ethical values. I am not thinking so much of the dangers with which technical progress has directly confronted mankind, as of the stifling of mutual human considerations by a "matter-of-fact" habit of thought which has come to lie like a killing frost upon human relations.[43]

A summary of Haldane's predictions in *Daedalus*[44]:

- "In the next war," he wrote, "no one will be behind the front line. It will be brought home to all that war is a very dirty business." Unfortunately, he was right on target!
- The first test-tube baby will be born in 1951. Birth control has done much more than ectogenesis to separate sex from reproduction.
- Coal and oil fields will be exhausted and we will have to tap those inexhaustible sources of power, the wind and the sunlight. He saw an England lined with windmills. Wind and solar energy have seen dramatic development in recent years.
- Surplus power will be used for cleavage of water to hydrogen and oxygen, the former to be liquefied and stored underground. The hydrogen fuel cell exists but is not yet a commercial entity.

- Sugar and starch, converted from cellulose, will become "as cheap as sawdust." New nitrogen-fixing organisms will double wheat yields and lead to abundant cheap food, collapsing agricultural economies.

Scientific Predictions in 1964

Haldane made some scientific predictions in 1964, especially what might be achieved by the year 2000, in an article in the *New York Times* shortly before his death in India on December 1 of the same year. It was titled, "A Scientific Revolution? Yes, Will We Be Happier? May Be."[45] He predicted that we would have a fair knowledge of the large-scale structure of objects within a billion light years, and in particular "we shall be able to rule out a lot of theories which now seem possible ... the theory that matter is being continuously created, so that though the galaxies move apart, matter does not thin out, will very likely be disproved. If it has *not* been disproved in its first 50 years, it will seem much more likely to be true than it does today, and may be generally accepted."

Haldane predicted that chemistry would be extremely accurate and the metachemistry of short-lived particles will be about as systematic as is classical chemistry in 1964. Nuclear fusion reactions based on the conversion of hydrogen to helium would be practicable on a laboratory scale, but not yet a source of industrial power by the year 2000. There would be a few people resident on the moon and Mars, but a voyage to the nearest star would not even be a possibility!

Most infectious diseases would be pretty rare, but new ones would appear often enough to warrant caution. Cancer would usually be curable, but some types would be preventable if diagnosed early. Most people would die between 80 and 100, and a few would live to be 120. Both eugenics and preventive medicine would be needed to extend life spans of most people to 100. Psychology and brain physiology would be advanced but not yet fused, as physics and chemistry had done. A serious problem would be that all advanced knowledge would be expressed only in terms of advanced mathematics. This would widen the gap between different sciences and also between scientists and others. We know about processes lasting up to 4 billion years—the age of the earth, but we cannot accurately study events lasting less than about a millionth of a second, let alone the interaction between some kinds of elementary particles that last only about one one-hundred-thousand-billionth-billionth of a second. Haldane thought we would be able to watch the rapid events that take place inside our cells and inside cellular organs such as mitochondria.

Haldane predicted further that the science of five centuries hence would be psychophysics, which would make it possible to describe any event in the language of either physics or psychology. By the year 2000 we would be a long way from that goal, and would be still further from the goal of a scientific understanding of human societies. Haldane wrote that it is the study of exceptional substances and processes—such as the alkali metals and halogens, which are not found uncombined in nature; electric conduction in metals; and radioactivity—that have led to our mastery of physics and chemistry. He wrote further that happiness is a byproduct, and if you aim at it directly you will not achieve it. He wrote, "It *can* be achieved, but I don't think most people will have achieved it by A.D. 2000."

"Biological Possibilities for the Next Ten Thousand Years"

In a remarkable address to the symposium on the "Biological Possibilities for the Human Species in the Next Ten Thousand Years," Haldane considered some alternative possibilities. They are:

(1) Man has no future, (2) A nuclear war will cause so much biological damage that civilization will have to be rebuilt from barbarism, (3) A nuclear war with such damage will lead to a highly authoritarian world state, (4) Rational animals of the human kind cannot achieve the wisdom needed to use nuclear energy, unless they live for several centuries, and (5) A nuclear war will not occur, but some kind of world organization will gradually emerge, after a general disarmament. Haldane also considered the possibility that mankind could very probably be destroyed by processes still more lethal than nuclear reactions. He considered further the likelihood that a fringe group could get hold of enough fissile material to force their own government to precipitate an international war.[46]

Biologically speaking, Haldane thought that qualitatively novel genetic changes such as mutations or other types are less likely to arise than what is already known from the effects of radioactivity and cosmic particles. He blamed "imaginative writers with a superficial knowledge of biology such as Aldous Huxley and John Wyndham," who have written of new types of mutations, for having done a considerable disservice to clear thinking."

Haldane pointed out that one of the most important tasks facing mankind is a complete revision of educational methods. Teaching methods appear to be aimed at developing children of a capacity a little below the median, and very great harm is done by wrong timing and wrong methods. We may have to wait for human clonal reproduction before scientific methods can be applied.

Referring to the human health situation, Haldane predicted that methods of prevention of many forms of malignant and cardiovascular diseases would soon be developed. An overwhelming desire to colonize Mars or some other planet without introducing terrestrial bacteria and viruses might well provide an incentive to render human beings completely aseptic. After eradicating many of the infectious diseases and deficiency diseases, the next stage in the struggle for health would be against congenital diseases and those that occur in later years. Haldane discussed what he called "rational geriatry." A congenitally weak organ may fail through chronic environmental stress. He wrote, "One reason why I have gone to India is to avoid chronic 'rheumatic' joint pains. I do not mind the heat, since I dress almost rationally, wearing as few clothes as decency permits. Infections such as amoebic dysentery, which are still hard to avoid, are no more trying than English respiratory infections. . . . Perhaps retirement may come to mean retirement to a congenial climate."

Haldane suggested that it is far more important to discover the capacities of young people, and guide them into suitable occupations. Recognition of rare capacities is even more important. Supernormal hearing is very rare, but supernormal vision is even rarer. Supernormal smelling may be quite common. Supernormal muscular skill is highly desired, especially in sports. Aptitude tests may eliminate the worst half or three-quarters, but they do not identify that rare individual who is exceptionally talented. Once poverty becomes a memory of a distant past, there will be much less interest in acquiring material objects, and more and more interest in our bodies and minds, and those of others in whom we are interested.

One Thousand Years

Haldane was inclined to be extremely cautious on the desirability of eugenic applications when many others were rushing to apply eugenics. He predicted further that our descendants would be more interested in their own biology than us, and would have far more knowledge and control of it. Haldane hoped that, by understanding and intellectualizing their normal pleasures, our successors would convert them into servants rather than masters. One of the human goals is emotional homeostasis, which would be achieved by the integration of the emotions, similar to the integration of the activities of antagonistic muscles.

Cloning

Haldane predicted that clones would be made from people aged at least fifty, except athletes or dancers, who would be cloned younger. These would be individuals who have excelled in a certain socially acceptable accomplishment. Other clones would be descendants of people with very rare capacities, such as permanent dark adaptation, lack of the pain sense, and special capacities for visceral perception and control. Centenarians, if reasonably healthy, would be useful subjects for cloning.

We have lost, in our evolutionary past, capacities that are valuable to us, such as our olfactory capacities and the capacity for healing. Hybridization with animals possessing these capacities is clearly impossible. But it is possible to introduce small fragments of the genome of one species into another. That type of intranuclear grafting might enable our descendants to incorporate many valuable capacities of other species without losing those that are specifically human. The most obvious abnormalities in extraterrestrial environments are concerned with gravitation, temperature, air pressure, air composition, and radiation. Haldane wrote, "Clearly a gibbon is better preadapted than a man for life in a low gravitational field, such as that of a space ship, an asteroid, or perhaps even the moon. A platyrhine with a prehensile tail is even more so. . . . The human legs and much of the pelvis are not wanted. Men who had lost their legs by accident or mutation would be specially qualified as astronauts."

Haldane wrote that a drug with an action like that of thalidomide, but on the leg rudiments only, not the arms, may be useful to prepare the crew of the first spaceship to the Alpha Centauri system, reducing their weight, and their food and oxygen requirements.

One sense that is better developed in other species, such as birds, than in our species, is that of time. We have largely lost this sense, mainly because of our dependency on the sun, and later on our watches and clocks. Haldane suggested that the negative aspect of time, of which death is the most significant aspect, might not concern us so much if we could realize human life as a finite pattern in time, capable of all degrees of perfection.

Other problems discussed by Haldane included extremes of temperature and pressure. At air pressures below about a quarter of an atmosphere, a pressure suit is needed. It was typical of Haldane that he considered once again populations that would be easier to train, such as individuals of an Andean or Tibetan ancestry who might be able to live at an external pressure of one-fifth of an atmosphere, wearing a suit that allows breathing at a pressure a few millimeters above that outside. This would be both safer and more comfortable than if the difference were greater. He commented that there is no prospect, in the next ten thousand years, of adapting human beings to breathe air in which the partial pressure of oxygen is less than 1 percent of a terrestrial atmosphere. However, previous experience has shown that given an artificial breathing mixture, humans can live quite happily at all pressures from 1/4 atmosphere to 20 atmospheres, and very likely at higher ones. Other dangers include radiation and high speed particles. Resistance to radiation is a desirable character in astronauts, and may or may not be inherited although it occurs rarely in bacteria.

Division of Species

One idea that Haldane pursued in his writings occasionally, is the real prospect of our species dividing into two or more branches, either through specialization for life on different stars or for the development of different human capacities. This could be a dangerous development, as such species may fail to understand each other, leading to quarrels or even wars. Haldane predicted that, in ten thousand years, the world community will still be polytypic, but much more polymorphic than today. He wrote,

> I hope that the lowest 50% of present mankind for any achievement will be represented by only 5% in our descendants. I do not believe there will be universal racial fusion. ... I do not believe in racial equality, though of course there is plenty of overlap; but I have no idea who surpasses whom in what. ... When opportunities are nearly equalized some races are found to produce far more superior people at some particular skill than others.[47]

Haldane referred to the fact that men and women of African ancestry in the United States, excel in sprinting. He predicted that tropical Africans would produce more potential biologists than potential physicists. However, the intellectual elite of the world would be of very mixed racial origins, perhaps with a median color about that of northern Indians today. The intellectual elite will be more polymorphic than the general population, partly because they will largely be products of assortative matings, that is, when individuals in the same profession tend to marry each other. The physiological and psychological polymorphisms will be far more important, leading to much more tolerance.

Notes

1. J. B. S. Haldane, *The Inequality of Man and other Essays* (1932), 1.
2. J. B. S. Haldane, *Daedalus, or Science and the Future* (London: Kegan Paul, 1923; and New York: E. P. Dutton, 1924).
3. Ectogenesis is the growth of an organism outside the body in which it would normally be found, such as the growth of a fetus in an artificial uterus outside the mother's body.
4. Freeman Dyson, "*Daedalus* after Seventy Years," in K. R. Dronamraju, ed., *Haldane's Daedalus Revisited* (Oxford: Oxford University Press, Oxford, 1995), 55–63.
5. Lord Reith (1889–1971) (John Charles Walsham Reith) was a Scottish broadcasting executive who established the tradition of independent public service broadcasting in the United Kingdom. He was the first director-general of the BBC during the years 1927–1938. His concept of broadcasting as a way of educating the masses marked the BBC and similar organizations around the world for a long time.
6. The BBC is a British public service broadcasting organization. The BBC is headquartered at Broadcasting House in London and smaller production centers throughout the UK. The BBC is the world's oldest national broadcasting organization and the largest broadcaster in the world by number of employees, with about 23,000 staff.
7. Margaret Sanger (1879–1966) opened the first birth control clinic in the United States in 1916, and established organizations that evolved into the Planned Parenthood Federation of America. She was born in Corning, New York. In 1910 she moved to Greenwich Village and started a publication promoting a woman's right to birth control (a term that she coined). Obscenity laws forced her to flee the country until 1915. Sanger fought for women's rights her entire life. From 1952 to 1959, Sanger served as president of the International Planned Parenthood Federation. She died in 1966, and is widely regarded as a founder of the modern birth control movement. Sanger's writings are curated by two universities: New York University's history department maintains the Margaret Sanger Papers Project, and Smith College's Sophia Smith Collection maintains the Margaret Sanger Papers collection.
8. Marie Stopes (1880–1958) was a British pioneer in the field of birth control, and also an author, paleobotanist, academic, and campaigner for women's rights. She was the first female academic on the faculty of the University of Manchester. Her first marriage to the botanist Ruggles Gates was dissolved because it was not consummated even after two years, mainly because Gates was impotent. With her second husband, Humphrey Verdon Roe, she founded the first birth control clinic in Britain. Stopes edited the newsletter *Birth Control News*, which gave explicit practical advice. Her sex manual *Married Love* was controversial and influential: It brought the subject of birth control into wide public discourse. She was never in favor of abortion, arguing that preventing conception was all that was needed.
9. Louisa K. Haldane (1863–1961), mother of JBS Haldane and Naomi Mitchison, wife of John Scott Haldane, the Oxford University physiologist.
10. In 1994, in a personal conversation, the eminent Harvard biologist Ernst Mayr, who was a good friend of Haldane, murmured with satisfaction, "sought after," to my comment that Haldane attracted or "sought after" controversies. Haldane and Mayr were involved in a controversy regarding the importance of mathematical population genetics. Their exchange was summarized in my book *Haldane, Mayr and Beanbag Genetics* (Oxford: Oxford University Press, 2011). In response to Mayr's criticism of his work, Haldane wrote his famous "Defense

of Beanbag Genetics," a classic example of Haldane's pugnacious and witty writing for which he was well known (*Perspectives in Biology and Medicine* 7 [1964]: 343-59).
11. Lord Birkenhead (1872–1930), best known to history as F. E. Smith, was a British Conservative statesman and lawyer who held the offices of lord chancellor and secretary of state for India. He was Winston Churchill's greatest personal and political friend until his death at the early age of fifty-eight from cirrhosis of the liver, which was brought on by alcoholism.
12. Krishna R. Dronamraju, ed., *Haldane and Modern Biology* (Baltimore: Johns Hopkins University Press, 1968).
13. Lady Ottoline Violet Anne Morrell (1873–1938) was an English socialite and hostess. She was influential in artistic and literary circles and enjoyed an intimate friendship with several writers including Bertrand Russell, Aldous Huxley, T. S. Eliot, and D. H. Lawrence, and artists such as Augustus John and Dora Carrington. Morrell was known to have many lovers. In 1902, she married the MP Philip Morrell, with whom she shared a passion for art and a strong interest in Liberal politics. They shared what would now be known as an open marriage for the rest of their lives. The hospitality offered by the Morrells was such that most of their guests had no suspicion that they were in financial difficulties. Many of them assumed that Ottoline was a wealthy woman. This was far from being the case, and during 1927 the Morrells were compelled to sell the manor house and its estate and move to more modest quarters in Gower Street. Lady Ottoline's most interesting literary legacy is the wealth of representations of her that appear in twentieth-century literature. She was the inspiration for Mrs. Bidlake in Aldous Huxley's *Point Counter Point*, for Hermione Roddice in D. H. Lawrence's *Women in Love*, for Lady Caroline Bury in Graham Greene's *It's a Battlefield*, and for Lady Sybilline Quarrell in Alan Bennett's *Forty Years On*. *The Coming Back* (1933), another novel that portrays her, was written by Constance Malleson, one of Ottoline's many rivals for the affection of Bertrand Russell. Some critics consider her the inspiration for Lawrence's Lady Chatterley. Huxley's roman à clef *Crome Yellow* depicts the life at a thinly veiled Garsington.
14. Julian S. Huxley, *Memories* (New York: Harper & Row, 1970), 74–75.
15. Naomi Mitchison, personal communication to the author, 1982.
16. D. H. Lawrence, *Lady Chatterley's Lover* (Florence: Giuseppe Orioli, 1928; first American ed., New York: Grosset & Dunlap, 1932).
17. Aldous Huxley, *Brave New World* (London: Penguin Books, 1932). Includes revolutionary ideas in reproductive and molecular biology, a fictionalized version of Haldane's ideas in *Daedalus*. Huxley was the younger brother of the biologist Julian Huxley.
18. J. B. S. Haldane, *The Inequality of Man and other Essays* (London: Chatto & Windus, 1932).
19. J. B. S. Haldane, "Human Biology and Politics" (The British Science Guild: 10th Norman Lockyer Lecture, 1934), 3.
20. Krishna R. Dronamraju, *Foundations of Human Genetics* (Springfield, IL: Charles C. Thomas, 1989).
21. Haldane, "Human Biology and Politics."
22. Letter dated August 18, 1947. Ethna Donnelly was a Los Angeles doctor who wanted to establish a eugenic program to sterilize subnormal individuals. She sought Haldane's support.
23. E. Carlson, "The Parallel Lives of H.J. Muller and J.B.S. Haldane—Geneticists, Eugenists, and Futurists," in Dronamraju, *Haldane's Daedalus Revisited*, 90–101. Carlson was a student of Herman J. Muller, who won the Nobel Prize for his discovery that atomic radiation causes genetic mutations. Muller's Nobel award was made in 1946, immediately following the atomic bombing of Hiroshima and Nagasaki, which underscores the importance of Muller's discovery, from a biological as well as a sociopolitical point of view.
24. J. B. S. Haldane, "The Implications of Genetics for Human Society," in *Genetics Today: Proceedings of the 11th International Congress of Genetics*, ed. S. J. Geerts (London: Pergamon Press, 1964), xci–cii. Haldane discussed a whole range of social, ethical, political, and scientific issues related to the applications of human genetic knowledge for the betterment of human race.
25. Tom Driberg (1905–1976), was a British journalist, politician, and High Anglican churchman who served as a Member of Parliament (MP) from 1942 to 1955 and from 1959 to 1974. He was first elected to parliament as an Independent, and joined the Labour Party in 1945. He

never held any high office, but rose to senior positions within the Labour Party and was a popular and influential figure in left-wing politics for many years.

26. J. B. S. Haldane, "The Implications of Genetics for Human Society," in *Genetics Today*, Proceedings of the XI International Congress of Genetics (London: Pergamon Press, 1965), xcv–xcvi.
27. J. B. S. Haldane, "Human Evolution: Past and Future," in *Genetics, Paleontology and Evolution*, ed. J. L. Jepson, G. G. Simpson, and E. Mayr (Princeton, NJ: Princeton University Press, 1949).
28. Haldane, "The Implications of Genetics for Human Society," xcviii–xcix.
29. Haldane, "Human Biology and Politics."
30. Haldane, "Human Biology and Politics," 34.
31. Haldane, "Human Biology and Politics."
32. Haldane, "Human Evolution: Past and Future."
33. Carlson, "The Parallel Lives of H.J. Muller and J.B.S. Haldane," in *Haldane's Daedalus Revisited*, ed. K. R. Dronamraju (Oxford: Oxford University Press, 1995), 90–101.
34. Haldane, "The Implications of Genetics for Human Society," xci–cii.
35. Haldane, "Human Evolution: Past and Future," 405–20.
36. J. B. S. Haldane, "Biological Possibilities for the Human Species," in *Man and His Future*, ed. G. E. W. Wolstenholme, Ciba Foundation symposium (London: J. and A. Churchill, 1963), 337–47.
37. Haldane, "The Implications of Genetics for Human Society."
38. Dyson, "*Daedalus* after Seventy Years"; Haldane, *Daedalus, or Science and the Future*.
39. Haldane, *Daedalus, or Science and the Future*.
40. Haldane, *The Inequality of Man and other Essays*, 98–100.
41. J. Lederberg, Foreword, in Dronamraju, *Haldane's Daedalus Revisited*, xii–ix.
42. Dyson, "*Daedalus* after Seventy Years."
43. A. Einstein, *Ideas and Opinions* (New York: Crown Publishers, 1954). Based on *Mein Weltbild*.
44. Haldane, *Daedalus, or Science and the Future*.
45. J. B. S. Haldane, "A Scientific Revolution? Yes, Will We Be Happier? May Be." *New York Times*, 1964.
46. J. B. S. Haldane, "Biological Possibilities for the Human Species in the Next Ten Thousand Years," in *Man and His Future*, ed. G. E. W. Wolstenholme (London: J. and A. Churchill, 1963).
47. Haldane, "Biological Possibilities for the Human Species."

4

Population Genetics

Haldane was a polymath, whose scientific contributions covered multiple disciplines. His most extensive and important work was in genetics, especially theoretical population genetics or mathematical genetics. His contributions have significantly impacted on every aspect of genetics.

In 1900, the physiologist John Scott Haldane took his eight-year-old son, the future "JBS," to a lecture at a Royal Society conversazione by A. D. Darbishire on Mendel's laws of inheritance, then newly rediscovered. It made a big impression on young Jack, who wanted to learn more. The term "genetics" itself was still to be invented by Bateson. While he was still a student at Eton, Jack began analyzing Darbishire's results and came up with an unexpected and exciting discovery—"linkage." The younger Haldane realized that his mathematical skill and intuition endowed him with a powerful tool that provided him a lifelong approach to solving scientific problems. In addition to his own research, JBS went on analyzing other scientists' results throughout his life, often discovering novel results which they themselves missed in their experiments. In 1911 he presented the results of his analysis of Darbishire's results in a seminar organized by Professor E. S. Goodrich of Oxford University. Had he published his findings at that time, he would have been the first to discover genetic linkage in vertebrates, simultaneous with the discoveries of the Morgan School, in fruit flies, at Columbia University in New York. However, when Jack consulted R. C. Punnett, he was advised to obtain his own experimental results and confirm Darbishire's work independently.

Haldane and his sister, Naomi, began breeding experiments with mice and guinea pigs at home. Visitors to the Haldane household saw cages with mice and guinea pigs instead of the more traditional croquet balls and tennis courts. In her chapter "Beginnings," which Naomi contributed to my book *Haldane and Modern Biology*,[1] she wrote,

> Our first joint scientific experiments began when I was about twelve. I had by that time a number of guinea pigs and was making my own observations on their lives and loves. . . . Then came Mendelism, which at that time was easily understood, even by someone such as myself with no scientific knowledge. . . . Early genetics was relatively unmathematical . . . We had the rats and mice in the animal house of the zoology department at Oxford. . . . Our first paper, on color inheritance in mice, was published in 1915.

The paper by Haldane, Sprunt, and Haldane, titled "Reduplication in Mice" was published in the *Journal of Genetics* in 1915.[2] It was his first paper in genetics, but not his first scientific publication, because he had already published two physiological papers, including one with his father, in 1912. In the introductory paragraph of his first genetical paper, Haldane wrote, "Darbishire's experiments indicated the existence of Reduplication[3] in mice, and this work was undertaken to verify and extend his results. Owing to the war it has been necessary to publish prematurely, as unfortunately one of us (A.D.S.) has already been killed in France. The reduplication occurs between the factors . . . C and E, the absence of C producing albinism, that of E (when C is present) pink eyes and a colored but pale coat."

In 1915 Haldane joined the family regiment, the Black Watch, and quickly acquired the reputation as a first-class regimental officer, as remembered by Eric all, in the days of trench warfare, brave but no taker of undue risks. He trained at Nigg, and we swam together in the cold North Sea and I kept taking omens and bargaining with whatever powers had our lives now in a choking hold."[4]

We have his sister's account of his savage life in that war:

> The third Battalion of the Black Watch went to France in early 1915, into the horrors of trench warfare, treading on the faces of one's own half-buried dead. My brother took to making his own bombs and going out with them

Figure 4.1 Haldane and a colleague in Black Watch uniforms, 1919.

along the mazes of old water-logged trenches, listening for German conversation, lobbing in his bomb and watching the resultant shower of enemy arms and legs. He became immersed in this savage life where in summer one wore only a kilt and boots; he must have killed a lot of people, though, of course, nothing compared with what the man who operates the bomb hatch in a modern plane has as his responsibility.

Naomi wondered whether her brother was troubled by this past. She wrote:

Perhaps it is easier to come to terms with those one has killed in this nearly hand-to-hand way, though it may involve one in almost Jainist attitude to life. Nearly half a century later, when I watched him letting a horsefly suck from his hand while he watched her beautiful eyes, I asked him whether *this* had anything to do *that*. He disclaimed it so fiercely that it may have been true.[5]

Returning from the war in 1919, Haldane immediately published two papers in the *Journal of Genetics*, which indicates that he must have been working on these scientific problems while serving in the front. Both papers dealt with the problem of estimating linkage values and the calculation of distances between loci on the chromosome maps. He derived a method of accurately estimating what was later called the "mapping function"[6] and the map distance between loci. Later, several genetic map functions have been proposed by others to infer the unobservable genetic distance between two loci from the observable recombination fraction between them, and more complex situations involving multiple loci were considered later. Haldane's approach introduced accuracy into the mapping methods that were then being followed by the American school under T. H. Morgan (1866–1945)[7] at Columbia University in New York. Haldane also introduced in the same paper the terms "morgan" and "centimorgan" for units of map distance, which were used for a long time afterward.

Population Genetics

Haldane's major contribution to science is the founding of "population genetics," a distinction he shared with two others, Ronald A. Fisher and Sewall Wright. Their main goal was to explain the evolutionary theory of Charles Darwin in the light of the laws of inheritance discovered by the Austrian monk Gregor Mendel. The foundations laid by Haldane, Fisher, and Wright were based on theoretical and mathematical investigations of the evolutionary process. Haldane's contributions to population genetics spanned about forty years, from the 1920s to his death in 1964. Some of his papers were published posthumously. The contributions of Fisher and Wright also spanned several decades. No other field of science was dominated for so long by so few. Although they were in general agreement, they differed in certain details concerning the relative importance of various evolutionary processes involved. Fisher and Haldane were interested in analyzing evolution in large populations, whereas Wright analyzed evolution in small isolated populations.

While the early population genetics, founded by Fisher, Haldane and Wright, was theoretical and mathematical, experimental population genetics came later and was mostly dominated by the experimental work with the fruit fly *Drosophila*, which played a big part in testing the theoretical models.

Benefits of Population Genetics

Both directly as well as indirectly, population genetics has revolutionized and benefited human society through its numerous applications in agriculture, public health, and clinical medicine. It has changed our fundamental outlook on life and society in a radical manner. Among the many spinoffs of the applications of population genetics are the epidemiology of various diseases and defects in populations, estimation of radiation-induced genetic damage, improved crop and animal breeding methods and higher food productivity, improved methods of DNA fingerprinting applications in recent years, better understanding of the role of genetic and environmental factors in IQ determination and various psychological and behavioral disorders, a vastly better understanding of racial diversity (and perhaps tolerance), and the genetic basis of kinship in human families as well as that between the human species and other species on earth, to name a few.

Haldane, Fisher, and Wright

Haldane was a polymath who contributed significantly to several branches of science. He was a gifted mathematician who applied his talent to the genetic analysis of evolution in populations and numerous other problems. Haldane was a public figure whose utterances, quips, and statements were frequently reported in the popular press. Long after his death, Haldane still remains one of the most quoted intellectuals today.

Haldane was most generous in acknowledging the assistance of others, no matter how slight it was, in his publications. He was equally quick at acknowledging his own mistakes, especially in his mathematics, and it was not unusual to come across the improved and corrected versions of his papers, written by himself.

Fisher and Wright received formal academic degrees in mathematics and zoology, respectively, while Haldane received no formal academic qualification in any branch of science. Fisher made profound contributions to statistics and genetics. Wright's early work was mostly in animal breeding and genetics, especially the application of *path coefficient analysis*.[8] Fisher and Wright published many scientific papers and books of professional nature, as did Haldane, but unlike Haldane they were not engaged in popular writing.

In comparing these three pioneers, James F. Crow, a well-known authority on population genetics, wrote,

> Fisher had nothing like the breadth and depth of knowledge in all areas that Haldane had. Neither did Wright, although by ordinary standards he was a

learned man. Haldane's restless curiosity and willingness to delve into any subject under the sun diluted his contributions to any single subject. He learned easily and remembered essentially everything, and he worked hard and wrote easily. I for one would not have wanted it otherwise. He would have produced more population genetics if he stayed to his last, but his total output would have been less varied and less interesting, and we would have been the losers.[9]

Their interrelationships are also interesting. Fisher and Wright differed strongly in their approach to population genetics, and they carried their differences to the personal level. They did not speak to each other. Wright and Haldane treated each other with great respect and mutual admiration. When Haldane lectured at the University of Wisconsin, Wright introduced him in glowing terms. And Haldane reciprocated in a similar fashion. They shared many interests and enjoyed talking to each other. Fisher and Haldane enjoyed arguing with each other. Their relationship was marked by swings. They were both on the faculty of University College London for several years. Their paths crossed often at the meetings of the Genetical Society of Great Britain during the 1930s and the 1940s. Again, in their last years they were both at the Indian Statistical Institute (ISI) in Calcutta, where Haldane was a research professor from 1957 to 1961. Fisher was a frequent visitor, stopping on his way from London to his new home in Australia.

The director of ISI, Prasanta Chandra Mahalanobis, was tactful enough to maintain friendly relations with both Fisher and Haldane, two notoriously difficult men—a remarkable feat that he achieved with much skill and diplomacy. Haldane, for his part, attempted to maintain friendly and civil relations with Fisher, partly because he did not want to embarrass the director, who was their host. On one occasion, when he was invited to address a meeting at the institute, which was convened to honor Fisher, Haldane made the following remarks:

> I am very glad to see Professor Fisher again. For a number of years we were colleagues at the University College London. Then he went away to Cambridge and I didn't see as much of him as I could have wished. When we were at the University College, we were chronically in slight disagreement. But the disagreement was not so great as to prevent collaboration, and that, after all, what matters. . . . But although formally we were each quarrelling with the other, the net result, I think, was fairly constructive. I would also like to say that some of my juniors, including my wife, said that Professor Fisher was much kinder than I when they brought statistical problems. . . . Although Professor Fisher may sometimes have been rather hard on his contemporaries or his seniors, he has, at least among those I know, a reputation of being extraordinarily kind to junior workers. And that, I think, is the reputation which matters, because the junior workers will be alive after all the others are safely disposed of and it is they who will hand out the tradition of what sort of a man Fisher was.[10]

It is obvious that Haldane was "bending" backward to be nice to Fisher. It would surely have surprised their former colleagues in England who were more accustomed to see much rancor between Fisher and Haldane!

What Is Population Genetics?

Population genetics began as an attempt to marry Darwin's theory of evolution with the science of genetics that was founded by the Austrian monk Gregor Mendel. The laws of inheritance, which Mendel published in 1866, were at first widely ignored by the scientific community. But his work was rediscovered and confirmed independently by three scientists in 1900, Carl Correns in Germany, Hugo de Vries in Holland, and Erich von Tschermak in Austria.[11]

Mendel's success was due to his clear conception of new scientific methods, which according to the Cambridge University biologist William Bateson, were "absolutely new in his day." Mendel focused his attention on drawing a simple relationship between single uncomplicated characters, such as flower color and seed coat texture, and the corresponding underlying inherited factors (later called genes).

After the rediscovery of Mendel's laws in 1900, the concept of the gene and its function advanced rapidly. In the early years after the "rediscovery," Bateson introduced terminology that helped the rapid growth of this new science. He introduced the term "genetics" in 1906 to describe the new science and also introduced several other terms, that are still used in genetics today.[12]

As Darwin pointed out, evolution occurs at the population level. Population genetics is the science that deals with an analysis of the genetic structure of populations under various phases of the evolutionary process.

Hardy-Weinberg Law

The next important step in population genetics was advanced in 1908 by two individuals who worked independently of each other in two countries, the mathematician Godfrey Harold Hardy[13] at Cambridge, England, and the physician Wilhelm Weinberg[14] in Stuttgart, Germany. Using simple mathematics, they showed how the relative proportions of certain genes in populations remain constant from one generation to next. Although they worked independently of each other, their discovery came to be called the "Hardy-Weinberg equilibrium" or the "Hardy-Weinberg law" or the "Hardy-Weinberg principle."

The Hardy-Weinberg law is strictly valid only if several conditions are fulfilled:

1. The population must be large enough so that sampling errors can be ignored.
2. There must be no mutation.
3. There must be no selective mating.
4. There must be no selection.

G. H. Hardy

Hardy, though a great mathematician, had no knowledge of genetics. A statistician, Udny Yule, who played cricket with Hardy, asked him a mathematical question, whether the dominant genes in a population would automatically increase in frequency in mixed populations. Hardy immediately came up with the solution, which came to be called "Hardy's law," later the "Hardy-Weinberg Law."

Interestingly, the mathematical part of this work was so simple that Hardy was embarrassed to show it to his mathematical colleagues in England. He published his short paper in a general scientific journal (not specializing in mathematics), far away from home, the *Science* journal in Washington, DC!

> To the Editor of Science: I am reluctant to intrude in a discussion concerning matters of which I have no expert knowledge, and I should have expected the very simple point which I wish to make to have been familiar to biologists. However, some remarks of Mr. Udny Yule, to which Mr. R. C. Punnett has called my attention, suggest that it may still be worth making. ... Suppose that *Aa* is a pair of Mendelian characters, being dominant, and that in any given generation the number of pure dominants (*AA*), heterozygotes (*Aa*), and pure recessives (*aa*) are as p:2q:r. Finally, suppose that the numbers are fairly large, so that mating may be regarded as random, that the sexes are evenly distributed among the three varieties, and that all are equally fertile. A little mathematics of the multiplication-table type is enough to show that in the next generation the numbers be as $(p + q)2:2(p + q)(q + r):(q + r)2$, or as $p1:2q1:r1$, say. The interesting question is—in what circumstances will this distribution be the same as that in the generation **before?** It is easy to see that the condition for this is $q2 = pr$. And since $q2 = p1r1$, whatever the values of *p*, *q*, and *r* may be, the distribution will in any case continue un-changed after the second generation.[15]

Wilhelm Weinberg

Wilhelm Weinberg (1862–1937) was a German physician who practiced in Stuttgart. In 1908 he published a paper in German, expressing the concept that later became the Hardy-Weinberg principle. He and Hardy worked separately and independently. James F. Crow writes, "Why was Weinberg's paper, published the same year as Hardy's, neglected for 35 years? The reason, I am sure, is that he wrote in German. At the time, genetics was largely dominated by English speakers and, sadly, work in other languages was often ignored."[16]

Weinberg was born in Stuttgart and studied medicine at Tübingen, Berlin, and Munich, receiving an MD in 1886. Much of his academic life was spent studying genetics, especially focusing on applying the laws of inheritance to populations. Weinberg made other contributions to statistical genetics, including the first estimate of the rate of twinning, and he pointed out that identical twins would have the same sex, while dizygotic twins could be either of the same or opposite sex. He derived the formula for estimating the frequency of monozygotic and dizygotic twins.

Evolutionary Synthesis

Evolutionary synthesis began as a process to strengthen the foundation of Darwinism by synthesizing biological evidence from multiple disciplines. The first phase of this process was founded by J. B. S. Haldane, R. A. Fisher, and S. Wright, who worked independently of each other. They employed mathematical methods to build an extensive framework that provided a satisfactory interpretation of Darwinism in terms of Mendelian genetics. Much of this foundation was laid down during the years 1918–1932, although Haldane and Fisher continued to publish extensions of their work until the 1960s, and Wright continued well into the 1980s.

What is important to remember is that the genetic foundations laid down by the three pioneers, Haldane, Fisher, and Wright, established Darwinism on a firm scientific footing. It revived and strengthened the role of natural selection in evolution at a time when interest in the role of selection had clearly declined or totally neglected. The two important books—Fisher's *The Genetical Theory of Natural Selection* and Haldane's *The Causes of Evolution*, as well as Wright's extensive paper played a decisive role in turning the tide in favor of the recognition of the important role of natural selection in the evolutionary process.[17]

Haldane's Contributions to Population Genetics

Haldane's contributions to population genetics differed significantly from those of Fisher and Wright. He was not a system builder, hence there was no system to defend. His approach was to solve problems as they arose. On the other hand, Fisher and Wright advanced models of population genetics that they defended for several decades. They were involved in bitter personal disputes that divided geneticists. In fact, some accounts of population genetics narrated its history solely in terms of the Fisher-Wright controversy, almost ignoring Haldane's numerous contributions to population genetics (see, for instance, Provine[18]). Clearly there is much more to population genetics than the mere dispute between Fisher and Wright.

Haldane, on the other hand, was an open-minded scientist, who often admitted his own mistakes and published papers correcting his previous work. But he also loved controversies. The issues he chose and cared deeply to argue were often much broader, even beyond science, extending into other domains such as theology, politics, ethics, warfare, and science fiction. With respect to Haldane's open-mindedness, I quote one of the most gifted mathematical population geneticists, James F. Crow of the University of Wisconsin:

> Where did Haldane stand? What did he do? And why is his name not associated with any global theory of evolution? The answer, I believe, lies in his

eclecticism and interest in the work of others. Haldane worked out in great detail the way selection acts on gene frequencies. He anticipated Fisher's work on the probability of fixation of a favorable mutant gene. His paper on metastable equilibria paralleled and anticipated much of Wright's work. But rather push any particular view, Haldane saw good in all of them. His open-mindedness, his detailed knowledge of the work of others, and his breadth of interest kept him from the staunch support of a particular view that characterized the writings of many others. So, in a way, the fact that Haldane's work has led to no school of thought is to his credit. As an example, the approach to inbreeding through identity by descent, which owes a great deal to Haldane, is replacing the correlation approach of Wright, although Wright's method completely dominated the field in the early days.[19]

Fisher received formal training in mathematics. The mathematics of Wright, on the other hand, was self-taught. Haldane formulated a problem and then ground out the answer, solving a number of important problems. He was not particularly interested in an elegant approach; Jim Crow called his style "brute-force mathematics."[20] He worked very fast and published his results, often without a second look. Consequently, some minor errors appeared in his papers, but he did not care because they did not affect the general conclusions of his papers. In later years, Haldane asked one of his students, Suresh Dinakar Jayakar,[21] to check the mathematics in his papers before publication.

Early Population Genetics

The 1959 symposium at the famous Cold Spring Harbor Laboratory in New York was held to mark the centennial of the publication of Charles Darwin's classic *On the Origin of Species*. The noted evolutionary biologist Ernst Mayr opened the symposium with the question, "But what, precisely has been the contribution of this mathematical school to the evolutionary theory, if I may be permitted to ask such a provocative question?"[22]

The early work in population genetics was characterized as "beanbag genetics" by Ernst Mayr because he likened the treatment of single genes to the work of early Mendelians who used beans of various colors in bags to study Mendel's laws of genetics.[23] Mayr emphasized that to consider genes as independent units is meaningless from the physiological as well as the evolutionary viewpoint, and emphasized that interactions between genes play a crucial role in evolution.

In their early work, both Haldane and Fisher treated genes as noninteracting independent units, whereas Wright considered their interaction (epistasis) as an essential part of evolution. In their later work, both Haldane and Fisher also considered more complicated situations including genic interaction. All three pioneers of population genetics took into account the complications arising from linkage, dominance, and epistasis in their calculations. In retrospect, Haldane[24] commented that Mayr overstated his case. Initially, Haldane treated genes as noninteracting independent units for the sake of simplifying mathematical analyses.

There is a lot more to "beanbag genetics" than was apparent from Mayr's critique. It goes far beyond counting Mendelian ratios or a simple phenotype-genotype relationship. Indeed, beanbag genetics today encompasses molecular clocks, nucleotide diversity, DNA-based phylogenetic trees, as well as the four major forces: mutation, selection, migration, and random genetic drift. Rates of evolution at the nucleotide level can be measured and compared among diverse populations and among species. But most important of all, genetic differences between populations could be measured only by hybridization. Today it is commonplace to compare DNA differences and similarities in diverse species, orders, and taxa. Finally, the most important evidence for the concept of "out of Africa" in human ancestry came from nucleotide diversity, in other words, beanbag genetics.

Early Population Genetics

Early developments in genetics could be categorized as "population genetics," much of which could also be characterized as "beanbag genetics" in today's terminology. Indeed, much of classical genetics is based on the assumption that the gene is the basic unit of inheritance for all practical purposes and that there is a direct relationship between a phenotype and its corresponding genotype. The independent nature of each gene and its specific function were generally accepted except where epistasis and linkage were considered. These apparently simple notions rapidly established classical genetics during the early decades of the twentieth century. One can say that the successful establishment of genetics as a scientific discipline is mainly due to the concepts and methods of "beanbag genetics."

The founder of genetics, Mendel himself pioneered "beanbag genetics." It is well known that Mendel's success was due to his clear conception of some procedures that, according to Bateson, were "absolutely new in his day." One of these was his treatment of individual characters and the underlying factors as discrete noninteracting entities. Among other reasons for Mendel's success was his clear vision of drawing a simple relationship between a phenotype and its genotype. These are the same concepts that were followed by Haldane in formulating his "mathematical theory of natural selection" and to which Mayr objected in his critique of "beanbag genetics." It must be added, however, that Haldane took into account the effect of other complications such as epistasis and inbreeding in his later papers in his mathematical series.

In response to Mayr's challenge, Haldane[25] wrote a remarkably spirited defense of his work, shortly before his death. It was titled "A Defense of Beanbag Genetics." Haldane pointed out that Mayr made a large number of enthusiastic statements about the biological advantages of large populations that are unproved and not very probable. Haldane explained that the genetic structure of a species depends largely on local selective pressures, on the one hand, and migration between different areas, on the other. We can judge the "success" of a species both from its present geographical distribution and numerical frequency and from its assumed capacity for surviving environmental changes and for further evolution.

Haldane[26] wrote that there is not enough knowledge to indicate whether a particular species (including humans) would benefit more by increased "gene cohesion" or gene flow or some other factor from one area to another, as suggested by Mayr. A geneticist can deduce the evolutionary consequences given further data on various other parameters if the types of interactions that may occur between a genotype and various environments are known, as discussed by Haldane.[27]

Sewall Wright contributed a chapter to the Haldane memorial tribute that I edited in 1968.[28] He provided a brief comparative account of the contributions of the three pioneers. Extensive biographic accounts of Fisher[29] and Wright[30] have been published, and the reader is referred to those publications for additional information.

The mathematics of all three pioneers agreed essentially despite some differences in their approach. Haldane expressed his results in terms of the ratio, u, of the frequency of the mutant gene (A') to that of its type allele (A), (distribution 1A: uA'). In 1922, Fisher used a function of the gene frequency distribution (1-p) A:pA', viz., = $\cos^{-1}(1-2p)$, which has the property that its sampling variance is constant, $1/2N$, but he shifted later to the gene frequency itself, which was used systematically by Wright from the beginning.

However, all three differed greatly in the application to evolution. Haldane assigned usually constant selective values, but occasionally variable, to each gene or, in some cases each genotype involving two or more interacting loci, deducing deterministically the number of generations required to bring about a specific change in the gene frequency ratio under different types of genetic situations. Fisher, on the other hand, proposed general theorems of evolution, which are applicable to situations of varying complexity. The most important of these was his fundamental theorem of natural selection: "The rate of increase in fitness of any organism at any time is equal to its genetic variance in fitness at that time."[31] Genetic variance was defined as the additive component of the total genotypic variance. This theorem assumes that selection always depends on the net effect of each gene and that there can be no selection among interaction systems between various genes. Both Fisher's and Haldane's approaches were similar in their deterministic character when treating large populations. Both authors recognized that the fate of a single mutation is decided by a stochastic process.

Wright's[32] approach, on the other hand, was directed toward ascertaining whether some way might exist in which selection could take advantage of the enormous number of interaction systems resulting from a limited number of unfixed loci. He proposed that such a system might exist in a large population that is subdivided into many, small local populations, isolated enough to facilitate considerable random differentiation of gene frequencies. At the same time, their isolation is limited enough to stop any diffusion of the more successful interaction systems.

Natural Selection

Haldane's interest in evolution began with studies of natural selection, which Charles Darwin postulated as the main agent of evolutionary change. In 1924, in his first paper in population genetics, Haldane wrote, "In order to establish the view that natural selection is capable of accounting for the known facts of evolution, we must show not only that it can cause a species to change but that it can cause it to change at a rate which will account for present and past transmutations."[33]

Haldane cited some examples of his work where mathematical methods yielded deeper insights regarding evolutionary problems, which would not have been possible otherwise. In response to Mayr, Haldane[34] wrote:

> Our mathematics may impress zoologists but do not greatly impress mathematicians. Let me give a simple example. We want to know how the frequency of a gene in a population changes under natural selection. I made the following simplifying assumptions:
>
> (1) The population is infinite, so the frequency in each generation is exactly that calculated, not just somewhere near it,
>
> (2) Generations are separate. This is true for a minority only of animal and plant species. Thus even in so-called annual plants a few seeds can survive for several years,
>
> (3) Mating is at random. In fact, it was not hard to allow for inbreeding once Wright had given a quantitative measure of it,
>
> (4) The gene is completely recessive as regards fitness. Again it is not hard to allow for incomplete dominance. Only two alleles at one locus are considered.
>
> (5) Mendelian segregation is perfect. There is no mutation, nondisjunction, gametic selection, or similar complications.
>
> (6) Selection acts so that the fraction of recessives breeding per dominant is constant from one generation to another. This fraction is the same in the two sexes. where q_n is the frequency of the recessive gene, and a fraction k of recessives is killed when the corresponding dominants survive.

Haldane acknowledged[35] that H. T. J. Norton[36] gave an equivalent equation in 1910, and Haldane (1924) produced a rough solution when selection is slow, that is, when k is small. However, he pointed out that even such a simple-looking equation would not yield a simple relation between q and n. Many years later, Haldane and Jayakar[37] solved this equation in terms of automorphic functions. Haldane noted that the mathematics are not much worse when inbreeding and incomplete dominance are taken into account. But they are much more complicated when selection varies from year to year and from place to place or when its intensity changes gradually with time. When such problems are solved, the mathematics employed would be truly impressive.

Haldane stated at the outset that he would be dealing only with the simplest possible cases, involving a single completely dominant Mendelian factor or its absence. In one instance, considering the effects of slow selection, Haldane[38] showed that for an autosomal factor (he was still using the term "factor" which was introduced by his mentor William Bateson, and was later replaced by "gene"), the number of generations required to change the frequency, under slow selection (k = .001), are:

Gene frequency change	Number of generations required
0.001% to 1.0%	6,921
1.0% to 50.0%	4,819
50.0% to 99.0%	11,664
99.0% to 99.999%	309,780

When external conditions (such as a change in the temperature or a new disease) change drastically, many genes that have been less favorable to an organism in the past may become more favorable in the new environment. The practical cases to which Haldane applied his theory were ones that involved such change. One of these cases was based on field data assembled by the ecologist Charles Elton on the steady decline in the proportion of silver fox pelts among fox pelts marketed in various parts of Canada in the century preceding 1933.[39] Silver is due to a simple recessive gene. Haldane calculated that this gene must have been at an average selective disadvantage of about 3% per year or 6% per generation if generation length is taken as two years, compared with its wild type in red.[40]

Betty Norton's Correspondence with Haldane

In 1951, Norton's sister, Betty, wrote to Haldane to complain that a biography of the economist, John Maynard Keynes, by Roy Harrod[41] contained a reference to her brother Harry that was unflattering and inadequate. She sought Haldane's help in redressing the injustice done to her late brother. She was aware that Haldane had acknowledged, in 1938, in a book called *Background to Modern Science*, her brother's pioneering contribution to the mathematical theory of natural selection. However, she did not seem to be aware of Haldane's earlier scientific papers, where he had already acknowledged Norton's contribution while stating that he reached his conclusions quite independent of Norton's work. In the correspondence that ensued, it is easy to see that Haldane responded promptly and fairly to Miss Betty Norton's request. In a letter to Miss Norton, dated February 2, 1951, Haldane wrote, "I have always felt that he [Harry Norton] may have thought that I 'jumped his claim', on the other hand the publication of his work was, if I remember, delayed for about seventeen years and as we only overlapped to a relatively small extent, I felt justified in going ahead with reference to 'the important unpublished work of H.T.J. Norton'." Although Harry Norton, as early as 1910, had communicated his preliminary results to Punnett, who reproduced them in his book *Mimicry in Butterflies*,[42] Norton did not publish his results fully until 1926, in a lengthy paper of forty-five pages in the *Transactions of the London Mathematical Society*.[43]

Haldane's paper, published in 1927, on the treatment of selection in overlapping generations is of special importance because he combined a demographic approach to the study of evolution in the context of Mendelian genetics. Considering a population where the intensity of selection is independent of its size, Haldane addressed the problem of measuring population growth. He calculated the birth rate and death rate, which are not functions of its density. He showed that the oscillations of the population about an exponential function of the time "are either damped or at least increase less rapidly than the population itself."[44] If population is in equilibrium, oscillations are damped and a stable equilibrium occurs. Haldane concluded that his result was a new proof of the great value of Lotka's theorem on the stability of the normal age distribution.[45]

Haldane then considered the mode of selection of a dominant autosomal factor in the population, only for female zygotes under the following assumptions: The sex ratio at birth is taken as fixed; the number of dominant and recessive female zygotes is fixed; random mating occurs; and selection and population growth are slow. Haldane then calculated the reproduction rates of the three female phenotypes at a time t, and he defined selection as the probability for a female zygote to reach age x.

Haldane's approach integrated the mode of selection in demographic structures, showing that selection, when generations overlap, produces an effect that is analogous to that obtained when generations are nonoverlapping. He also showed that the mode of selection was the same for the sex-linked factor. Furthermore, Haldane integrated his work with the population growth model of Dublin and Lotka, related to the real structure of the United States population in the 1920 census.[46] Haldane computed the differential reproduction rate of different phenotypes at a given time, under the assumption when recessives are rare, when dominants are rare, and when the death rate and the reproduction rate are the same in the two sexes.

Demographic Structures

Of the three great pioneers of population genetics, Haldane's attempt was the first to find the solution to genetic problems by integrating them into demographic structures. He was the first to show that a synthesis was necessary between demography and genetics in order to establish the real mode of selection and its processes in a human population. However, Haldane's brilliant interpretation met with no following at that time. The French demographer Jean Sutter commented that the integration of genetic problems into demographic structures progressed very slowly for several decades, because of the domination of the deterministic models.[47] After Haldane's death, a breakthrough occurred when Malecot[48] made use of the chain processes of modern probabilities, which enabled him to set up interesting stochastic models. However, the pioneering contribution of Haldane is still remembered.

Interestingly, one of his correspondents after World War II was the well-known Italian demographer/statistician Corrado Gini (1884–1965), who developed the "Gini coefficient" for measuring the disparity in the national income of a country. He asked Haldane to send any books and papers that he may have published in the war years. Scientific research was disrupted in Italy during the war years. Communication with foreign countries was difficult. In a letter dated April 30, 1946, Gini wrote, "Here the academic and Scientific life goes on very slowly and with great difficulties. All the public institutions are in very bad financial conditions." Many years later, Gini visited the Indian Statistical Institute in Calcutta at the invitation of the director Mahalanobis, and wanted to see Haldane as well.

Someone wrote of him:

> Oh, listen to Corrado Gini
> He calls Vitamins, Vitamini.

Roles of Selection and Mutation in Evolution

One of the interesting problems considered by Haldane was the dynamics of mutation-selection balance, to which he returned repeatedly in several later papers. Haldane wrote that if selection acts against mutation, it remains ineffective unless the rate of mutation is greater than the coefficient of selection. Furthermore, it is not selection but mutation that is quite effective in causing an increase of recessives where these are rare in the population. Usually, mutation is also more effective than selection in weeding out rare recessives in a population. Haldane concluded: "Mutation therefore determines the course of evolution as regards factors of negligible advantage or disadvantage to the species. It can only lead to results of importance when its frequency becomes large."[49]

Gene Fixation

Haldane took up the stochastic treatment of investigating the probability of fixation of mutant genes. When a new mutant appears in a finite population, it either gets lost or becomes established (fixed) in the population after a certain number of generations. This is a fundamental aspect of population genetics that Haldane considered early in his series of pioneering papers in population genetics. In 1927 Haldane[50] investigated the probability of fixation of mutant genes, using the method of generating function suggested by Fisher in 1922.[51] Haldane showed for the first time that a dominant mutant gene having a small selective advantage k in a large random-mating population has a probability of about $2k$ of ultimately becoming established in the population. The calculation of the probability of fixation is more complicated if the advantageous mutation is completely recessive, but Haldane ingeniously showed that it is of the order of k/N, where k is the selective advantage of the recessives and N is the population number. Haldane's results were confirmed by later investigations. He also discussed the situation involving an equilibrium between recurrent mutation and selective elimination and the mechanism by which that equilibrium is reached in the case of complete dominance. Thirty years later, in 1957, Motoo Kimura confirmed Haldane's calculations and extended them to situations involving different levels of dominance.[52] Later, in 1962, Kimura developed a general formula for the probability of eventual fixation, $u(p)$, which was expressed in terms of the initial frequency, p, but also including random fluctuations in selection intensity and random genetic drift in small populations.[53]

When the population is entirely self-fertilized or inbred by brother-sister mating, both dominant and recessive factors have about the same chance as a recessive. Haldane further stated that in situations with partial self-fertilization or inbreeding an advantageous recessive factor has a finite chance of establishment after one appearance, however large be the population.

Metastable Populations

In Part VIII of his series of papers on the mathematical theory of natural selection, Haldane's concept of "Metastable populations" stands out. He wrote, "Almost every species is, to a first approximation, in genetic equilibrium; that is to say no very drastic changes are occurring rapidly in its composition. It is a necessary condition for equilibrium that all new genes which arise at all frequently by mutation should be disadvantageous, otherwise they will spread through the population. Now each of two or more genes may be disadvantageous, but all together may be advantageous."[54] Haldane cited the observation of Gonsalez, in 1923, who found that, in purple-eyed *Drosophila melanogaster*, arc wing or axillary speck (both due to recessive genes) shortened life, but together they lengthened it.

For instance, in the case of a system involving two dominant genes *A* and *B*, he expressed the relative fitnesses of the four phenotypes as follows:

$$AB1, aaB1 - k_1, Abb1 - k_2, aabb1 + K.$$

In an elegantly presented model of the two-locus case, assuming no linkage complications, Haldane plotted the trajectories of points representing the genetic composition of a population in a two-dimensional coordinate system. For a system containing *m* genes, he showed that a population can be represented by a point in m-dimensional space. He suggested further that related species represent stable types of the kind described in his model and that the process of speciation may be the result of a rupture of the metastable equilibrium. He added that such a rupture will be especially likely where small communities are isolated. Independently, Sewall Wright, in 1931, described his theory that evolution is a trial-and-error process that occurs in multidimensional adaptive surfaces, and that it tends to occur rapidly in small isolated populations.[55] There was, however, an important difference between Haldane and Wright. While Wright spent much of his life discussing and advancing his theory, Haldane, having made the suggestion, moved on to other subjects.

Types of Selection

In 1956, Haldane contributed to our understanding of the various types of natural selection that can impact on the process of evolution.[56] Selection that alters the mean of any character may be called *linear*. If it reduces the variance of a character, which involves elimination of extreme phenotypes, it is called *centripetal*. An example is the effect of selection on human birth weight, which eliminates babies with extremely low and high birth weights. If it increases the variance, it is called *centrifugal*. Kettlewell's observations on selection for melanism is both *linear* and *centripetal*.

Haldane distinguished between the two types of selection: *effective* selection, which changes the gene frequencies, and *ineffective* selection, which does not.

If the mean of a character changes in a certain expected direction it is called *directional* evolution. When the variance is reduced, by eliminating extreme genotypes, then it is called *normalizing* evolution. If its reduction involves elimination of genotypes that vary greatly in diverse environments, it is called *stabilizing* selection. But if the variance is increased, it is called *disruptive* evolution. However, Haldane did not consider a process "disruptive" if it leads to the establishment of a stable polymorphism. The term "normalizing evolution" was proposed by Haldane in this discussion to replace "normalizing selection" because the elimination of phenotypically extreme genotypes may be ineffective and does not necessarily lead to normalization.

It is in fact not possible to know whether selection will be effective without the knowledge of principles of genetics. It is known that at equilibrium (when the intensity of selection is balanced by the rate of mutation) evolution is neither *stabilizing* nor *normalizing*. Haldane emphasized that a new vocabulary is noted for the different types of *phenotypic* and *genotypic* selection, for example, selection at the human Rhesus (Rh) locus due to neonatal jaundice caused by the D antigen, which is phenotypically *disruptive* but genotypically *directional*.

Measurement of Natural Selection in Man

In 1954, Haldane proposed to measure the intensity of natural selection by $I = \ln s - \ln S$, where s is the fitness of the optimum phenotype and S is that of the whole population.[57] When he applied this to Karn and Penrose's data on human birth weight distribution, Haldane[58] found that while the death rate in all the babies in their sample was about 4.5 percent, that in the group whose weights lay between 7.5 and 8.5 pounds was only 1.5 percent. Two-thirds of these deaths were phenotypically selective, and the total intensity of selection was 3 percent. But this was phenotypic selection, not genotypic selection.

Haldane wrote,[59] "a great deal of phenotypic selection occurs before birth and is hard to measure. Genotypic selection can be measured with some degree of accuracy in certain situations as in the case of hemophilia or certain blood group genes."

Haldane suggested that we should try to argue from phenotypic to genotypic selection if adequate data are available. If an environmental influence simultaneously alters a metrical character and raises fitness, there is phenotypic selection in the direction in which that character is altered. For example, if an adequate diet promotes growth in length there is phenotypic selection for length. Even when a metrical character is wholly determined genetically, phenotypic and genotypic selections could be in opposite directions. If we consider a pair of alleles segregating in a population under random mating, Haldane showed that in a population at equilibrium phenotypic selection may be in the direction of the more dominant character. However, when the population is not in equilibrium, phenotypic and genetic selections may be in opposite directions. Haldane concluded that phenotypic and genotypic selections are usually in the same direction when measured over long periods.

Notes

1. Naomi Mitchison, "Beginnings," in Krishna R. Dronamraju, ed., *Haldane and Modern Biology* (Baltimore, MD: Johns Hopkins University Press, 1968), 299–305.
2. J. B. S. Haldane, A. D. Sprunt, and N. M. Haldane, "Reduplication in Mice," *Journal of Genetics* 5 (1915): 133–35.
3. Reduplication is the association of genes being on the same chromosome, which was later called "linkage."
4. Mitchison, "Beginnings," 304.
5. Mitchison, "Beginnings," 304.
6. Mapping function: A formula expressing the relation between distance in a linkage map and recombinant frequencies to calculate map distances corrected for multiple crossover products.
7. Thomas Hunt Morgan (1866–1945) was an American embryologist and geneticist. He was the great-grandson of Francis Scott Key, who wrote the "Star Spangled Banner." He won the Nobel Prize in Physiology or Medicine in 1933 for discovering the role of chromosomes and genes in understanding genetic inheritance. At the famous "fly room" at Columbia University in New York, Morgan led a vigorous team of young scientists including Alfred Sturtevant, Herman Muller, Calvin Bridges, and others, all of whom went on to enjoy illustrious careers in later years. The Morgan School, together with the Bateson School at Cambridge University in the UK, the Wilhelm Johannsen School in Copenhagen, and several other European scientific centers, contributed to the founding of genetics as a scientific discipline on a firm footing. Later, Morgan moved to the California Institute of Technology (Caltech), establishing its Division of Biology, which has produced seven Nobel laureates over the years.
8. Path coefficient analysis was a method developed by Sewall Wright in the 1920s to determine whether or not a multivariate set of nonexperimental data fits well with a particular causal model.
9. J. F. Crow, "Haldane, Fisher, and Wright," in Indian Statistical Institute, *J.B.S. Haldane: A Tribute*, Foreword by B. L. S. Prakasa Rao (Calcutta: Indian Statistical Institute, 1992), 61–67.
10. J. B. S. Haldane, "Farewell to R.A. Fisher (16 March 1959; Indian Statistical Institute)," in Indian Statistical Institute, *J.B.S. Haldane: A Tribute*, 44–46.
11. Gregor Johann Mendel (1822–1884) was an Augustinian friar who bred pea plants and discovered the fundamental principles of genetics. He demonstrated that the inheritance of certain characteristics in pea plants follows particular patterns. The profound significance of Mendel's work was not recognized until 1900, when the independent rediscovery of these laws by three European botanists initiated the modern science of genetics. These botanists are Hugo de Vries (1848–1935), Carl Correns (1864–1933), and Erich von Tschermak (1871–1962).
12. William Bateson (1861–1926) was an English biologist who coined the term "genetics" in 1906 for the new science of inheritance. Bateson introduced several other terms in genetics that are commonly used today, including *zygote*, for the individual that develops from the fertilized egg, *homozygote* and *heterozygote*, and F_1 and F_2 for first and second filial generations following hybrid crosses, as well as the term "allelomorph" (later changed to "allele"). In his book, *Materials for the Study of Variation*, Bateson showed, in 1894, that biological variation exists both continuously for some characters and discontinuously for others, and he coined the terms "meristic" and "substantive" for the two types.
    First mention of the term "genetics" by Bateson in a letter to Adam Sedgwick in 1905.

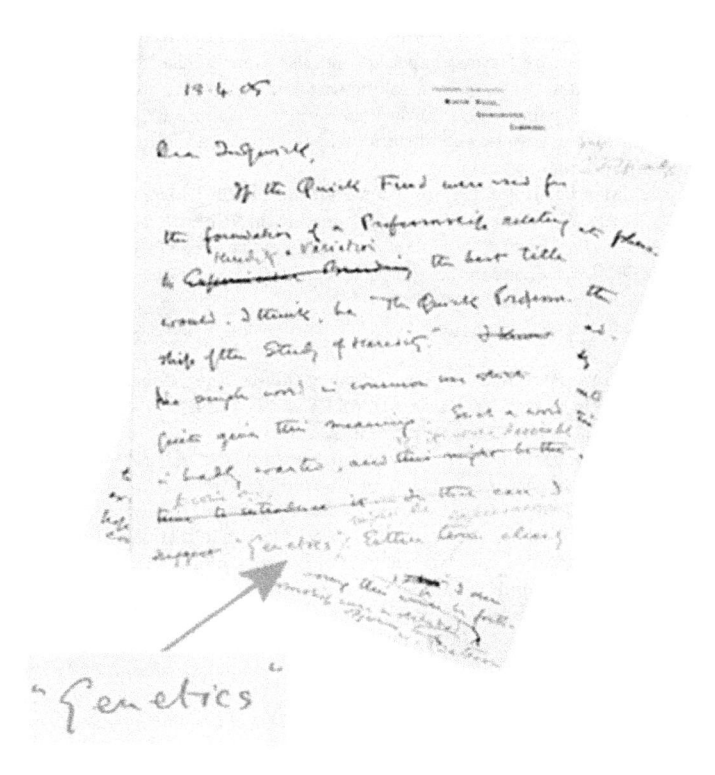

13. Godfrey Harold Hardy (1877–1947) was born in Cranleigh, Surrey, England. Both his parents were greatly interested in mathematics. Hardy was influenced by the French mathematician Camille Jordan and became well acquainted with European mathematical methods of his time. Hardy was appointed to the Savilian Chair of Geometry at Oxford in 1919. In 1931 he returned to Cambridge as the Sadleirian Professor until his retirement in 1942. Hardy's name is closely tied to the science of genetics because of the Hardy-Weinberg equilibrium, although he considered it a very minor aspect of his career. In 1911, Hardy started his long collaboration with fellow English mathematician John Littlewood in developing mathematical analysis and analytic number theory. Hardy preferred pure mathematics without any traditional applications of it in social devastations because he objected to the use of mathematics and its applications in warfare. Hardy's most famous collaboration started in 1914 with the Indian mathematician Srinivasa Ramanujan. He became Ramanujan's mentor and they developed the Hardy-Ramanujan asymptotic formula, which has been greatly used in physics required for finding quantum partition functions of atomic nuclei and also to derive thermodynamic functions of noninteracting Bose-Einstein systems. Hardy claimed that his greatest contribution in mathematics was his discovery of Ramanujan.
14. Wilhelm Weinberg (1862–1937), a German physician and obstetrician-gynecologist practicing in Stuttgart, published a paper in German that later came to be a part of the Hardy-Weinberg equilibrium. Weinberg also contributed other concepts that are routinely used in human

genetics. Weinberg developed the principle of genetic equilibrium independently of the British mathematician G. H. Hardy. He expressed his ideas in a lecture on January 13, 1908, about six months before Hardy's paper was published in English. Weinberg's contributions remained unknown to the English-speaking world for more than thirty-five years until Curt Stern wrote a brief paper ("The Hardy-Weinberg law." *Science* 97 [1943]: 137–38). Before 1943, the concepts in genetic equilibrium had been known as "Hardy's law" or "Hardy's formula" in English-language texts.

15. G. H. Hardy, "Mendelian Proportions in a Mixed Population," *Science* 28 (1908): 49–50.
16. J. F. Crow, "Hardy, Weinberg and Language Impediment," *Genetics* 152 (1999): 821–25.
17. R. A. Fisher, *Genetical Theory of Natural Selection* (Oxford: Clarendon Press, 1930); J. B. S. Haldane, *The Causes of Evolution* (London: Longmans, Green, 1932); S. Wright, "Evolution in Mendelian Populations," *Genetics* 16 (1931): 97–159.
18. W. B. Provine, *Sewall Wright and Evolutionary Biology* (Chicago: University of Chicago Press, 1986).
19. J. F. Crow, "The Founders of Population Genetics," in *Human Population Genetics*, ed. A. Chakravarthi (New York: Van Nostrand Reinhold, 1984), 177–94.
20. J. F. Crow, personal conversation with the author, 1965.
21. Suresh Dinakar Jayakar (1937–1988) was an Indian biologist who pioneered the use of quantitative approaches in genetics and biology. He studied mathematical statistics, physics, and mathematics at the University of Lucknow and in 1959 joined the Indian Statistical Institute in Calcutta, where he attended lectures by J. B. S. Haldane in mathematical genetics. At that institute, Jayakar received early instruction in genetics in a course taught by this author (Krishna Dronamraju) and additional training with Helen Spurway (Mrs. Haldane). He moved to Orissa when Haldane moved and became the director of the Genetics and Biometry Laboratory after Haldane's death. He made studies of the yellow-wattled-lapwing in collaboration with Helen Spurway. He collaborated with Helen Spurway and Krishna Dronamraju in studies of the nest-building activity of the wasp *Sceliphron madraspatanum* (Fabr.). Spurway Helen, S. D. Jayakar, and K. R. Dronamraju, "One Nest of *Sceliphron madraspatanum* (Fabr.). (Sphecidae: Hynemoptera)." *Journal of the Bombay Natural History Society* 61 (1964): 1–42.

He moved to work with L. L. Cavalli-Sforza at the University of Pavia and continued his studies on quantitative genetics and sex-determination mechanisms.
22. E. Mayr, "Where Are We?" *Cold Spring Harbor Symposia on Quantitative Biology* 24 (1959): 1–14.
23. E. Mayr, *Animal Species and Evolution* (Cambridge, MA: Harvard University Press, 1963). "Beanbag genetics," an initially derogatory term, was popularized by the evolutionary biologist Ernst Mayr, to describe the early mathematical population genetics that was founded by R. A. Fisher, J. B. S. Haldane, and S. Wright. Mayr wrote that the Mendelian was apt to compare the genetic contents of a population to a bag full of colored beans. Mutation was the exchange of one kind of bean for another. This subject is discussed in greater detail in Krishna R. Dronamraju, *Haldane, Mayr and Beanbag Genetics* (New York: Oxford University Press, 2009).
24. J. B. S. Haldane, "A Defense of Beanbag Genetics," *Perspectives in Biology and Medicine* 7 (1964): 343–59.
25. Haldane, "A Defense of Beanbag Genetics."
26. Haldane, "A Defense of Beanbag Genetics."
27. J. B. S. Haldane, "The Interaction of Nature and Nurture," *Annals of Eugenics* 13 (1946): 197–204.
28. Sewall Wright, "Contributions to Genetics," in Dronamraju, *Haldane and Modern Biology*, 1–12.
29. Joan Fisher Box, *R.A. Fisher: The Life of a Scientist* (New York: Wiley, 1978).
30. William B. Provine, *Sewall Wright and Evolutionary Biology* (Chicago: University of Chicago Press, 1986).
31. R. A. Fisher, *Genetical Theory of Natural Selection* (Oxford: Clarendon Press, 1930).
32. Wright, "Contributions to Genetics."
33. J. B. S. Haldane, "A Mathematical Theory of Natural and Artificial Selection: Part I." *Transactions of the Cambridge Philosophical Society* 23 (1924): 19–41.

34. J. B. S. Haldane, "A Defense of Beanbag Genetics," *Perspectives in Biology & Medicine* 7 (1964): 343–359.
35. J. B. S. Haldane, "A Mathematical Theory of Natural and Artificial Selection: Part IV," *Proceedings of the Cambridge Philosophical Society* 23 (1927): 607–15.
36. H. T. J. (Harry) Norton was a mathematical population geneticist who studied under G. H. Hardy at Cambridge. He showed his early work to Haldane in 1910 but published very little until 1928. Some of his early work on natural selection was included in Punnett's book *Butterflies*. Norton suffered from mental health problems. His sister Betty Norton asked Haldane to make sure that Norton's early contribution to population genetics was duly recognized.
37. J. B. S. Haldane and S. D. Jayakar, "The Solution of Some Equations Occurring in Population Genetics," *Journal of Genetics* 58 (1963): 291–317.
38. Haldane, 1924 "A Mathematical Theory of Natural and Artificial Selection: Part I."
39. C. Elton, *Voles, Mice and Lemmings: Problems in Population Dynamics* (Oxford: Oxford University Press, 1942).
40. J. B. S. Haldane, "The Selective Elimination of Silver Foxes in Eastern Canada," *Journal of Genetics* 44 (1942): 296–304.
41. Roy Harrod, *The Life of John Maynard Keynes* (London: Macmillan, 1951).
42. R. C. Punnett, ed. *Mimicry in Butterflies* (Cambridge: Cambridge University Press, 1915), 154–55. Includes a table on natural selection by Harry Norton.
43. H. T. J. Norton, "Natural Selection and Mendelian Variation," *Proceedings of the London Mathematical Society* 28 (1928): 1–45.
44. Haldane, "A Mathematical Theory of Natural and Artificial Selection: Part IV."
45. A. J. Lotka, "The Stability of the Normal Age Distribution," *Proceedings of the National Academy of Sciences, Washington* 8 (1922): 339–45.
46. L. I. Dublin and A. J. Lotka, "On the True Rate of Natural Increase," *Journal of the American Statistical Association* 20 (1925): 305–39.
47. J. Sutter, "Haldane and Demographic Genetics," in Dronamraju, *Haldane and Modern Biology*, 73–77.
48. G. Malecot, *Le Mathematiques de l'heredite* (Paris: Masson, 1948).
49. J. B. S. Haldane, "A Mathematical Theory of Natural and Artificial Selection: Part V. Selection and Mutation," *Proceedings of the Cambridge Philosophical Society* 23 (1927): 838–44. Gene fixation.
50. Haldane, "A Mathematical Theory of Natural and Artificial Selection: Part V."
51. R. A. Fisher, "On the Dominance Ratio," *Proceedings of the Royal Society of Edinburgh* 42 (1922): 321–41.
52. M. Kimura, "Some Problems of Stochastic Processes in Genetics," *Annals of Mathematical Statistics* 28 (1957): 882–901.
53. M. Kimura, "On the Probability of Fixation of Mutant Genes in a Population," *Genetics* 47 (1962): 713–19.
54. J. B. S. Haldane, "A Mathematical Theory of Natural Selection: Part VIII. Metastable Populations," *Proceedings of the Cambridge Philosophical Society* 27 (1931): 137–42.
55. S. Wright, "Evolution in Mendelian Populations," *Genetics* 16 (1931): 97–159.
56. J. B. S. Haldane, "Natural Selection in Man," *Acta Genetica et Statistica Medica* 6 (1956): 321–32.
57. Haldane, "Natural Selection in Man."
58. Haldane, "Natural Selection in Man."
59. J. B. S. Haldane, "The Measurement of Natural Selection," *Caryologia* (Turin), Suppl. 6 (1954): 480–87.

5

Evolutionary Biology

Haldane's mathematical investigations of the evolutionary process began in 1924 and continued until his death in 1964, some publications appearing posthumously. During his last years, S. D. Jayakar was his coauthor in several papers dealing with various conditions for polymorphism and related topics. Haldane's first paper, in 1924, was stimulated by the early study of natural selection by H. T. J. (Harry) Norton, which was presented briefly by R. C. Punnett in his book *Mimicry in Butterflies* (1915). Haldane's investigations during the first eight years (1924–1932) were summarized in his book *The Causes of Evolution* (1932).

Haldane's contributions to evolutionary theory spanned four decades, from the 1920s until his death in 1964. While many of his papers dealt with the dynamics of the evolutionary process, he wrote on "The Statics of Evolution" in a paper he wrote for a festschrift for Julian Huxley, *Evolution as a Process*.[1] Haldane discussed evolution as an almost unobservable process in a lifetime, considering the processes that bring it about as being almost in equilibrium. In reality, however, evolution is probably proceeding with unparalleled speed. Haldane suggested that, after the series of violent climatic changes of the Pleistocene, "man has altered the ecological conditions of most animal and plant species, including his own."[2] One situation that was investigated by Haldane involved conflict between mutation and selection. Many of the rare human congenital abnormalities are dominants, the homozygous mutant being unknown or lethal, and the heterozygote abnormal. Some others are sex-linked recessives. Natural selection rapidly eliminates the genes in question, while new ones arise by fresh mutation. Equilibrium is reached soon unless the mutation rate changes rapidly. The same process occurs for autosomal recessives, only more slowly. It is quite possible that much of the "normal" variation in a species is due to this cause.

Another situation prevails when selection occurs by labile pathogens. For instance, when rusts attack wheat crops, one genotype can be immune to one strain of a rust but susceptible to another. The difference may often depend on a single gene substitution. In a natural population of hosts and pathogens composed of many genotypes, there will be selection for pathogens that are adapted to common biochemical variants. In fact, both host and pathogen will constantly alter their prevailing genotype, as it relates to the host–parasite relationship. The same is true of animals and their parasites. This process will encourage a diversity of types in the host, and also in the pathogens, resulting in certain evolutionary consequences. Haldane discussed the topic of adaptation with reference to different types of "clines,"[3] which were introduced into biology by Julian Huxley.

Recurrent Mutation

In a series of papers, which dealt with the evolutionary consequences of the occurrence of mutations on a population, Haldane considered two kinds of effects: primary and secondary.[4] The former include the gradual spread of the new mutant in the population, replacing the original gene. The secondary effects include a number of situations arising from the spread of the new mutant in a population. Secondary effects of frequent disadvantageous mutations include (1) an increase of dominance due to mutation of dominant alleles, (2) increase of dominance due to spread of modifying genes, (3) selective value of certain situations such as polyploidy and duplication, (4) heterogametism of male rather than female sex, (5) concentration of mutable genes in the X chromosome, and (6) development of internal balance in the X chromosome (including possibly X/autosome balance), to name a few.

But, surprisingly, Haldane concluded that mutation is a necessary but not sufficient cause of evolution.[5] While mutation provides the raw material for natural selection to act on, the actual evolutionary trend itself would be determined by selection. Haldane then showed that recurrent mutation cannot overcome the impact of selection of even quite moderate intensity. He mentioned the case of Stadler's work in maize,[6] where the largest value for the probability of a gene mutating to a less fit mutant was 4×10^4, the lowest being less than 10^6. Haldane argued that even if we take 10^3 as an upper limit in ordinary cases, a coefficient of selection k of 10^3 would prevent them from spreading very far.

In *The Causes of Evolution*, Haldane also considered two specific instances where very high mutation rates, induced by heat, have played an important role in evolution.[7] First, it may have been responsible for the orthogenetic evolution of species near the tropical limit of their range, and may have been at least partially responsible for the occurrence of greater diversity of species found in tropical as compared with temperate and arctic habitats. Second, the gonads of mammals and birds are permanently at a higher temperature than is usual in other organisms. Haldane speculated that when this temperature first evolved, it increased the mutation rate of their genes. Simultaneously, many new ecological niches were open, offering many opportunities for selective advantage for the many new types of mutation. Therefore, he suggested that these facts may have played a significant part in the very rapid evolution of mammals during the Eocene.[8] However, Haldane concluded that confirmatory evidence on the evolution of populations of *Drosophila* or other species under the influence of high temperature was lacking at that time. He suggested further that similar outbursts of mutation might have been caused by natural radioactivity or by certain chemical substances in the past. He doubted, however, that the amount of radioactive substances near the earth's surface 10^9 years ago was sufficient even to double the present mutation rates!

As in many publications of Haldane, there was a good deal of speculation in his 1933 paper on recurrent mutation.[9] But Haldane was both honest and bold about his speculations. He was not afraid to go out on the limb, when so many other scientists are reluctant to speak out. There was refreshing intellectual honesty in Haldane's writings. He invoked mutation pressure to explain the loss of useless organs, such as eyes in cave animals. However, it is selection that is invoked today to explain such situations. For example, cave animals that can see are free to migrate, and do so, out of the caves. Furthermore, as emphasized by H. J. Muller, an organ that is useless is not selectively neutral. It exerts a strain on the organism, by depleting valuable resources that could be useful elsewhere.

Haldane suggested in 1932, when he was forty, that when there are excessive numbers of young males in a population, who can get no mates, "and may be a nuisance to the herd," a few sex-linked lethal genes may be a positive advantage to the species.[10] He argued further that the presence of genes conducing to cancer, which kills off superfluous old men and women, would confer a similar advantage to the species. This turned out to be a prophetic statement, as he himself died of cancer at the age of 72!

Among other ideas of Haldane are explanations, in terms of mutation and natural selection, for *tachygenesis*[11] and *recapitulation*.[12] Genes will come into action rather late, but will extend their sphere of action in time. With respect to *neoteny*,[13] genes that originally determined temporary embryonic or larval characters will tend to extend their action forward into adult life. Haldane cited the example of the embryonic cranial flexure, which later culminates in the human head. Haldane attributed this to mutation pressure, not natural selection. He wrote that the process has been aided by the tendency of the genes concerned to develop a "factor of safety," protecting them against mutant genes that would otherwise pose danger in the heterozygous condition.

Mutation versus Natural Selection

Haldane wrote, "In general, mutation is a necessary but not sufficient cause of evolution. Without mutation there would be no gene differences for natural selection to act upon. But the actual evolutionary trend would seem usually to be determined by selection."[14] He argued that mutations may give rise to primary and secondary effects, the former due to the accumulation of mutant genes, the latter to the selective value of conditions that protect the organism against lethal genes. He suggested that the consequences might include the disappearance of useless organs, recapitulation, and the fact that the heterogametic sex is usually male.

In several papers and books, Haldane dealt with the quantitative treatment of mutations in populations and families. His investigations included both the statics and the dynamics of the mutation process. Much of his discussion centered on the impact of mutation on the evolutionary process, especially its role in relation to the impact of selection in populations. He was the first among the great biologists to show that by balancing mutation rate against selection intensity, one can arrive at an understanding of the role of mutation in the evolutionary process. By using this principle, Haldane showed how the mutation rates of various human diseases and other traits could be estimated. The first of these was mutation rate for hemophilia,[15] which was first mentioned by Haldane in his classic work *The Causes of Evolution* in 1932.

In 1935, Haldane's initial research on human mutation rates[16] followed some earlier work on the mutation rates of the fruit fly (*Drosophila*) by Muller[17] and maize (*Zea*) by Stadler.[18] As he put it, "Satisfactory data on rates of spontaneous mutation exist for *Zea* and *Drosophila*, but not for vertebrates. However, such data may be determined for man by indirect methods. Clearly the rate at which new autosomal genes recessive to the normal type appear can only be accurately estimated where either inbreeding or very extensive back-crossing to recessives is possible. But under ordinary conditions new dominants or sex-linked recessives can be detected more readily."[19]

In his investigation of the effect of natural selection against the gene for hemophilia on the human X chromosome, Haldane pointed out that the loss of this gene in each generation must be balanced by recurrent mutation. If this were not so, the disease would quickly die out. Thus, natural selection would not act on female heterozygous carriers of the gene (XX) who are normal but on affected hemizygous males (XY). Accordingly, Haldane argued that nearly one-third of the known cases of hemophilia must arise by mutation of the gene in each generation. In *The Causes of Evolution*, Haldane stated that the mutation rate for hemophilia is of the order of once in a hundred thousand generations or somewhat more. Later, he gave the general formula for sex-linked genes as $2\mu + v = (1-f)x$, where μ and v are the mutation rates in females and males respectively, x is the frequency of hemophilia among males at birth, and f is the fitness of hemophilic males as a fraction of the normal. He calculated further that the mutation rate was much higher in males than in females, and made similar calculations for sex-linked muscular dystrophy.

Haldane's estimate in 1935 of a human mutation rate of about one spontaneous mutation in about 50,000 life cycles has been revised since. He suggested further that for other cases, such as neurofibromatosis, the mutation rate might well exceed 1 per 100,000. It is important to note that Haldane's investigation of hemophilia was conducted before the recognition of hemophilia A and B (and Christmas disease). With remarkable intuition, he speculated at that time that there are probably two distinct allelomorphs at the same locus, the milder type arising less frequently by mutation than the severe type.

In his Royal Society's Croonian Lecture, Haldane pointed out that the mutation rate is probably more or less adaptive; "Too high a mutation rate would flood a species with undesirable mutations, too low a one would probably slow down evolution."[20] Man and *Drosophila melanogaster* have about the same rate per generation, and if this were increased ten times it would result in a very great loss in fitness. Haldane doubted whether the human mutation rate could be lowered much further, because a "substantial fraction" of it is caused by natural radiation. In fact, with remarkable foresight, Haldane predicted that a very great prolongation of human life, or at least the reproductive period, might be incompatible with the "survival" of the human species.

The Strange Case of Peppered Moth: Industrial Melanism

Haldane enjoyed applying his theoretical and mathematical analysis to practical situations that involved observations and data collected in the field by other scientists. This was in fact a major part of his contribution to science. His analysis and insights often provided valuable information that was not foreseen by the investigators themselves. One such analysis involved industrial melanism[21] in the industrial districts of England.

Early in the nineteenth century, the peppered moth was known to most naturalists, including Charles Darwin, as a predominantly white-winged moth liberally speckled with black. While the moths were resting on tree trunks, the lighter form was protected from predators because it blended with the light-colored lichens on the trunks. Later, with increasing industrialization in the nineteenth century, soot, smoke, and other industrial pollutants from factories darkened the landscape, and the tree trunks where the moths rested were no exception. This sudden change in their environment made the moths highly vulnerable to birds. However, by the turn of the twentieth century, the black (*carbonaria*) variant of the moth had largely if not entirely replaced the light-speckled (typical) form in the most polluted parts of UK. Experiments by Bernard Kettlewell showed conclusively that the darker melanic forms were eaten much more frequently by birds in the rural areas, where there is less pollution, as compared with the melanic forms in the urban areas, where the tree barks are darker because of greater pollution.

Many years later, with the passage of the 1956 Clean Air Act, the black variant of the moth began to decline and the white form, which was better camouflaged on lichen, reemerged on the trees. Peppered moth evolution is commonly used as an example of industrial melanism. Yet, despite its importance within evolutionary biology, the molecular genetic and developmental control of the *carbonaria*-typical polymorphism is unknown. All that is known on this topic is that the trait is controlled by a dominant gene at a single Mendelian locus.

The following conclusions were reached after extensive experiments and field observations.[22]

	Number of moths eaten by birds	
	Pepper	Melanics
Urban (more pollution)	15	43
Rural (less pollution)	164	26

1. Industrial melanism is genetically controlled by a single locus in *B. betularia*.
2. Populations have undergone evolutionary change in color pattern.
3. That change is consistent with the interpretation that it was due to natural selection, in that there is differential survival of the genotypes caused by differential predation on a particular background.
4. Results confirm qualitative prediction of equation for gene frequency change.

Haldane on Industrial Melanism

Long before the studies of Kettlewell,[23] Haldane calculated that the gene for melanism conferred a selective advantage of about 50% on its carriers.[24] It is so high that few biologists accepted his findings at that time. Haldane wrote: "Few or no biologists accepted this conclusion. They were accustomed to think, if they thought quantitatively at all, of advantages of the order of 1% or less. Kettlewell ... has now made it probable that, in one particular wood, the melanics have at least double the fitness of the original type."[25]

Haldane also considered analysis of the nearly complete replacement of light-colored moths, *Biston betularia*, by a semidominant dark mutant form that in the end became completely dominant, presumably by direct selection of modifiers. In a paper published in the *Proceedings of the Royal Society of London* in 1956, Haldane wrote, "Dr. Kettlewell's proof that the dark form *carbonaria* of *Biston betularia* has replaced type, at least in part as the result of selection by bird predators, gives me the right to bring my calculation (Haldane 1924) on this matter up to date."[26] Haldane's analysis of Kettlewell's data led to new and unexpected conclusions: Selection against the melanic form was generally much less intense than was found by Kettlewell, there may have been immigration of the lighter form from unpolluted areas into polluted areas, and selection has slowed down (selective advantage) due to some special reason. One such possibility is due to what geneticists call "balanced polymorphism," which involves the existence of two kinds of melanic forms. Although they differ in their underlying genetic basis, they are indistinguishable externally.

Evolution of Dominance

In population genetics, one of the key issues was explaining the evolution of inheritance. Fisher's theory of the evolution of dominance[27] was concerned with the evolution of modifier genes that act on other genes to make them dominant or recessive, and that these other genes are then themselves subject to natural selection. It was expanded in his book *The Genetical Theory of Natural Selection*. However, Sewall Wright and J. B. S. Haldane believed that the main explanation for dominance should be based on physiological factors, and that selection for modifiers was not a primary force. Subsequent works, particularly in molecular biology and biochemistry, have tended to favor the view of Wright and Haldane. At present, the prevalent view is that dominance cannot evolve as a direct result of selection. Furthermore, it has been argued that due to inherent constraints in biochemical systems, the manifestation of dominance is a default expectation and hence evolutionary explanations are not necessary. However, there are also several studies indicating that dominance levels can be modified as a result of changes in the genetic background. Furthermore, other studies have indicated that dominance selection is possible in certain circumstances. The controversy remains unresolved.

The Causes of Evolution

Haldane's early work in genetics was summed up in his book *The Causes of Evolution*. He synthesized the latest advances of that period in several branches of biology. That book stands as a pillar of the foundations of population genetics and evolutionary biology. He brought together evidence in support of Darwinian evolution from multiple disciplines including population genetics, cytogenetics, paleontology, and biochemistry, among others. The biologist Hampton Carson put it succinctly: "Haldane neatly conjoins Darwin and Mendel, Fisher and Wright, Newton and Kihara. In the evolutionary context, Haldane deals for the first time with inversions and translocations polyploidy and hybridization. The paleontological record is woven into the argument."[28]

Referring to the interrelationships between various species, Haldane argued that a biochemist who finds the same quite complex molecules in all plants and animals can hardly doubt their common origin; "There may be some reason in the chemical nature of things why all living creatures must contain glucose."[29] As early as 1920, he was referring to the gene in physicochemical terms: "The chemist may regard them as large nucleoprotein molecules, but the biologist will perhaps remind him that they exhibit one of the most fundamental characteristics of a living organism: they reproduce themselves without any perceptible change in various different environments."[30]

After summing up paleontological and geological evidence, Haldane considered the following questions:

1. What is the nature of heritable differences within a species?
2. Are the differences between species of the same or of a different character?
3. Does natural selection occur in nature?
4. If so, will it account for the formation of species?
5. Must we allow for other causes of evolutionary change?
6. A value judgment and "not scientific" (in his words), How do we judge the process of evolution: Is it good or bad, beautiful or ugly, directed or undirected?

Haldane spent much of his life attempting to find answers to these questions.

The Causes of Evolution contains some of Haldane's most famous statements, which are still quoted today. For instance, with respect to the role of divine power in guiding the evolutionary process, Haldane wrote, "There are two objections to this hypothesis. Most lines of descent end in extinction, and commonly the end is reached by a number of different lines evolving in parallel. This does not suggest the work of an intelligent designer, still less of an almighty one."[31]

According to Haldane, the second objection—on moral grounds, is a more serious one. He was referring to the evolution of several parasitic species, which have lost the use of certain organs but cause considerable pain and suffering to man and other species. Haldane equated this process to moral breakdown in a human being. He wrote, "We have now to ask whether God made the tape-worm. And it is questionable whether an affirmative answer fits in either with what we know about the process of evolution or what many of us believe about the moral perfection of God."

Among various possibilities, Haldane suggested that we could regard the dark as well as the bright side of evolution as a manifestation of divine ingenuity. He quoted *Isaiah*: "I make peace, and create evil: I the Lord do all these things." With tongue firmly in cheek, Haldane wrote that the tapeworm presents just as much ingenuity as does the rose. He added that perhaps we should give the devil credit for a large share in evolution. Or, we can say that at present it does not seem necessary to postulate divine or diabolical intervention in the evolutionary process. He concluded, "The question whether we can draw theological conclusions from the fact that the universe is such that evolution has occurred in it is quite different, and very interesting."

With respect to the so-called progress in evolution, Haldane commented that the change from monkey to man might well seem a change for the worse to a monkey, "But it might also seem so to an angel." He wrote further, "we must remember that when we speak of progress in evolution we are already leaving the relatively firm ground of scientific objectivity for the shifting morass of human values."[32]

It has been speculated that Haldane's motive for studying evolution may have been due to his lively interest in religion. However, Haldane's task was to convince the nonbiological lay readers, not his fellow biologists, most of whom were already convinced of the fact of Darwinian evolution. Even more important, it would be too simplistic to say that Haldane's interest in evolution was motivated by a single cause, especially an antireligious bias. It would be more accurate to say that his interest in evolution was due to several factors. Foremost among these was his interest in biology, especially in genetics, which was evident from an early age.

Reasons for Ignoring *The Causes of Evolution*

Why was Haldane's *Causes* less often quoted than the works of his cofounders of population genetics (those of R. A. Fisher and S. Wright) in the scientific literature? There are at least three reasons:

1. Some writers have interpreted the development of mathematical population genetics primarily as an argument between Fisher and Wright, with Haldane playing a quiet role in the background.
2. The Harvard University biologist Ernst Mayr[33] offered the following explanation: Haldane's *Causes* may be considered by some as a somewhat popular or semipopular work, was less often referred to in the United States in the technical evolutionary literature, and was missing in many libraries.
3. At least one reviewer of my book of Haldane's collected papers[34] commented that many fellow scientists in his generation were not referred to study Haldane's books and papers, perhaps because of Haldane's communist and Marxist connections.

In the years preceding the publication of *The Causes of Evolution* there was a general decline in support for Darwin's theory of natural selection. Haldane's defense of Darwinism in the *Causes* was a necessary prelude to his exposition of the Darwinian theory of evolution. Strange as it may seem from today's point of view, Haldane, in the 1920s, had to refute at first several misconceptions and false beliefs in the public mind about the theory of evolution. Haldane was also concerned with correcting Darwin's errors, such as his belief in the theory of blending inheritance. He pointed out that blending inheritance requires a huge reduction in variation in each generation followed by the production of an equally huge amount of new genetic variation in the next generation. Darwin erroneously believed that this could result from environmental interaction. Haldane further wrote that Darwin observed blending (i.e., diminution of variation) because the mating system of his domestic animals and plants was suddenly changed, such as mating together two races of pigeons that had been bred separately for many years.

What was remarkable about Darwin was that even though he was ignorant of the underlying causes of variation in a population, namely, mutation, recombination, and chromosomal aberrations, he proposed a theory that was essentially correct in its major features.

Haldane attempted at first to show that all claims of Lamarckian inheritance, especially the inheritance of acquired characters that was still popular in France, were not supported by evidence. He explained that certain claims of experimental evidence were in fact due to unconscious selection. Haldane refuted such claims so completely that there was no mention of that topic in Dobzhansky's *Genetics and the Origin of Species*,[35] which was published five years later.

Beanbag Genetics

The evolutionary biologist Ernst Mayr, in his address to a Cold Spring Harbor Symposium in 1959,[36] questioned the three founders of population genetics, Haldane, Fisher, and Wright: "But what, precisely has been the contribution of this mathematical school to the evolutionary theory, if I may be permitted to ask such a provocative question?" He was particularly concerned with the mathematical school because one of his favorite subjects, namely, speciation, was not addressed by them. He addressed the problem of what he called "beanbag genetics," in which "evolutionary genetics" was presented as an input or output of genes, similar to adding or removing certain beans from a beanbag. Elsewhere, Mayr wrote that the Mendelian was apt to compare the genetic contents of a population to a bag full of colored beans. Mutation was the exchange of one kind of bean for another. Mayr objected to the concept that genes were either dominant or recessive and they have constant selective values, as well as the tendency to equate genes and characters, implying a simple gene–character relationship. Mayr commented that the treatment of genes in early population genetics as noninteracting independent units is misleading from an evolutionary and biochemical viewpoint. However, Mayr praised Haldane in his review of the reprint of *The Causes of Evolution* in 1992. Mayr wrote, "I have always admired Haldane for the frankness with which he acknowledged his ignorance about certain problems.... Haldane was the most open-minded of the Fisher-Haldane-Wright trio. He perceived aspects of evolution that are often ignored."[37]

It is not clear who could be the target of Mayr's criticism. Among the three founders of population genetics, Wright had always used epistatic and interacting systems in his formulation of population genetics. Haldane and Fisher considered simpler models in their early papers but included complicated and interacting systems in their later papers. In response to Mayr's criticism, Haldane wrote "A Defense of Beanbag Genetics,"[38] in which he defended the early mathematical foundations of Fisher, Wright, and himself, which he regarded as an essential framework that needed to be tested, but adequate experimental data were not forthcoming.

Mathematical Analysis

Haldane defended the use of mathematical analysis. He wrote that a mathematical analysis of the effects of selection is necessary and valuable to study the effects of natural selection. "Many statements which are constantly made, e.g. 'Natural selection cannot account for the origin of a highly complex character' will not bear analysis." He added that the conclusions drawn by common sense on this topic are often very doubtful.[39] Haldane wrote that unaided common sense may indicate an equilibrium, but rarely tells us if it is stable. Haldane predicted that the permeation of biology by mathematics had only just begun and would continue in the future. In "A Defense of Beanbag Genetics" Haldane wrote, "In the consideration of evolution, a mathematical theory may be regarded as a kind of scaffolding within which a reasonably secure theory expressible in words may be built up ... without such a scaffolding verbal arguments are insecure."[40] In the same work, Haldane cited an example from astronomy, one of the sciences that often interested him. Some critics thought that if the sun attracted the planets, as Newton had suggested, they would fall into it. Newton had to show further that the inverse-square law led to stable elliptic motion, and that spheres attracted one another as if they were particles.

Rates of Evolution: Hard Parts

Another important idea of Haldane was concerned with the measurement of rates of evolution and his suggestion that a unit called *darwin*[41] can be used as a measure of the rate of change in evolution. For instance, the rate of change in the size of a certain organ during the course of evolution can be measured in terms of a certain change per generation or in a given time, such as one year. In some species of insects, the generation time is exactly one year. However, we can never do so for fossil forms that have become extinct.

In the case of a metrical or a measurable character, the simplest and most direct measurements are linear. Even with a sample of ten specimens, considering the change in an organ, such as the adult mammalian tooth, we can calculate their mean value with a standard error of about 2 percent. On the other hand, a ratio of lengths can be stated with a somewhat greater precision. When comparing rates of change of different organs, it is more accurate to compare changes on a percentage basis rather than absolute length. If, on the other hand, we are concerned with changes in vertebrate bones, we should choose a measure of shape, which does not change much with the age of the animal.

Another method suggested by Haldane defined the unit of change (a Darwin) as the increase or decrease of a mean value through one standard deviation of the character in question as determined from a population found at a single horizon. The use of the standard deviation as a yardstick is interesting because, according to the theory of evolution, the variation within a population at any time constitutes the raw material available for evolution. However, this variance has already been diminished to some extent because the fossil record consists mainly of the hard parts of adults, such as teeth and bones.

Haldane's analysis included an analysis of fossil records of certain tooth measurements in ancestors of horses. He suggested that the evolutionary rate could be measured in a unit such as a *darwin*, which represents an increase or decrease of size by a factor of e per million years, or an increase or decrease of 1/1000 per 1,000 years. According to this measure, the evolutionary rates in the fossil ancestors of horses would range around 40 millidarwins. Rates under a *millidarwin* would be hard to measure, and a rate of one *darwin* would be very rare. However, domesticated animals and plants have undergone changes that can be measured in *kilodarwins*, but not *megadarwins*.

Haldane estimated that the median figure of 40 *millidarwins* (or 4×10^8) for the horse evolution implies that the means were increased by a standard deviation in half a million years. To Haldane, such minute changes indicate the extreme slowness of evolutionary changes under natural selection. Hence, other processes also must be taken into account. Although certain mutation rates may reach quite substantial rates such as 7×10^9 per year, it is natural selection that ultimately decides the direction of evolution.

For Haldane, the rate of evolution is set by the number of loci in a genome, and the number of stages through which they can mutate. For instance, if pre-Cambrian organisms had much fewer loci than their descendants, they may have evolved much quicker, but the possibilities open to them would be limited.

Cost of Natural Selection (Evolution)

Toward the end of his life, Haldane's fertile mind produced yet another important idea on the study of evolution. In his paper "The Cost of Natural Selection,"[42] Haldane explained his idea as follows. Suppose a population is in equilibrium under selection and mutation; that is to say, any change due to natural selection is balanced by fresh mutations. A sudden change occurs in the environment, such as a change of climate, the appearance of a new predator or parasite, the sudden availability of a new food source, or pollution or contamination of the environment, or migration to a new habitat, and so on. The species is less adapted to the new environment, and its reproductive capacity is lowered. However, it is gradually improved as a consequence of natural selection. The process involves a number of deaths, or their equivalents in lowered fertility. Kimura has referred to this loss as the substitutional (or evolutional) load. It is the "cost" paid by a species for evolving.[43]

Haldane outlined the goal of his paper in the following terms: "In this paper I shall try to make quantitative the fairly obvious statement that natural selection cannot occur with great intensity for a number of characters at once unless they happen to be controlled by the same genes. ... The principal unit process in evolution is the substitution of one gene for another at the same locus. ... I shall show that the number of deaths needed to carry out this unit process by selective survival is independent of the intensity of selection over a wide range."

Following Haldane, if selection at the ith selected locus is responsible for d_i of these deaths in any generation, the reproductive capacity of the species will be $\Pi(1 - d_i)$ of that of the optimal genotype, or $\exp(-\Sigma d_i)$ nearly, if every d_i is small. Thus the intensity of selection approximates to Σd_i. With reference to a particular locus, Haldane showed that the total number of selective deaths or the equivalent in lowered fertility required depends mainly on the initial frequency of the gene that was later favored by natural selection.

Haldane's conclusions are best described in the discussion of his paper. The unit process of evolution, the substitution of one allele by another, usually involves a number of deaths equal to about 10 or 20 times the number in a generation, always exceeding this number, but rarely approaching 100 times this number. He took 30 as a mean to allow for occasional high values. If two species differ at 1,000 loci, and the mean rate of gene substitution is one per 300 generations, then Haldane estimated that it would take at least 300,000 generations to generate an interspecific difference. It might take much longer. Haldane wrote, "To conclude, I am quite aware that my conclusions will probably need drastic revision. But I am convinced that quantitative arguments of the kind here put forward should play a part in all future discussions of evolution."

James F. Crow and his colleagues have suggested that the observed rates of amino acid changes in proteins such as hemoglobin, which average approximately one amino acid change per 10 million years in recent mammalian history, are consistent with a moderate amount of natural selection during this period; 10,000 gene loci, each with a nucleotide substitution every two million years, would require a reproductive excess of 15 percent (assuming 30 as the average cost per substitution).[44] Sewall Wright had suggested that evolution may largely depend not on substitution of initially rare mutants but on minor shifts in gene frequencies, especially of those with intermediate frequencies.[45] Such a model would require less reproductive excess than Haldane's approach. Haldane summed up his investigation as follows: "On the whole it seems that the rate of evolution is set by the number of loci in a genome, and the number of stages through which they can mutate. If pre-Cambrian organisms had much fewer loci than their descendants, they may have evolved much quicker, though the possibilities open to them were more limited."[46]

Haldane's discussion was only concerned with that part of mortality and lowered fertility that was selective. In a later paper, Haldane warned the reader against extending the theory to situations to which it does not apply. For instance, a genotype that was initially unfavorable becomes gradually neutral and then favorable. Such facts may lead to some modification in the application of the theory of cost of selection. However, it is abundantly clear that any serious discussion of the evolutionary process must include the questions raised by Haldane.

In contrast to Haldane's estimate, calculations by the Japanese geneticist Motoo Kimura[47] of the rate of evolution in terms of nucleotide substitutions gave a value so high that Kimura postulated that many of the mutations involved must be neutral or almost neutral ones. This gave rise subsequently to the neutral theory of evolution, to which several scientists have contributed in later years. However, the "cost of selection" concept of Haldane has been criticized, especially by Warren Ewens.[48]

Haldane's Dilemma

The paleontologist Leigh Van Valen introduced the term "Haldane's Dilemma" in 1963. Van Valen wrote,

> Haldane ... drew attention to the fact that in the process of the evolutionary substitution of one allele for another, at any intensity of selection ... a substantial number of individuals would usually be lost because they did not already possess the new allele ... because it necessarily involves either a completely new mutation or (more usually) previous change in the environment or the genome, I like to think of it as a dilemma for the population: for most organisms, rapid turnover in a few genes precludes rapid turnover in the others. A corollary of this is that, if an environmental change occurs that necessitates the rather rapid replacement of several genes if a population is to survive, the population becomes extinct.[49]

That is, since a high number of deaths are required to fix one gene rapidly, and dead organisms do not reproduce, fixation of more than one gene simultaneously would conflict. Note that Haldane's model assumes independence of genes at different loci; if the selection intensity is 0.1 for each gene moving toward fixation, and there are N such genes, then the reproductive capacity of the species will be lowered to 0.9^N times the original capacity. Therefore, if it is necessary for the population to fix more than one gene, it may not have reproductive capacity to counter the deaths.

Notes

1. J. S. Huxley, ed., *Evolution as a Process* (London: Longmans, Green, 1954).
2. J. B. S. Haldane, "The Statics of Evolution," in *Evolution as a Process*, ed. Huxley, J. S. et al. (London: Allen & Unwin, 1954), 111.
3. J. S. Huxley, "Clines: An Auxiliary Taxonomic Principle," *Nature* 142 (1938): 219–20.
4. J. B. S. Haldane, *The Causes of Evolution* (London: Longmans, Green, 1932).
5. J. B. S. Haldane, "The Part Played by Recurrent Mutation in Evolution," *American Naturalist* 67 (1933): 5–19.
6. L. J. Stadler, "Mutations in Barley Induced by X-Rays and Radium," *Science* 68 (1928): 186–87.
7. Haldane, *Causes of Evolution*, 32.
8. Haldane, *Causes of Evolution*, 80.
9. Haldane, *The Part Played by Recurrent Mutation in Evolution*, 5.
10. Haldane, *Causes of Evolution*, excess males.

11. Tachygenesis: Acceleration of development by the shortening of ancestral stages during embryonic development.
12. Recapitulation: Repetition of the ancestral evolutionary stages in the development of an organism.
13. Neoteny: The preservation, in the adult stage, of what were embryonic characters in the ancestor (as in some amphibians).
14. Haldane, *Causes of Evolution*, 60–61.
15. Haldane, *Causes of Evolution*, 31–32.
16. J. B. S. Haldane, "The Rate of Spontaneous Mutation of a Human Gene," *Journal of Genetics* 31 (1935): 317–26.
17. H. J. Muller, "Artificial Transmutation of the Gene," *Science* 66 (1927): 84–87.
18. L. J. Stadler, "Mutations in Barley," 186–87.
19. Haldane, "The Rate of Spontaneous Mutation of a Human Gene," 317.
20. J. B. S. Haldane, "The Formal Genetics of Man: Croonian Lecture," *Proceedings of the Royal Society of London, B* 135 (1948): 147–70.
21. Industrial melanism: During the industrialization of England, when the number of coal-burning factories increased, it was found that the number of melanic (darker) forms of the peppered moth (*Biston betularia*) also increased. The darker form, which was rare in the original population of normally light-colored moths, increased greatly in numbers in the polluted areas. This change of increase in the darker forms is called "industrial melanism."
22. H. B. D. Kettlewell, *Industrial Melanism in* Biston betularia (Oxford: Oxford University Press, 1973).
23. H. B. D. Kettlewell, *The Evolution of Melanism: The Study of a Recurring Necessity; With Special Reference to Industrial Melanism in the Lepidoptera* (Oxford: Clarendon Press, 1973); and H. B. D. Kettlewell, "A Résumé of Investigations on the Evolution of Melanism in the Lepidoptera," *Proceedings of the Royal Society, B* 145 (1956): 297–303.
24. J. B. S. Haldane, "A Mathematical Theory of Natural and Artificial Selection: Part I." *Transactions of the Cambridge Philosophical Society* 23 (1924): 19–41.
25. J. B. S. Haldane, "A Defense of Beanbag Genetics," *Perspectives in Biology and Medicine* 7 (1964): 343–59.
26. Haldane, "Natural Selection in Man."
27. R. A. Fisher, "On the Dominance Ratio," *Proceedings of the Royal Society of Edinburgh* 42 (1922): 321–41.
28. Hampton Carson, Personal communication to the author (2002).
29. Haldane, *Causes of Evolution*, 79–80.
30. J. B. S. Haldane, "Some Recent Work on Heredity," *Transactions of the Oxford University Junior Scientific Club* 1 (1920): 3–11.
31. Haldane, *Causes of Evolution*, 86.
32. Haldane, *Causes of Evolution*, 83.
33. E. Mayr, "Haldane's *Causes of Evolution* after 60 Years," *Quarterly Review of Biology* 67 (1992): 175–86.
34. K. R. Dronamraju, ed. *Selected Genetic Papers of J.B.S. Haldane* (New York: Garland, 1990).
35. T. Dobzhansky, *Genetics and the Origin of Species* (New York: Columbia University Press, 1937).
36. E. Mayr, "Where Are We?" *Cold Spring Harbor Symposia on Quantitative Biology* 24 (1959): 1–14.
37. Mayr, "Haldane's *Causes of Evolution* after 60 Years," 179.
38. Haldane, "A Defense of Beanbag Genetics."
39. Haldane, "A Defense of Beanbag Genetics," 125.
40. Haldane, "A Defense of Beanbag Genetics," 350.
41. J. B. S. Haldane, "Suggestions as to Quantitative Measurement of Rates of Evolution," *Evolution* 3 (1949): 51–56.
42. J. B. S. Haldane, "The Cost of Natural Selection," *Journal of Genetics* 55 (1957): 511–24.
43. M. Kimura, "Evolutionary Rate at the Molecular Level," *Nature* 217 (1968): 624–26.
44. J. W. Drake, B. Charlesworth, D. Charlesworth, and J. F. Crow, "Rates of Spontaneous Mutation," *Genetics* 148 (1998): 1667–86. Also see N. Takahata, "Molecular Clock: An Anti-Neodarwinian Legacy," *Genetics* 176 (2007): 1–6.

45. S. Wright, "Evolution in Mendelian Populations," *Genetics* 16 (1931): 97–159.
46. Haldane, "The Cost of Natural Selection," 523.
47. Kimura, "Evolutionary Rate at the Molecular Level."
48. W. J. Ewens, *Mathematical Population Genetics* (Berlin: Springer-Verlag, 1979).
49. Haldane's dilemma is due to the paleontologist Leigh Van Valen's "Haldane's Dilemma, Evolutionary Rates and Heterosis," *American Naturalist* 97 (1963): 185–90. In 1957, Haldane, in his now famous paper, "The Cost of Natural Selection," pointed out that in the process of the evolutionary substitution of one allele for another, at any intensity of selection, a substantial number of individuals would usually be lost because they did not already possess the new allele. Kimura (1960) has referred to this loss as the substitutional load, because it necessarily involves either a completely new mutation or (more usually) previous change in the environment or the genome. Van Valen (1963) considered it a dilemma for the population: For most organisms, rapid turnover in a few genes precludes rapid turnover in the others.

6

On Being a Guinea Pig

> Life without danger would be like beef without mustard. But since my life is useful it would be wrong to risk it for the mere sake of risk, as by mountaineering or motor racing.
> —J. B. S. Haldane, in *Keeping Cool and Other Essays*

Rear Admiral Albert R. Behnke[1] of the US Navy once recalled J. B. S. Haldane's penchant for self-experimentation under painful conditions:

> Perhaps no scientist has ever been better endowed by heritage, natural gifts, and training than J.B.S. Haldane to pursue studies in diving physiology and the effects of inhalation of gases. His investigations in applied and basic physiology, although only a small part of his published scientific endeavor, have had worldwide influence, because of the qualities of leadership and certain attributes centering in his ability to resolve a problem into essentials, to separate relevant from irrelevant, to employ simple means in brilliantly conceived tests, to make critical observations based upon an intimate knowledge of experimental conditions, and to express his results with mathematical precision.
>
> Those of us privileged to know him cherish his memory, and all of us who have been or are engaged in naval investigation have been sustained by his dedicated, profound, and courageous contributions to our basic knowledge and to the solution of our practical problems. We shall continue to practice his Golden Rule of medical experimentation, "To test on ourselves first that which we would have others do."[2]

Behnke and Haldane had remarkably parallel beginnings with the US and British navies, respectively. Haldane was already trained in some diving experiments from his teenage years by his father. However, the long series of diving experiments with the navies conducted by both Haldane and Behnke were abruptly initiated by submarine accidents involving HMS *Thetis* in the case of Haldane and USS *Squalus* in the case of Behnke. Both occurred just before the onset of World War II, in 1939. Both were involved in diving research for their respective navies under rules of wartime secrecy. It was only after the war they were able to exchange information about their wartime research in diving.

Haldane held Albert R. Behnke in great respect, and it was reciprocated. They shared an active interest in diving physiology. Shortly after Haldane's death, I was visiting the Department of International Health at the San Francisco Medical Center, where I delivered a lecture on Haldane, including his work on diving physiology. To my great surprise, a white-haired and well-dressed gentleman approached me afterward and introduced himself, "I am Albert Behnke." I invited him at once to contribute an article for the Haldane memorial volume that I was then editing. He wrote an excellent article on Haldane's physiological experiments, which was included in the book *Haldane and Modern Biology*.[3]

One reviewer of *Haldane and Modern Biology*, Professor Lionel Penrose, who was a former colleague of Haldane at University College London, commented on Behnke's contribution,

> Outside genetics, Haldane was perhaps best known for his work on the effects of the inhalation of gases at high pressure. The article by A.R. Behnke and R.W. Brauer is extremely informative and, for many people, explains for the first time what was attempted and what was accomplished by these investigations. The preoccupation of J.B.S. with the physiology of CO, CO_2, and oxygen was clearly a result of his persistent interest in the work of his father, J.S., for, as a boy, he had already been exposed to abnormal conditions for experimental purposes concerning problems of coal miners. The experiments were almost always extremely uncomfortable and were justified by J.B.S. in the rule which he gave for medical experimentation, to "test on ourselves first that which we would have others do." This line of thought can be followed up by the reader who turns to the appendix by Naomi Mitchison, which interestingly supplements the story of the early years of J.B.S. given by his mother in her autobiography. ... The assembly of material by K.R. Dronamraju, the editor, is however, impressive.[4]

Subject-Investigator

Haldane's self-experimentation largely falls under two groups; the first group of experiments, which were conducted in the 1920s, which he initiated to extend his father's experiments on the toxic effects of carbon monoxide and carbon dioxide, and the second and much larger group of experiments, involving diving and inhalation of various gaseous mixtures and escape from submarines under water, which he undertook for the British Navy during World War II. He undertook some experiments shortly before that war to investigate a submarine accident in which many sailors lost their lives.

JBS explained that he conducted experiments on himself because a guinea pig or a rabbit "made no serious attempt to co-operate with one." And, "to do the same sort of things to a dog ... requires a license signed in triplicate by two archbishops."

Haldane's father was the Oxford University physiologist, John Scott Haldane, who was noted for his painful and courageous self-experimentation in human physiology. He preferred to do experiments on people, including himself and his son, rather than other animals. He believed animals would be more afraid, not understanding the purpose or process of what was happening. Humans could be interested in the results, thus ignoring fear and pain.

True to the family motto "Suffer," John Scott Haldane conducted painful physiological experiments. He laughed at pain that would be unbearable for others. However, JBS explained that his father's attitude was much more like that of a good soldier who will risk his life to defeat the enemy than that of an ascetic who deliberately undergoes pain. His father did not seek pain in his work, but greeted it with much laughter when it occurred. JBS started his research career by helping his father, mainly as a "bottle-washer and a calculator, for a good many years." Young Jack Haldane accompanied his father down mine shafts to assist in the collection of samples of "bad air." To demonstrate the effects of breathing firedamp, his father told Jack to stand up and recite Mark Antony's speech from Shakespeare's *Julius Caesar*, beginning "Friends, Romans, countrymen." Jack began to pant, and somewhere about "the noble Brutus" his legs gave way and he collapsed onto the floor, where the air was fine. Haldane later commented, "In this way I learnt that firedamp is lighter than air and not dangerous to breathe."[5]

Haldane wrote a great deal about his father's accomplishments in physiological research. In a BBC radio broadcast, "Some Adventures of a Physiologist,"[6] Haldane spoke of his father: "Intellectual adventure is the hardest kind to share. However, a few scientists manage to combine intellectual and physical adventure. My father, Dr. J.S. Haldane, who died in March 1936, was one of this kind. As I helped him with his work, mainly as a bottle-washer and a calculator, for a good many years, I am going to talk mainly about his work rather than my own." As Haldane put it, his father's research aims could be summed up in three questions: What is bad air? What makes air dangerous to breathe? And how can its bad effects be prevented? His father found that he could stay indefinitely in air containing one volume of carbon monoxide in 2,000. When he breathed one part in 500, which knocked a mouse out in 4 minutes, he noticed no effect in the first half-hour. But after 71 minutes, when the experiment was stopped, his notes read, "Vision dim, limbs weak, difficulty in getting up or walking without assistance. Movements very uncertain."[7]

Haldane wrote that from these experiments, a simple principle emerged. A mouse is, in fact, no more sensitive to carbon monoxide than a man in the long term. This is because all warm-blooded animals produce about the same amount of heat per unit of surface. Three thousand mice weigh as much as one man, but have twenty times his surface area. They therefore produce twenty times as much heat per minute, and need twenty times as much oxygen. Similarly they take up carbon monoxide at twenty times the rate.

So Haldane's father, John Scott Haldane, concluded that a mouse or a small bird could be used as an indicator of the presence of carbon monoxide. An indicator is badly needed, because carbon monoxide has practically no smell, nor does it cause any irritation, like the poisonous war gases. A man may feel nothing till he falls down. John Scott Haldane was responsible for initiating the use of canaries as indicators of lethal gases in British coal mines. The reason is clear. When he went down a mine to investigate after a colliery explosion in South Wales, he found that only five men had been killed by the force of the explosion. And the remaining fifty-two men were killed by carbon monoxide poisoning! The same situation was found in other explosions.

The physiological experiments of both father and son have had worldwide influence because of the qualities of leadership. The impact of JBS, in particular, exceeded his father's because of his mathematical and analytical skills, enabling him to perform systematic analyses, to use his biochemical knowledge, and to express his results with quantitative rigor.

Several experiments were conducted under controlled conditions in specially constructed chambers. As these were often painful experiments for the subject-investigator (JBS) it is not surprising to find a sentence such as this one in one of the reports prepared by JBS: "Since these experiments are extremely uncomfortable, I have unfortunately not been able to confirm my observations by those of other experimenters."[8] This propensity for conducting heroic experiments that combined subject with investigator proved highly rewarding, because Haldane's rigid scientific discipline enabled him to record detailed and accurate observations according to protocol under what were obviously difficult circumstances. Several of his observations have formed the foundations for important investigations in later years.

Experiments on Blood Acidity (1920s)

Following his father's work on the amount of carbon dioxide dissolved in the blood, JBS realized that the carbon dioxide dissolved in the blood stimulated the brain to send messages to the breathing muscles. However, JBS wanted to find out whether this was due to the acidity of the blood or due to some other reason. JBS then started experiments of his own to find the answer.

If breathing is regulated by the acidity of the blood, JBS argued that an increase in the amount of the alkaline bicarbonate in the blood would slow down the breathing to retain more carbon dioxide to balance it. Conversely, if the amount of bicarbonate is diminished, the breathing will speed up to lower the amount of carbon dioxide.

JBS ate about an ounce and a half of bicarbonate of soda, and, sure enough, it slowed down their breathing and the carbon dioxide in the blood rose to balance it. However, increasing the acidity of the blood proved problematic. Drinking hydrochloric acid was almost impossible because it is a corrosive poison and burns your mouth and throat. So they diluted it greatly and drank it, but it was not enough to produce much effect. So he tried to disguise it in the form of ammonium chloride and drank moderate amounts of it. An ounce of ammonium chloride liberated enough acid to cause shortness of breadth, and Haldane panted for several days afterward.[9]

Strangely, this discovery later proved to be of some practical use. A form of tetany in babies was found to be due to the blood being too alkaline, and was cured with ammonium chloride. Haldane used to cite this example to show how fundamental research may lead to unforeseen practical solutions in some cases.

JBS never considered his experiments as being too risky. He started in low doses and small steps and gradually worked up to larger steps that would produce dramatic results. His knowledge of biochemistry and physiology assured him of the limits of safety in his experiments, although they may seem dangerous to onlookers. Of course, he suffered much discomfort and pain during the experiments but survived to tell the story.

Diving and Inhalation of Gases (1930s and 1940s)

Haldane's early discoveries in physiological research prepared him well for his later distinguished work in underwater physiology,[10] which he undertook for the British Navy. Haldane, as subject-investigator, investigated the role of carbon monoxide as a tissue poison in 1927. He found, as did his father, that mammals could live on oxygen dissolved in their blood at high pressures when almost all of their hemoglobin was combined with carbon monoxide at a partial pressure that did not exceed one atmosphere. However, Haldane further demonstrated that the addition of more than one atmosphere of carbon monoxide affected some substance in their tissues, resulting in the death of animals, even though there was adequate oxygen in solution. Haldane's research and his mathematical skills enabled him to describe the kinetics of carbon monoxide poisoning of enzymes within the context of basic formulations of the German biochemists Michaelis and Warburg.

A generation earlier, John Scott Haldane calculated a system of staged decompression, in which the diver came up rapidly to the depth at which the pressure was halved, then ascended in stages designed to avoid the "bends." These early decompression tables, prepared by John Scott Haldane in 1907, were universally accepted and used until 1956. The concepts framed by JSH remain the foundation for all tables devised since. No other development in the history of diving has saved so many lives. The gratitude of the British Navy was expressed in the form of a gift to Louisa Kathleen Haldane, which included a silver coffee pot, candlesticks, and an engraved salver!

Underwater Physiology for the British Navy

In 1941, in a brief review of the field of underwater physiology,[11] J. B. S. Haldane (with E. M. Case) listed six main problem areas: mechanical effects, nitrogen intoxication, oxygen intoxication, aftereffects of carbon dioxide, bubble formation during decompression, and cold. These topics have largely remained the key areas of concern.

HMS *Thetis*

J. B. S. Haldane's advice was sought when the submarine HMS *Thetis* sank during trials in Liverpool Bay in 1939, three months before World War II erupted. Even though the vessel was intact with its stern protruding from the water, after almost eighteen hours of being confined in the sub, only four of the 108 men on board survived. Almost half of the victims were civilian mechanics. Nineteen were members of the Amalgamated Engineering Union, and others were members of the Electrical Trades Union. Haldane was asked by these trade unions to represent their interests in the public enquiry. Many years earlier, his father investigated the causes of the high death rate in the slums of Dundee, and JBS now claimed that this investigation was merely an extension of his father's experiments. However, this later investigation went much further. Haldane emphasized that the special qualities required for a volunteer in his project were unique. One obvious requirement was not to panic under stress or under conditions of claustrophobia. Even men of great fighting record and bravery in the war easily panicked under experimental conditions. The tests were conducted in a Siebe Gorman[12] metal cylinder, which is similar to a boiler lying on its side. It is eight feet long, four feet in diameter, and with a door at one end that closed on to a rubber sealing ring. The chamber had no lights and no telephone link with those outside. Communication was maintained by a tapping code or by small slips of paper with messages that were held up to the small glass windows.

The air in a submerged and damaged submarine deteriorates in many ways, for instance, through the escape of acid fumes from batteries or from engines, and also any carbon dioxide exhaled by the surviving crew. In experiments, the subjects were exposed to gas mixtures rich in carbon dioxide at pressures ranging from one to ten atmospheres approximately corresponding to depths from 0 to 90 meters.

Haldane recruited four former soldiers, W. Alexander, Patrick Duff, George Ives, and Donald Renton. Their attitude during the experiments was summed up by Duff: "I felt bad, but trusted the Professor." They locked themselves in a steel chamber at the Siebe Gorman factory in London to simulate the environment inside a disabled submarine. This experiment showed Haldane how different men would behave under experimental conditions. At the end of an hour everyone was panting severely. As the carbon dioxide concentration rose, all had headaches, and they became sick and incapacitated. Some vomited. Then they came out of the chamber and attempted to put on the Davis submarine escape gear, simulating the conditions in the submarine, but were unable to do so. After two to three minutes, Haldane removed the mouthpiece to vomit repeatedly, bringing up a pint of clear fluid. He had not eaten or ingested fluids during the previous sixteen hours in order to parallel conditions in the submarine.

This experiment was carried out after Haldane himself had been sealed up in the chamber for fourteen and a half hours, to test if the impact of a gradual rise in carbon dioxide pressure is same as that experienced by him in previous experiments when the rises were abrupt. Haldane's investigation showed that certain physiological conditions were not taken into consideration by the Admiralty, which led to further official investigations. The result was that the Government finally invited Haldane to continue his investigations on a much broader scale. No doubt these developments received an urgent impetus by the deepening war in which the allies were actively involved.

The main aim of Haldane's experiments was to investigate the safety conditions under which men might escape from sunken submarines, the effects of breathing various gases involved under such circumstances, and the range of psychological responses of men under different combinations of temperature and pressure while breathing various gaseous mixtures.

The real danger of this work was summed up by Haldane: "The work is of a very severe character. For example, I was on one occasion immersed in melting ice for thirty-five minutes, breathing air containing six and a half percent of carbon dioxide, and during the latter part of the period also under ten atmospheres' pressure. I became unconscious. One of our subjects has burst a lung, but is recovering; six have been unconscious on one or more occasions; one has had convulsions."[13]

Figure 6.1 JBS Haldane in chamber of horrors.

Investigation of Oxygen Tolerance Levels

One of the important problems of World War II involved oxygen inhalation by underwater demolition and salvage teams. Medical personnel of the US Navy had earlier determined tolerance limits for the inhalation of oxygen in the dry chamber and by divers during decompression at rest. More than five hundred decompressions on oxygen beginning at the 60-foot level had been made prior to 1943. A routine procedure had been established for inhalation of oxygen as the prime therapeutic adjuvant in the treatment of decompression sickness.

Haldane provided vital information, as advisor and also subject, to the distinguished group of diving and medical officers under the direction of the surgeon Lieutenant-Commander Donald. Under Haldane's direction, that group conducted one of the most exhaustive programs of diving experiments ever attempted in which more than one thousand dives in water were carried out at toxic levels to water depths of 90 feet. Some idea of the range of these critical physiological experiments was indicated by Haldane in the following letter to Albert R. Behnke of the U.S. Navy, dated May 7, 1943.

Dear Dr. Behnke:

I had hoped to send you a full report on our own work but I am told that this had better go through official channels. So will you please get moving at your end to expedite proper contact. Meanwhile, I comment on some points in your memorandum for which I thank you very warmly.

Regarding sensitivity to oxygen, we have tested 28 naval subjects in air [i.e., in the dry tank] at 90 feet, breathing from a "Salvus" apparatus with a flow of about 2 litres per minute at one atmosphere, carbon dioxide being absorbed by soda lime. The times varied from 6 to 90 minutes. . . . Only the best quarter lasted over 45 minutes. The tests rarely ended in a convulsion, but always with face twitching or some well marked symptom. We cannot predict who will be tough. Thus, the second place in our records . . . is held by a rather unathletic woman [Dr. Helen Spurway, who later became his wife]. We opine that your two subjects would last about an hour at 90 feet and thus fall into our toughest quarter.

We are inclined to think that you are erring on the dangerous side regarding the breathing pure oxygen underwater. We have had several convulsions in under 30 minutes at 40 feet, and one at 35 feet . . . in the course of a year, Dr. Spurway [future Mrs. Haldane] and I have had the following experiences. We both started with tolerances of over 35 minutes at 90 feet in the dry. We both had convulsions in other tests. Her tolerance fell so low that she convulsed after 13 minutes in the dry at 90 feet, while I had symptoms after ten and a half minutes. Since then her tolerance has gradually risen to over 80 minutes, while mine went up to 32, but has fallen to a steady level around ten.

We therefore think that for 100% safety from convulsions you must aim at a pretty low partial pressure[14] of oxygen, not more than two atmospheres[15] absolute, though many people can last 30 or even 60 minutes in diving suits at 3 atmospheres absolute. [Haldane's recommendation was confirmed and accepted in several later studies and represents today the upper limit for work under water during the course of oxygen inhalation.]

It was on this basis that we chose 45% oxygen mixtures for the special suit which Lt. Hoffman will describe to you. The Tables worked out for it are based on actual experience. . . . The British decompression Tables for depths up to 120 feet and times up to one hour err, if anything, on the safe side (for greater depths or longer times are not safe).

Yours sincerely,
J.B.S. Haldane

Gaseous Mixtures

Haldane and Martin Case investigated the narcotic action of nitrogen in several studies, often employing several volunteers. They studied the narcotic effects of other gases and gaseous mixtures including helium, hydrogen, nitrogen, and argon mixed with oxygen were investigated under conditions that called for arithmetic and other tests of cerebral function. All facets of the pressurized environment were explored. Such diverse cases as odontalagia were recognized as effects of hyperbaric pressure variation, and Haldane himself lost a tooth in the process.

Haldane's notes presented some interesting facts. On one occasion, he wrote, "EMC and JBSH lay inside the pressure chamber in a bath containing water and large amounts of broken ice."[16] EMC was Martin Case, who had earlier lived with the Haldanes during his student years at Cambridge University. He was variously described by colleagues as able, amusing, eccentric, an intrepid diver, and a strange philosophic person. Besides Case, Haldane recruited his colleagues at University College London, including his secretary, Elizabeth Jermyn, the geneticists Hans Kalmus[17] and James Rendle, and also a former prime minister of Spain, Dr. Juan Negrin.

The civilian scientists and naval ratings reacted rather differently to the stresses of experimentation. As long as the conditions remained tolerable, the sailors were more disciplined and complained less, but when conditions became critical, the scientists, especially the women scientists, were less inclined to panic, possibly because they understood better the nature of the research.

Research activities of that period were confidential and were subjected to wartime secrecy and regulations. The following letter from Haldane, which was addressed to Hans Kalmus was typical of that period:

> Oct. 15th, 1942
> Dear H. K.
>
> Can you come down here for a series of experiments from October 27th to 30th? You would stay with us, with all expenses met, and a guinea a day for other expenses. You would be expected not to talk about the work even though it may not appear to have anything secret about it. I hope that it may be possible to get you into further work later on, but this depends on the navy, who won't be here during the week in question.
>
> Don't answer this. But see me next Monday.
>
> Yrs sinc
> J.B.S. H.
>
> You would come upon the 26th.

Mental Tests

There were occasional amusing moments when the mental capacities of the volunteers were put to test. Illusions and strange behavior were fairly common. Some subjects felt "awful." Others felt elated. One said repeatedly, "I'm going to die, I'm going to die."

One laughed uproariously when told he was cheating in a test. Some described themselves variously as feeling drunk or faint. Irishmen and Americans generally got very excited and quarrelsome. So did some Englishmen. Helen Spurway's arithmetic actually improved under experimental conditions, whereas other distinguished volunteers failed even in doing simple calculations or additions!

Haldane himself had hallucinations under oxygen poisoning, accusing one of his colleagues of trying to murder him deliberately. The young naval officer-in-charge, Lieutenant Kenneth Donald, cautioned Haldane not to subject himself to too much oxygen poisoning, as it might impair his capacity to continue the experiments or might cause a lasting damage of some kind. Donald recorded on one occasion, "Haldane convulsed several times in my arms in the wet pressure pot where he was in a diving suit underwater and I was on a platform above them."[18]

Haldane's state of mind under these circumstances was best indicated by the notes he had recorded while sitting inside the "chamber of horrors":

> I am breathing rapidly and deeply and my pulse is 110. . . . Soon I feel much better, though perhaps my writing is a little wobbly. But why cannot my companion behave himself? He is making silly jokes and trying to sing. His lips are rather purple, the colour of haemoglobin when uncombined with oxygen. I feel quite unaffected; in fact I have just thought of a very funny story. It is true I can't stand without support. My companion suggests some oxygen from the cylinder. . . . To humour him I take a few breaths. The result is startling. The electric light becomes so much brighter that I fear the fuse will melt. The noise of the pump increases fourfold. My note-book, which should have contained records of my pulse rate, turns out to be filled with the often repeated but seldom legible statement that I am feeling *much* better, and remarks about my colleague, of which the least libelous is that he is drunk. I put down the oxygen tube and relapse into a not unpleasant state of mental confusion.[19]

Because of these experiments, Haldane was able to claim that he probably held the world record of one and a half hours' continuous spasm of the hands and face.

A particular danger was the unpredictability of the seizures. On one occasion Helen Spurway lasted 85 minutes at a pressure of three atmospheres, and on another only 13 minutes. Haldane was at first more resistant than most, but after going through one hundred experiments, he became so highly sensitive that he began to twitch violently after only five minutes' exposure to oxygen.

On another occasion, Haldane suffered an injury to his spinal cord that was due to a bubble of helium formed in that organ while being decompressed during testing in 1940, on behalf of the British Admiralty. During my association with Haldane during his last years in India, it was a common sight to see him carrying a cushion to lecture rooms.

Haldane discovered that although oxygen is a tasteless gas at atmospheric pressure it begins to acquire at about five or six atmospheres the taste of rather stale ginger beer, "a trivial discovery which, for some reason, pleases me greatly," as he later described in his Personal Note for the Royal Society.[20]

With respect to life in higher atmospheres, Haldane wrote, "You feel pretty queer at 10 atmospheres. The air is so thick that you feel quite a resistance when you move your hand, and your voice sounds very odd, as if you were trying to imitate a Yankee twang and overdoing it very badly."[21]

Effects of Carbon Dioxide

The controversial role of carbon dioxide in the etiology of oxygen poisoning and inert gas narcosis has remained a problematic area for investigation. Some investigators with no firsthand experience have previously attributed various symptoms and reactions to carbon dioxide, but it is the subject-investigator who resolved conflicting ideas on the basis of consistent observation and accurate observation. In a letter to the *British Medical Journal* in 1947,[22] Haldane summed up his position, which is also in accord with that of Captain Behnke:

> Sir:
>
> May I venture to doubt Sir Leonard Hill's theory (June 21) that oxygen convulsions and nitrogen and oxygen narcosis are both due to carbon dioxide accumulation? Subjectively and objectively the symptoms of the two conditions are utterly different. I have never experienced or observed confusion or euphoria before an oxygen convulsion. Confusion is universal and euphoria common in nitrogen narcosis. Carbon dioxide enhances both oxygen and nitrogen poisoning. The obvious explanation of this fact is that it causes cerebral vasodilation, and thus facilitates absorption of other gases by the brain. I hope shortly to publish data which may help to clear this question up.
>
> With regard to questions of priority raised in his letter, the Official Secrets Act makes a full discussion illegal. . . . So long as publication is restricted it is impossible to assign credit objectively, and controversy can lead nowhere.
>
> Yours sincerely,
J.B.S. Haldane

One of the volunteers in Haldane's experiments during wartime was his colleague Hans Kalmus,[23] who wrote of his impressions in his autobiography, *Odyssey of a Scientist, An Autobiography*, published long after Haldane's death. Kalmus wrote that he was delighted to join Haldane's group:

> Exposing myself to mixed atmospheres at high pressures and to the hazards of decompression was both more dangerous and more interesting than hospital service. . . . We worked parallel with a naval team, which did the "wet" experiments, using diving gear and water tanks, while we worked with respirators and in compression cylinders. Our brief was to replicate the conditions which might arise in a damaged submarine, and to discover an ideal gas mixture for the respirators to facilitate the escape of those trapped underwater.

Kalmus wrote further:

> The sudden decompression of a diver causes "bends," severe pains caused by nitrogen bubbles in small blood vessels. Under high pressure an excess of nitrogen dissolves in the blood, and this is released when the pressure is suddenly eased. One of our tasks was to take measurements for a decompression table, which would relate the severity and duration of pressure to safe decompression procedures. "Bends" can be avoided if nitrogen is removed from what the victim breathes while under pressure; either he should breathe pure oxygen or a mixture of oxygen and the inert gas, helium.

Several volunteers who participated in the experiments suffered convulsions. Kalmus wrote that he never had an oxygen convulsion but he lost consciousness on several occasions when he was exposed to high concentrations of carbon dioxide in the air. But the high-pressure experiments were only a part of the team's war effort. They were also required to perform some mathematical work for the army. Kalmus spent hours sitting at an electromechanical calculator and solved some intricate differential equations by iteration. But they were not told the true purpose of these calculations.

Kalmus wrote of his personal impressions of Haldane: "Official and even private celebrations in his honor he did not easily bear. When a farewell dinner was organized by his colleagues on the eve of his departure for India, he refused to sit in the place of honour, next to his wife, and managed to go the whole evening without a drink, except for a glass of water." Haldane, according to Kalmus, said that he owed much of his department to two men: Hitler, who had created so many refugees, and Rockefeller, whose foundation paid them. Kalmus wrote, "Haldane went out of his way to help displaced scholars. He knew very well, I am sure, how deeply in his debt we all were."[24]

Other Sensory Experiences

Toward the end of his life, Haldane recalled the "thrilling" experiences of his life:

> I should find the prospect of death annoying if I had not had a very full experience mainly stemming from my work. . . . One thing which I am really sorry to have missed is walking to France on the sea bottom, which incidentally would have involved some interesting physiological research beforehand. . . . I have tried morphine, heroine, and *bhang* and *ganja* [hemp prepared for eating and smoking]. The alterations of my consciousness due to these drugs were trivial compared with those produced in the course of my work. I once dreamed that I was reading a life of Christ written and illustrated by Edward Lear. But I can only remember Pontius Pilate's moustache. If you want a dream as original as that, don't take opium, but eat sixty grams of hexahydrated strontium chloride. I have had some of the standard adventurous experiences such as being pulled out of the crevasse in a glacier, and more which are unusual. For example, I was one of the two people to pass forty-eight hours in a miniature submarine.

He added to his thrills, "Most of my joyful experiences have been by-products. Thus I have enjoyed the embraces of two notoriously beautiful women. In neither case was there any wooing. After knowing one another for some time we felt like that, and no words or gifts were needed. I later married a colleague. We had known one another for some years, and our love was largely based on respect for each others' work."[25]

Honored at the US Naval Medical Center in Bethesda, Maryland

In 1949, Haldane was honored as a distinguished lecturer at the United States Medical Research Institute and the Naval Medical Center in Bethesda, Maryland.[26] When

Haldane arrived, he was greeted and seated at the banquet table before dinner by the hosts. In the small talk that followed it seems that Haldane mumbled mostly and spoke so softly that he was barely audible. I was told that at that point that the hosts gave up any idea of listening to Haldane's inspiring after-dinner speech. They were preparing themselves to say a quiet goodbye to him soon. However, when Haldane began speaking they were in for a treat! That was one of the most inspiring and informative lectures they had ever heard on the subject of diving physiology.

The following response[27] was received when I contacted the British Government to suggest an appropriate recognition of Haldane's wartime heroic experiments on diving and inhalation of poisonous gases. There was no other individual who performed a service of that kind. This response from the government was disappointing.

CABINET OFFICE
70 Whitehall, London SW1A 2AS
Telephone: 071-270 0400

Chancellor of the Duchy of Lancaster
Minister of Public Service and Science

WW/94/M/490

Dr Krishna R Dronamraju
President
Foundation for Genetic Research
PO Box 27701-497
Houston
Texas 77227
USA

15 June 1994

Dear Dr Dronamraju,

Thank you for your letter of 14 April to Malcolm Rifkind about the Wartime contributions of the British Scientist, J B S Haldane. This subject is the responsibility of my Department and the letter has been passed to me for a reply.

I am most grateful for your letter and specifically for raising the question of Professor J B S Haldane's contributions including his warning about the developments in pre-war Germany.

As your paper indicates, the Haldane family were an influential one, and J B S Haldane himself was highly and widely regarded in the UK. He was awarded the Fellowship of the prestigious Royal Society in 1932 which he much valued. He was also awarded the Royal Society Darwin Medal in 1952 and gave the Cronian Lecture in 1946.

As a former associate of the Professor you will be aware of the regard in which he was held as a person and as a scientist.

WILLIAM WALDEGRAVE

Notes

1. Captain Albert R. Behnke (USN Retd.) (1903–1992) was an American physician who was principally responsible for developing the US Naval Medical Research Institute (NMRI). Behnke separated the symptoms of arterial gas embolism (AGE) from those of decompression sickness and suggested the use of oxygen in recompression therapy. Behnke is also known as the "modern-day father" of human body composition for his work in developing the hydrodensitometry method of measuring body density, his standard man and woman models, and a somatogram based on anthropometric measurements. When the submarine USS *Squalus* sank in 1939, Behnke responded with fellow divers. They were able to rescue all thirty-three surviving crew members from the sunken submarine. The salvage divers used recently developed *heliox* diving schedules and successfully avoided the cognitive impairment symptoms associated with deep dives, thereby confirming Behnke's theory of nitrogen narcosis. Later in 1939, Behnke and Yarborough demonstrated that gases other than nitrogen also could cause narcosis. Taking advantage of the positive public support for US Navy diving following the *Squalus* rescue, Behnke contacted President Franklin D. Roosevelt, and received approval for the construction of his research laboratory (NMRI). Upon retiring from the US Navy in 1959, Behnke became a professor of preventive medicine at the University of California and director of the Institute of Applied Biology, Presbyterian Medical Center, San Francisco, California.
2. A. R. Behnke and R. W. Brauer, "Physiologic Investigations in Diving and Inhalation of Gases," in *Haldane and Modern Biology*, ed. K. R. Dronamraju (Baltimore: Johns Hopkins University Press, 1968), 267–75; 274.
3. A. R. Behnke and R. W. Brauer, "Physiologic Investigations in Diving and Inhalation of Gases," in Krishna R. Dronamraju, ed., *Haldane and Modern Biology*(Baltimore, MD: Johns Hopkins University Press, 1968), 267–75.
4. L. S. Penrose, "Review of Haldane and Modern Biology," *Annals of Human Genetics* 33 (1969): 225–26.
5. J. B. S. Haldane, Personal communication to the author.
6. J. B. S. Haldane, "Some Adventures of a Physiologist: A BBC Radio Broadcast," in *Keeping Cool and other Essays* (London: Chatto & Windus, 1940), 87–96.
7. J. B. S. Haldane Archives, University College London.
8. K. R. Dronamraju, ed., *Haldane and Modern Biology* (Baltimore: Johns Hopkins University Press, 1968), 267.
9. M. M. Baird, C. G. Douglas, J. B. S. Haldane, and J. G. Priestley, "Ammonium Chloride Acidosis," *Journal of Physiology* 57 (1923): xli.
10. Behnke and Brauer, "Physiologic Investigations in Diving."
11. J. B. S. Haldane, "New Deep-Sea Diving Method: The Case for Helium," *Fairplay* 163 (1944): 740; E. M. Case and J. B. S. Haldane, "Human Physiology under High Pressure: I. Effects of Nitrogen, Carbon Dioxide, and Cold," *Journal of Hygiene* 41 (1941): 225.
12. During his last years in India (1957–1964), I used to see Haldane frequently puffing at his "Siebe Gorman" pipe, which has a bowl that is exquisitely carved to resemble the head of a diving suit.
13. J. B. S. Haldane, personal communication to the author, 1960.
14. In a mixture of gases, each gas has a partial pressure, which is the hypothetical pressure of that gas if it alone occupied the volume of the mixture at the same temperature. The total pressure of an ideal gas mixture is the sum of the partial pressures of each individual gas in the mixture. It relies on the following isotherm relation:

$$V_x \times p_{tot} = V_{tot} \times p_x$$

- V_x is the partial volume of any individual gas component (X)
- V_{tot} is the total volume in gas mixture

- p_x is the partial pressure of gas X
- p_{tot} is the total pressure of gas mixture
- n_x is the amount of substance of a gas (X)
- n_{tot} is the total amount of substance in gas mixture

The partial pressure of a gas is a measure of thermodynamic activity of the gas's molecules. Gases dissolve, diffuse, and react according to their partial pressures, and not according to their concentrations in gas mixtures or liquids. This general property of gases is also true of chemical reactions of gases in biology. For example, the necessary amount of oxygen for human respiration, and the amount that is toxic, is set by the partial pressure of oxygen alone. This is true across a very wide range of different concentrations of oxygen present in various inhaled breathing gases or dissolved in blood.

15. Atmospheric pressure is the force per unit area exerted against a surface by the weight of air above that surface in the Earth's atmosphere.
16. Behnke and Brauer, "Physiologic Investigations in Diving and Inhalation of Gases," 271.
17. H. Kalmus, *Odyssey of a Scientist: An Autobiography* (London: Widenfeld and Nicolson, 1991), 58–86.
18. J. B. S. Haldane Archives, University College, London.
19. J. B. S. Haldane Archives, University College London.
20. J. B. S. Haldane, Personal Communication to the author.
21. J. B. S. Haldane, Personal Communication to the author.
22. J. B. S. Haldane, "Letter," *British Medical Journal* 2 (1947): 226.
23. Kalmus, *Odyssey of a Scientist*, 67–71.
24. Kalmus, *Odyssey of a Scientist*, 82.
25. J. B. S. Haldane, "On Being Finite," in *Science and Life, Essays of a Rationalist*, (London: Pemberton Publishing Co., 1968), 203.
26. Haldane's visit to the Bethesda Naval Medical Center, 1949. Albert R. Behnke, personal communication to the author.
27. Response from British Government to an inquiry.

7

Chemical Genetics*

During the 1920s and the 1930s, Haldane was at the most productive years of his life. In 1923 he published his most controversial and daring scientific predictions in his first book *Daedalus, or Science and the Future*. In 1924 he published the first of his many papers on the mathematical theory of natural selection, which contributed to the founding of population genetics. In 1927 he married Charlotte Franken, after her long sensational divorce. It was also in 1927 that Haldane accepted the part-time position as the officer-in-charge of the genetic program at the John Innes Horticultural Institution (JIHI)[1] in Merton. His first book of popular scientific essays, *Possible Worlds and Other Essays* (including his famous essay "On Being the Right Size") was published in 1927, establishing him as a skilled and versatile popular writer. His major work, *The Causes of Evolution*, was published in 1932, and was hailed instantly as a pillar of population genetics. In the same year, he was elected a Fellow of the Royal Society of London.

Haldane's part-time appointment at JIHI, an institution founded by his mentor William Bateson, led to a research program in biochemical genetics involving plant pigments. For ten years, Haldane visited JIHI twice a week, at a modest salary, to analyze Mendel's ratios in plant-breeding experiments. Furthermore, hiring Rose Scott-Moncrieff, Haldane initiated research into the biochemical genetics of plant pigments, pioneering research that preceded the biochemical genetic work of Beadle and Tatum with *Neurospora*.

Haldane's younger colleague, C. D. Darlington, wrote,

> Until near the end of his ten years at the John Innes Haldane had behaved with discretion. He had indeed done good service by holding at bay the criticisms of certain professional members of the governing body. These men thought that genetics was an over-rated or ill-founded science the development of which ought to be stopped. ... Professor E.W. MacBride was the leader of this school of thought. For the fifteen years he had been a council member, he had been urging the claims of Kammerer and a succession of other bogus Lamarckian experimenters. The columns of *Nature* record these repeated claims, long anticipating Lysenko, and also their repeated exposure. Against the last of these attacks Haldane's letter to *The Times* of October 3, 1936, had been a reasoned defence, a defence of genetics and of the John Innes Institution.[2]

Haldane's Ideas in Biochemical Genetics

Haldane formed his ideas in the chemistry of gene action quite early in his career. It was quite natural for him to develop an early interest in biochemistry because of his familiarity with his father's research in physiology at a time when biochemistry (or physiological chemistry, as it was called) was still considered a branch of physiology. Later, Haldane's ten-year work in the laboratory of the great biochemist Frederick Gowland Hopkins, taught him to appreciate that the biochemical interpretation of gene action is more fundamental than the morphological and the embryological interpretations of gene action that were popular in the 1920s and 1930s.

As early as 1920, Haldane[3] wrote that genes may be regarded as large nucleoprotein molecules from a chemical point of view, however, from a biological point of view, they exhibit one important characteristic of a living organism, namely, their capacity to reproduce themselves in different environments. Haldane further hypothesized that genes produce definite quantities of enzymes and that the members of a series of multiple allelomorphs produce the same enzyme in different quantities. He prophesied further that for the chemist the action of ferments in producing plant and animal pigments offers a fascinating field for investigation, and he himself initiated a research program on the biochemical genetics of plant pigments.

Three of his books include discussions of at least some aspects of biochemistry of gene action: *Enzymes*,[4] *New Paths in Genetics*,[5] and *The Biochemistry of Genetics*.[6] When he joined JIHI, Haldane had intended to develop an experimental research program in the biochemical genetics of plant pigments. He initiated and stimulated research on the genetics of plant pigments by Rose Scott-Moncrieff[7] and others. As was often the case with Haldane, he stimulated research by suggesting projects, methods, and analyses. His contribution was theoretical and influential. Simultaneously, research on the biochemical genetics of insect eye pigments was being carried out by Caspari on *Ephestia*[8] and by Beadle and Ephrussi on *Drosophila*.[9] These experiments led to the recognition of the biochemical basis of gene action and finally to the elucidation of the complex biochemical problems in terms of enzyme chemistry.

The research of Haldane and his collaborators was from the beginning aimed at an elucidation of the biochemical basis of gene action in a systematic fashion. The colors of the petals were chosen because a large array of genetically determined variation was available for several species. Plant pigments, contrary to the melanin pigments of vertebrates, are easily soluble and obtainable in large quantities, which makes them amenable to the preparative methods of organic chemistry. Haldane was fully aware of these facts.

By the late 1930s, the structure and variations of the main classes of pigments had been established. Genes controlling the relative amounts of the different pigments, the formation and inhibition of co-pigments, the state of oxidation and methylation of the pigments, and the pH of the cell sap had been identified in several species, as summarized by Caspari.[10]

Haldane's future plans at JIHI included identifying the synthetic pathways by which the different substances are related to each other, and the enzymes and catalytic agents involved in the formation of the plant pigments. These investigations were not pursued at JIHI for at least four reasons: (1) Events preceding World War II and the war itself made it difficult to continue research; (2) because of internal politics, involving the director of JIHI, Sir Daniel Hall,[11] who blocked Haldane's succession to the directorship, Haldane began making plans to resign from his position at JIHI; (3) his acceptance of the professorship in genetics at University College London (UCL) in 1933, followed by his election to the Weldon Chair of Biometry at UCL in 1937 effectively terminated any remaining connection Haldane had with JIHI; and (4) finally, after Beadle and Tatum[12] working with *Neurospora*, which has a shorter generation time, established the principles of biochemical genetics, there seemed no point in continuing the slow process of biochemical genetics using plant pigments. Haldane, being keenly aware of this point, devoted a good portion of his book *The Biochemistry of Genetics* (1956) to *Neurospora* and other fungi.

Gene-Enzyme Hypothesis

Before Haldane, several others, including Archibald Garrod,[13] Richard Goldschmidt, and Sewall Wright, had arrived at the gene-enzyme concept. However, Haldane stated that he was stimulated by a paper by the French biologist Lucien Cuenot, who first investigated the genetic basis of the pigmentation of the mouse, concluding that the interaction of three gene-controlled substances must be involved in forming the gray coat of the wild mouse. Cuénot spent two years working on mice, and found that three "mnemons" (genes) are responsible for the production of one pigment and two enzymes. The pigment (if present) is acted on by the enzymes to produce black or yellow color. In the absence of a pigment, an albino mouse results. After studying various crosses between mice, he concluded that these "mnemons," or genes, were inherited in a Mendelian fashion. Cuénot was the first person to describe multiple allelism at a genetic locus.

Haldane was not dogmatic about his formulation of the gene-enzyme theory. He made numerous suggestions in the following years, especially his recognition that the primary gene product must differ from the gene itself. Haldane also rejected his own earlier notion that the differences between alleles of the same gene are quantitative in nature, which was mainly due to the fact that when several pleiotropic effects of the multiple alleles of the same gene are considered, it is not possible to arrange alleles in a series of quantitatively decreasing effects valid for all characters. Hence, Haldane concluded that the differences between the gene products and the genes of different alleles at the same locus must be qualitative rather than quantitative.

The assumption of qualitative differences between the primary gene products at one locus led Haldane to propose that under suitable conditions and methods we should be able to demonstrate in the same cell the two gene products of a heterozygote. He argued further that in situations of codominance the antigens of the red blood cells of humans and animals should show this feature. He was the first to point out a direct relationship between antigens and their determining genes, and that these antigens may be primary gene products. When Irwin[14] discovered hybrid antigens in 1932, Haldane commented, "This observation is a conclusive disproof of the hypothesis that a particular type of gene always makes a particular type of antigen, and that no antigens are made otherwise." Later, he proposed an experimental design to test the frequency of the occurrence of hybrid antigens in heterozygous mice. He also considered the possibility that small molecules, such as vitamins and hormones, might be primary gene products.

Human Biochemical Genetics

Haldane[15] drew attention to the work of Garrod in human biochemical genetics, which was summarized in his book *Inborn Errors of Metabolism*, published in 1909. This was an important step from a historical point of view, because Garrod's pioneering work was ignored not only by his contemporaries but also by many others in the subsequent generation. Besides drawing attention to Garrod's work, Haldane emphasized the biochemical identity of the individual. With remarkable insight, Garrod suggested that the homogentisic acid excreted by alkaptonurics must be largely derived from the amino acid tyrosine. He described it as a congenital metabolic disorder, and, after consulting William Bateson, recognized its autosomal recessive nature. Later, he added other disorders such as cystinuria, porphyria, and pentosuria. Quite early in his investigations, Haldane regarded genes as organs of the cell, and regarded their function as a legitimate part of the discussion. He was not dogmatic about his original formulation of the gene-enzyme theory, but made several suggestions to improve the theory in later years.

The fate of Haldane's work in biochemical genetics at JIHI was closely tied to certain personalities at the institution. Prominent among these were the director, Sir Daniel Hall, and the young cytologist Cyril Dean Darlington.

Sir Daniel Hall (1864–1942)

When he hired Haldane, Director Hall promised him that he would retire soon and recommend to the board to appoint Haldane as his successor. However, as time passed, Hall changed his mind and decided to continue his service further. He continued until Haldane got tired of waiting and accepted the position of professor of genetics and later of biometry at University College London. In the meantime, Hall promoted the young cytologist Cyril Darlington and groomed him to become the next director of JIHI.

Cyril Dean Darlington (1903–1981)

Haldane's association with cytologist C. D. Darlington[16] at JIHI is of great interest because of its impact on both Darlington and Haldane during the crucial years of their lives. Clearly it affected the course of history of biochemical genetics. Because of the conflict between Haldane and the director of JIHI (and the role played by Darlington in fanning that conflict), the development of biochemical genetics was delayed in Great Britain. These events disrupted Haldane's research program in biochemical genetics at JIHI.

The Beginning

Darlington recorded in a draft manuscript that he first heard of Haldane in 1925 at the JIHI. Haldane had just been dismissed by Cambridge University for alleged sexual misconduct as Haldane had been co-respondent in a divorce case.[17] Darlington overheard William Bateson, director of JIHI at that time, remark, "I am not a prude but I don't approve of a man running about the streets like a dog."[18] It soon became clear that Haldane was clearly the object of an increasing gossip in a small university town. Bateson, it seemed, had reached an unwarranted conclusion about Haldane's moral character. Darlington commented that Haldane had apparently made an unfavorable impression on the older generation of Cambridge men.

Curiously, Darlington's private notes as well as his publications make no mention of the immense *personal* debt which he owed to Haldane's intellectual support and guidance. But Darlington readily admitted the value of Haldane's service to British genetics and especially to the JIHI. In his notes, he wrote,

> We were saved. For ten years Haldane came half the year at a modest salary, toiling away with those Mendelian ratios which he had described in his first letter to Bateson. He sat at the next desk to me. I was drawing chromosomes. He was doing sums. No microscopes for him. No calculating machines, no slide rules, no logarithms even. All long division sums based upon data collected and recorded by Bateson's assistants from plants grown by Bateson's gardeners. The sheets would pile up.[19]

With respect to Haldane's contribution to the JIHI, Darlington wrote,

> For ten years he had held the John Innes Institution together. His massive figure, his name, well known in the world outside, had protected the Institution and the young geneticists it sheltered from being gobbled up by the hungry wolves outside. For now in addition to those who would have liked to seize the place and destroy it as they eventually did, there were others who wanted to turn it into something like an ordinary research station where people do what they are told, or an ordinary department where they do what their predecessors have done. Haldane was able, by a letter to the press or by a broadcast, to hold these people at bay. And it was thanks to him that we remained alive in a little world of our own dangerously balanced between scientific theory and practical horticulture, badly paid without professional security but momentarily free.

This eloquent statement by Darlington, who knew firsthand what Haldane had contributed, is testimony to Haldane's character and the method by which he was able to serve the cause of science in a profound manner. This was evident not only in his work at the JIHI but also at Cambridge, London, and later in India. It was as much as a public figure as through his work in the basic sciences that Haldane contributed to the advancement of science.

Later, in the same notes, Darlington summed up Haldane's contribution to genetics in the following words: "Haldane had filled the gap Bateson left. And he was still filling the gap, a gap in Bateson's genetics. . . . He had picked up also one of the main threads of past and future genetics—the chemical study of gene action initiated by Garrod in man and by Onslow in plants."

Darlington went even further: "He (Haldane) had saved the John Innes Institution. . . . He had even largely saved genetics in England. He had done it, however, not by discovery or innovation but by immovable conservatism, by the authority his physical and intellectual presence carried with it."

Darlington was also eloquent on what Haldane meant to him personally. Darlington's biographer Oren Solomon Harman presents a detailed picture of the early relationship between Haldane and Darlington:[20]

> At thirty-four, Haldane was Olympian in form and manner, the idol of the forward-looking, emancipated, protesting intellectual youth of the time, and the model of the scientist of the future. "For about seven years", Darlington would remember, "I regarded him as my infallible mentor. He took the place of my father and of Newton in my still immature mind." . . . Haldane was gradually reconciling his firm belief in the innate inequality of man with his growing attraction to revolutionary politics. Concomitantly, he was moving further away from the Kantian idealism, anti-mechanism, and organicism he had imbibed at home through his father, the physiologist John Scott Haldane. This was having an effect on his protege. Struggling to unite moral and mental engagements with both the political and scientific, Darlington embraced Haldane's scientism and his confident pronouncements that "the progress of society depends . . . on the progressive application of science." From this sentiment sprang an equally fervent antireligious bias.[21]

The correspondence in the Darlington archives indicates cordial relations initially between Darlington and Haldane. With reference to his story book for children, *My Friend, Mr. Leakey*,[22] which was published by the Cresset Press in 1937, Haldane wrote:

> Dear Darlington: If you would like to try your hand at some more illustrations here is one of the stories in the book. I suggest illustrations where X is marked. I could call for one on Tuesday for submission to my publisher who is dining with me that night.
> Additional data:
> 1. Mr. Leakey had long ears (not monstrously long) with small tufts of hair at the top.
> 2. Pompey was 1 ft. long, his tail being as long as the rest of him.
> 3. Abdu'l Makkar (= servant of the crafty one) had a long straight nose.
> If you are fed up with the whole thing, drop it. But there might be money in it, and I am against the professional whimsical illustration.

Other instances of their cooperation during the ten long years of Haldane's association with the JIHI are found in the archives at University College London, at the Bodleian at Oxford, and also at the JIHI, which is now located at Norwich. Haldane's advice on a number of scientific papers was actively sought by Darlington and numerous others. The correspondence in the Haldane archives at University College London indicates that Haldane answered a great number of letters on numerous subjects. Occasionally, the following printed card was mailed by his secretary:

16 Park Village East, N.W.1
Professor Haldane thanks you for your letter, but regrets that owing to pressure of work he is unable to Undertake the work which you suggest,
He therefore begs to be excused.

The following letter of Haldane gives some idea of how effective he was as a helpful critic.

June 13, 1933

Dear Darlington:

I looked over your Zea paper, which has been sent to Baur. I do not understand Table VI, (max. and min. variance). What is the variance? I think you ought to put in a further explanation when you go over the proofs, explaining how you calculate. In your summary, para 3, I have altered the mapped length from "46%" to over "over 46%," to allow for double crossing over. You can delete "over" in proof, if you like.

Mather and Stone have confirmed Navashin about effects of X-rays. Primula sinensis turns out to have so much less interference than Drosophila that its longest chromosome (map) seems to be about 60 units long. You have heard about the ring of 8 in Campanula. White (in the Zool. lab. at UCL) has found a big effect of temperature on Xma frequency in Locusta. He is going on to study the effect of CO_2. On your theory that terminalization is due to electrostatic repulsion, CO_2 should reduce the charge on the chromosomes and thus diminish terminalization. I cannot predict its effect on Xma frequency, though I should rather expect it to diminish interference and thus increase frequency.

I have spent most of my time in the last 2 months dealing with German academic refugees. We taken on Brieger here, with the aid of a fund for such people. I hope to take two of them in my lab at UCL. Stern has got a job in America. Goldschmidt and Jollos are not yet sacked. As a result of my activities I am already getting unwelcome attentions from British Nazis.

I very much hope that you won't go to the German genetical congress. A number of German "geneticists" such as Fischer, are rabid Nazis, and they are doing their best to make capital out of any foreigner who even appears to sympathize with them. I may add that Goldschmidt has had to resign his presidency of the German genetical society, and Jollos is almost certainly going to be sacked soon. That only leaves Renner and Wettstein as serious geneticists in the country. I don't think it is our business as scientists to protest in any way against national policy which does not concern us directly ... but when there is wholesale interference with scientific work I do think we should sit up and take notice. When the rector of Franfurt university says, as he did on May 23rd, "The chief characteristic of this rebirth is the replacement of the humanistic idea by the national and political. Nowadays the task of the universities is not to cultivate objective science but soldierlike, militant science, and their foremost task is to form the will and character of their students," I don't feel inclined to cooperate with him.

I wish you would let me know what you actually found about Drosophila. You will doubtless communicate the results of your researches on the Japanese orally. I wonder if we shall go to Innsbruck again. It is at present the scene of streetfighting between Hahnschwanzlers and Nazis.

Yrs ever,
J.B.S. Haldane

Haldane's Foreword

Haldane wrote a highly favorable Foreword to the first edition of Darlington's *Recent Advances in Cytology*:

> I am fortunate in being allowed to write a foreword to this book. It should properly, I think, be termed a treatise on karyology, the study of the nucleus, rather than on cytology, of which karyology, is a part. . . . The average cytologist is primarily an observer, and unaccustomed to long chains of deductive reasoning. He will find this book extremely difficult. Unfortunately it is indispensible. For Dr. Darlington's deductive principles stand the test of practical application. (pp. 127–28)

However, as time progressed, Haldane was beginning to complain. As Harman wrote in his biography of Darlington,

> By 1936 Hall had yet to show signs of retiring, and Haldane was beginning to become Impatient. An intense dislike between the wives of the two men, and a barrage of Jewish refugee scientists from Germany, and Haldane putting up at a cramped Merton, contributed to an increasingly strained, fractious environment. Referring to JIHI, Haldane complained that "this institution is in a state of hopeless indiscipline. I do not, for example, think it desirable that ping-pong should be played in a laboratory containing valuable microscopes" and further complained that "Lady Hall has on at least one occasion eaten large quantities of fruit from plants of genetical importance."[23]

Final phase of the Haldane–Darlington Relationship

It is unfortunate that the story of Haldane–Darlington relationship does not end there. As time passed, Haldane saw the JIHI situation quite differently. It seemed that Darlington, who benefited greatly from Haldane's intellect and public stature, took advantage of the rift between Haldane and the director, Sir Daniel Hall. In 1934 Hall reached the age of seventy but showed no interest in retiring. He reneged on the promise he made to let Haldane succeed him as the director. Furthermore, Hall, it appears, orchestrated the Council of JIHI's decision not to accept Haldane's plans for the institution. Darlington, presumably with Hall's encouragement, simultaneously offered counterproposals, without Haldane's knowledge, which were found to be more acceptable.

Darlington's biographer, Harman, described the deteriorating relationship between Haldane and Darlington. The younger man, Darlington, who described his relationship with Haldane initially in glowing terms ("infallible mentor" and "father") now began to see his hero's intellectual shortcomings. As Harman put it, "The directorship bid came, after all, precisely at the moment Haldane was moving closer to Communism, and Darlington, of the same obstinate, obdurate, and ruthlessly critical nature as his teacher, was becoming increasingly disillusioned: disillusioned with Haldane, his science, and his politics."[24] When Haldane drafted his own proposals, in his bid for the directorship, he consulted Darlington, *as a friend*, and took care to mention that "Darlington was a 'first rate man' who could be relied upon to direct cytological research." Darlington, on the other hand, submitted his competing proposals in secret, "behind the back of his benefactor," while Haldane was in Spain.

Darlington's proposals led eventually to his own appointment as director of JIHI in 1939. Haldane, who had given so much support and help to that institution, especially to Darlington, for several years, now felt that Darlington's appointment confirmed his worst suspicions of collaboration between Hall and Darlington. Recalling those events, during his last years in India, Haldane commented that gratitude was a difficult emotion to bear for Darlington![25]

Norman Pirie

In 1965, Norman Pirie,[26] biochemist and former pupil and colleague of Haldane at Cambridge University, was asked by the Royal Society of London to write the obituary of Haldane for the Biographical Memoirs of the Royal Society. Pirie contacted several former colleagues of Haldane, including Darlington, for biographical information. In his reply to Pirie, Darlington wrote, "I am not clear how far he bore me ill-will for having taken the post that had been promised to him but I am afraid there was such deep resentment against everybody who remained at the 'John Innes' that this coloured much of what he said and did in the following years." Darlington wrote further, "This is my true opinion although inevitably it sounds ungrateful for he always praised my own work very highly and although I perhaps did not need his encouragement at least I found it exceedingly comforting in those ten years he gave it to me; notably in his introduction to the two editions of my *Recent Advances in Cytology*."

In the same letter (to Pirie), Darlington mentioned that the only letter that he ever received from Haldane (after World War II) was in 1964, but it was really intended for P. J. Darlington Jr., the American zoogeographer and was misaddressed by Haldane's secretary. Haldane wrote,

> Allow me to congratulate you on your election to the National Academy, simultaneous with my own. Zoogeography and phytogeography need recognition. . . . We do a bit of zoogeography here, and I think have added a few

species to the fauna of Orissa. More interesting, perhaps, is our zoogeochronography, to coin a monstrous word. This has three aspects. (1) Times of critical events such as nests, in different parts of India. (2) Times of arrival and departure of migrants. (3) Annual journeys of resident species.

Darlington responded: "Alas! I fear however that the bulk of the letter was intended for my distinguished American namesake, the zoogeochronographer, P.J. Darlington, who must have been elected to the National Academy which is not likely to come my way unless I live to a very great age. . . . May I congratulate you on having attained to a sufficient repute and also reputibility to be accepted in Washington."

Darlington offered to meet with Haldane during his visit to India in January 1965, but received no reply. In any case, he could not have done so, because Haldane died of cancer on December 1, 1964. In retrospect, was it by mistake that Haldane's letter about his election to the National Academy was addressed to the wrong Darlington? Or was Haldane enjoying some perverse pleasure in letting Darlington know that he attained a distinction which Darlington was not likely to achieve? We will never know.

Correspondence

A curious footnote to the Haldane–Darlington relationship was their correspondence in 1941, long after Haldane moved on to University College London. Darlington was concerned about the lack of funds for research at the JIHI during wartime and was soliciting Haldane's help to obtain the Royal Society's support. Haldane was then a member of the Council of the Royal Society. Haldane was concerned about the continuation of the breeding program of his former colleague at JIHI, Miss de Winton, much of which had been directed by Haldane himself during his years at the institute. Furthermore, Haldane was willing to overlook his past differences with Darlington in the interests of saving valuable research, a decency that Darlington did not show Haldane when he interrupted the important research program in the biochemical genetics of plant pigments that was being directed by Haldane at the institute.

Haldane wrote, "Miss de Winton's more important material . . . represents the results of some 35 years' work on *Primula sinensis*, and if lost, will be quite irreplaceable. . . . For certain purposes *Primula sinensis* is a particularly favorable plant, since it has a lot of genes, and being grown in pots, can readily be subjected to fairly well controlled environmental conditions." In his response, Darlington assured Haldane that Miss de Winton would continue her work.

What is surprising about this correspondence is that it is not consistent with Darlington's statement in his letter to N. W. Pirie, dated July 9, 1965: "I am not clear he bore me ill-will for having taken the post that had been promised to him but I am afraid there was such deep resentment against everybody who remained at the 'John Innes' that this coloured much of what he said and did in the following years." The above cited letters were written two years after Darlington's appointment to the directorship of JIHI. Indeed, as late as 1945 (six years after Darlington's appointment) Darlington and Haldane were exchanging polite letters, Darlington seeking (and receiving) Haldane's support in such matters as the election of D. G. Catcheside to the Fellowship of the Royal Society. This is not consistent with Darlington's later claim that Haldane bore him some "grudge" because of Darlington's appointment as the director of JIHI in 1939.

Although Darlington's accounts lack consistency, the report of a special committee appointed by the Council of JIHI in 1936 paid tribute to Haldane's services: "Mr. Haldane's advice has been freely sought, freely given and very highly appreciated, and apart from difficulties with the Director, we have no evidence that his present relations with members of the staff are otherwise than friendly. In the interests of the Institution it is, in our view, highly desirable that after Mr. Haldane has taken up his new duties these amiable relations should continue."

S. C. Harland

The Darlington archives at the Bodleian contain letters from Sydney Cross Harland, a cotton geneticist who lived and worked in Peru for several years. Darlington informed Harland that he had been asked by *Nature*[27] to review the biography of Haldane by Ronald Clark and requested Harland's opinion on certain aspects of Haldane's life and work. Darlington dragged in various details of Haldane's personal life, such as his inability to have children and his failed marriage to Charlotte, suggesting that these events are related in some fashion to Haldane's desire for public acclaim. I am not competent to make a psychological evaluation of these outrageous claims of Darlington (nor was Darlington) regarding Haldane's private life. However, Harland dealt with this matter quite appropriately in his reply to Darlington: "I note your points which seem to me somewhat controversial. . . . I do think, however, that your review [of Clark's biography of Haldane] would be first rate if you evaluated his contributions to genetics, and soft-pedalled on personalities. I don't think he [Haldane] was a jealous man. . . . To my mind his main attribute was his gallantry—he tended to overestimate the contributions of those whom he wanted to back."

In a later communication to Darlington, Harland wrote, "It is rather remarkable that Medawar's views on Haldane were sought. It is well known that Medawar was not one of Haldane's buddies. Penrose's opinions on the other hand were worth eliciting as he was closely associated with Haldane for many years." Harland was referring to Medawar's Foreword for Ronald Clark's biography of Haldane. When Medawar's Nobel Prize was announced in 1960, I was sitting next to Haldane in India, and I heard him say that Medawar did not deserve it! Haldane thought Peter Gorer, who discovered the histocompatibility locus (H-2) in the mouse, was the man who deserved it, but unfortunately Gorer had died in 1949.

Another colleague of Haldane and Darlington at the JIHI was W. J. C. Lawrence. In response to a draft article on Haldane by Darlington, Lawrence sent him the following comments:

> Haldane has claimed in print that probably his greatest contribution to genetics was his initiation of biochemical genetics at JI [John Innes] and the securing of Scott-Moncrieff to undertake the chemical side, a claim which I think is a just one and for which he should be given full credit. I believe you may remember that in his book on biochemical genetics he plainly stated his disappointment at not being made director of JI (John Innes)—if my memory is correct.

Perhaps the most intriguing aspect of Haldane's personality concerns his non-experimental attitude to genetics ... he felt there was a gap in genetical research in that the use of statistics and mathematics was ignored by geneticists in what, after all, is a science basically concerned with quantitative/qualitative phenomena. ... Finally, it would be fair to say that despite his avoidance of personal involvement in experimentation he nevertheless did make a continuous contribution throughout his life since without the benefit of his criticisms and suggestions a number of genetical papers would have been less satisfactory! He was never an experimenter [in genetics]. But a good mentor.

These views were in marked contrast with those of Darlington.

Notes

* All information about the correspondence and private notes of C. D. Darlington is from the Darlington Archive at the Bodleian Library in Oxford.
1. The Johns Innes Horticultural Institution is referred to as JIHI or the institution in the text.
2. Darlington archives Bodleian Library, Oxford; Oren Solomon Harman, *The Man Who Invented the Chromosome* (Cambridge, MA: Harvard University Press, 2004).
3. J. B. S. Haldane, "Some Recent Work on Heredity," *Transactions of the Oxford University Junior Scientific Club* 1 (1920): 3–11.
4. J. B. S. Haldane, *Enzymes* (London: Longmans, Green, 1930; repr. Cambridge, MA: MIT Press, 1965).
5. J. B. S. Haldane, *New Paths in Genetics* (New York: Harper, 1942).
6. J. B. S. Haldane, *The Biochemistry of Genetics* (London: Allen & Unwin, 1954).
7. R. Scott-Moncrieff, "The Classical Period in Chemical Genetics: Recollections of Muriel Wheldale Onslow, Robert and Gertrude Robinson and J.B.S. Haldane," *Notes & Records of the Royal Society of London* 36 (1981): 125–54.

    The research described by Rose Scott-Moncrieff in biochemical genetics was largely due to the stimulus and direction provided by Haldane at JIHI. She was hired by Haldane for this purpose. The biochemical part was advised by Sir Robert Robinson. Scott-Moncrieff commented on the question of priority: "G.W. Beadle and E.L. Tatum are believed by many of the present generation to be the originators of chemical genetics; they do not seem to be aware of this very extensive early work which Beadle himself clearly recognized, p. 125."
8. *Ephestia* is a small moth belonging to the family Pyralidae. Some species are significant pests of dry plant produce, such as seeds and cereals. Best known among these are probably the cacao moth and the Mediterranean flour moth. Ernst Caspari, "Uber die Wirkung eines pleiotropen Gens bei der Mehlmotte *Ephestia Kuhniella*," *Zetschrift fur Archives Entwicklungs Mechanist* 130 (1933): 353–81.
9. *Drosophila*: The common fruit fly. Because of its short generation time, low cost of maintenance, and great genetic variability, it is useful for teaching and research in genetics, as was first demonstrated by Thomas Hunt Morgan at Columbia University in New York.
10. E. Caspari, "Haldane's Place in the Growth of Biochemical Genetics," in Krishna R. Dronamraju, ed., *Haldane and Modern Biology* (Baltimore, MD: Johns Hopkins University Press, 1968), 43–50.
11. Sir Daniel Hall (1864–1942) was a British agricultural educationist and researcher. He was born in Rochdale, Lancashire. He was principal of Wye College and director of Rothamsted Experimental Station and the John Innes Horticultural Institution. He was elected as a Fellow of the Royal Society in 1909.

12. G. W. Beadle and E. L. Tatum, "Genetic Control of Biochemical Reactions in Neurospora," *Proceedings of the National Academy of Sciences, Washington* 27 (1941): 499–506.
13. A. E. Garrod, *Inborn Errors of Metabolism* (Oxford: Oxford University Press, 1909).
14. Garrod, *Inborn Errors of Metabolism*.
15. J. B. S. Haldane, "The Biochemistry of the Individual," in *Perspectives in Biochemistry*, ed. J. Needham and D. E. Green (Cambridge: Cambridge University Press, 1937), 1–7.
16. Cyril Dean Darlington (1903–1981) was an English biologist who discovered the mechanics of chromosomal crossing over, its role in inheritance, and its importance to evolution. His early research at John Innes Horticultural Institution was guided and encouraged by Haldane.
17. Harman, *The Man Who Invented the Chromosome*, 56.
18. Harman, *The Man Who Invented the Chromosome*, 56.
19. Darlington archives Bodleian Library, Oxford; Harman, *The Man Who Invented the Chromosome*.
20. Harman, *The Man Who Invented the Chromosome*, 55–56.
21. Harman, *The Man Who Invented the Chromosome*, 159.
22. J. B. S. Haldane, *My Friend, Mr. Leakey* (London: Cresset Press, 1937).
23. Harman, *The Man Who Invented the Chromosome*, 130–31.
24. Harman, *The Man Who Invented the Chromosome*, 56.
25. In the course of a private conversation with his wife, Helen Spurway, and the author in 1960, Haldane mentioned the behavior of Darlington and commented, "Some people like Darlington find 'gratitude' a difficult emotion to bear."
26. Norman Wingate (Bill) Pirie (1907–1997) was a British biochemist who, along with Frederick Bawden, discovered that a virus can be crystallized by isolating tobacco mosaic virus in 1936. He studied biochemistry under F. G. Hopkins and J. B. S. Haldane at Cambridge University. He was elected a Fellow of the Royal Society in 1949, delivered its Leeuwenhoek Lecture in 1963, and won its Copley Medal in 1971 for his virology work.
27. Two reviews by C. D. Darlington: "Review of *Haldane and Modern Biology*, (ed) K.R. Dronamraju," *Nature* 222 (1969): 56–57; "Review of *J.B.S., The Life of J.B.S. Haldane*, by Ronald Clark," *Nature* 220 (1968): 933–34.

8

Origin of Life

Haldane's essay "The Origin of Life" in *The Rationalist Annual* in 1929 was the first serious attempt in the twentieth century to explore a scientific explanation for the origin of life on planet earth. Classical notions of spontaneous generation that living beings are generated by decaying organic substances were championed by Aristotle, among others. There were many others who suggested a supernatural origin of life, but such assumptions began to be questioned in the seventeenth century. Increasing knowledge of the structure and the stages of development of microorganisms, quite possibly protozoa and bacteria, cast doubt on the old assumption that maggots and putrefaction appeared spontaneously in aging organic matter. Microscopy contributed to this stage of development significantly. The scientific basis for the origin of life began to interest some prominent scientists in the nineteenth century including Louis Pasteur and T. H. Huxley. Charles Darwin, in a letter dated March 29, 1863, to his friend Joseph Dalton Hooker, expressed frustration: "It is mere rubbish thinking, at present, of origin of life; one might as well think of origin of matter." However, eight years later, in another letter to Hooker, dated February 1, 1871, he wrote:

> It is often said that all the conditions for the first production of a living being are now present, which could ever have been present. But if (and oh what a big if) we could conceive in some warm little pond with all sort of ammonia and phosphoric salts—light, heat, electricity present, that a protein compound was chemically formed, ready to undergo still more complex changes, at the present such matter would be instantly devoured, or absorbed, which would not have been the case before living creatures were formed.[1]

Thus, Darwin, with remarkable prescience, advanced the subject closer to what later became the accepted doctrine in early twentieth century.

In 1929 Haldane wrote: "It was possible either to suppose that life had been supernaturally created on earth some millions of years ago, or that it had been brought to earth by a meteorite or by micro-organisms floating through interstellar space. But a large number, perhaps the majority, of biologists believed in spite of Pasteur, that at some time in the remote past life had originated on earth from dead matter as the result of natural causes."[2]

Even as late as the early twentieth century, inaccurate statements were made about life's origins in many textbooks. The first major advance was made when

Haldane (1929), with remarkable insight, proposed a plausible mechanism that any complex molecules that were generated in the primitive earth would be more stable if the prebiotic world were anaerobic than they would be in the presence of oxygen with no ozone in the atmosphere. Ozone absorbs ultraviolet light strongly, hence only a small fraction of that in the sunlight reaches us today. However, in an anaerobic world, a great deal more ultraviolet light would have gotten through, and it is known to be a powerful synthesizing agent.[3] Haldane wrote, "when ultra-violet light acts on a mixture of water, carbon dioxide, and ammonia, a vast variety of organic substances are made, including sugars and apparently some of the materials from which proteins are built up. ... In this present world such substances, if left about, decay—that is to say, they are destroyed by micro-organisms. But before the origin of life they must have accumulated till the primitive oceans reached the consistency of *hot dilute soup*."[4] As Haldane pointed out, the first precursors of life found plenty of food as they had no competitors in the struggle for existence. As the primitive atmosphere contained little or no oxygen, Haldane suggested that they must have obtained the energy needed for their growth by some other process than oxidation, by fermentation, or, as Pasteur[5] put it, "fermentation is life without oxygen." This is a necessity for the origin of life in an anaerobic environment. Haldane wrote, "If this was so, we should expect that high organisms like ourselves would start life as anaerobic beings, just as we start as single cells. This is the case. Embryo chicks for the first two or three days after fertilization use very little oxygen, but obtain the energy which they need for growth by fermenting sugar into lactic acid, ... in all probability you and I lived mainly by fermentation during the first week of our prenatal life."[6]

J.D. Bernal called this idea *biopoiesis* or *biopoesis*,[7] the process of living matter evolving from self-replicating but nonliving molecules, and proposed that biopoiesis passes through a number of intermediate stages. Such a process was briefly hinted by Haldane in his 1929 article, and was discussed in greater detail in his later papers on this subject.

Microspheres

Following Bernal's suggestion, Sidney W. Fox[8] studied the spontaneous formation of peptide structures under conditions that might have existed in the primitive earth. He showed that when amino acids dried out in a manner similar to the warm dry ponds in prebiotic conditions they formed long thread-like submicroscopic polypeptide molecules later called *proteinoid microspheres*, which showed many of the basic features of "life."

Fox discovered that the temperature was over 100 °C (212 °F) just 4 inches (100 mm) beneath the surface of a cinder cone from Hawaii. He suggested that this might have been the environment in which life was created, any molecules that formed would be washed into the sea.

Later, he placed lumps of volcanic lava over amino acids derived from methane, ammonia, and water, sterilizing all materials and baking the lava over the amino acids for a few hours. He obtained a brown, sticky substance that contained proteinoids from a combination of amino acids, which then combined to form small, cell-like spheres. These were named *microspheres*, but were later renamed *protobionts*.[9] They formed clumps and chains that were similar to cyanobacteria. They multiplied asexually, forming double membranes similar to cell membranes.

Half-Living Things

Haldane hypothesized that the first living or half-living things were probably large molecules that were synthesized under the influence of the sun's radiation. They were presumably able to reproduce in the particularly unique medium in which they originated. Each may have required a variety of specialized molecules before it could reproduce itself, and depended on chance for the occurrence of such situations. Haldane acknowledged that his idea that the intermediate stage on the way to the first organism was a self-reproducing molecule was influenced by the discovery of the viruses at the beginning of the twentieth century. He was also impressed by the discovery of the bacteriophage, by Felix D'Herelle[10] in 1917, which is capable of reproducing itself inside the cell. He was in sympathy with the hypothesis proposed by the geneticist Herman J. Muller, who compared the bacteriophage to a gene that copies itself within the cell.[11] Haldane regarded the virus as something more than a functional model; he saw it also as the true phylogenetic "missing link" between inanimate matter and life.[12]

Haldane's views on the origin of life changed with time throughout his career. At first, his position was similar to that of Oparin. In 1954 he considered that the first decisive step is the formation of a cell. Haldane supported his view by pointing out that the viruses are surrounded by a membrane-like structure. At the International Conference on the Origin of Life in 1963, he suggested that the first organism might have consisted of an RNA molecule that functioned as a single gene. At the same conference, Haldane discussed the possibility that life may have remained in the virus stage for many millions of years before a suitable assemblage of elementary units was brought together in the first cell.[13]

Although recognizing the first molecules as steppingstones to life, Haldane emphasized the organic dynamic nature of life, recognizing the first reproducing molecules as only half-living. Referring to the unicellular organisms, such as bacteria, Haldane speculated that they are made up of parts that cooperated, each part being specialized to a specific chemical function, preparing chemical molecules suitable for the growth of the other parts. Consequently, the organism can subsist on a few types of molecules, which lead to the more complex transformation needed for the growth of the cell as a whole. Haldane suggested that the hypothetical cell consists of numerous half-living chemical molecules that are suspended in water and enclosed in an oily film. The conditions for that situation were far better when the whole sea was a vast chemical laboratory. And life may have remained in that virus stage for many millions of years before the accidental assemblage of elementary units that formed the first cell. Although there must have been many failures, the first successful cell had plenty of food, a definite advantage over any competitors that may have appeared.

Asymmetry

Another point made by Haldane was that all organisms appear to have descended from a common ancestor because, of the two possibilities, most of our structural molecules show only one type of asymmetry throughout living nature, as indicated by the fact that they rotate the plane of polarized light and often asymmetrical crystals. He added further that there is nothing to prevent the existence of looking-glass organisms built from molecules that are the mirror images of our own bodies. However, as they do not exist, this event probably occurred only once, or, more likely, the descendants of the first living organism rapidly evolved to overwhelm any later competitors that may have appeared on the scene.

Haldane speculated further that, as the primitive organisms exhausted the food available in the sea, some of them began to synthesize in their own bodies, thus making sure that abundant food is available. The first plants may have come into existence, living near the surface and using the sunlight for making food. And the first animals may have descended from them and retained molecules descended from chlorophyll. However, the oxygen released by the first plants probably killed most of the other organisms.

As Haldane closed his 1929 article, he wrote, "The above conclusions are speculative. They will remain so until living creatures have been synthesized in the biochemical laboratory. We are a long way from that goal. . . . I do not think I shall behold the synthesis of anything so nearly alive as a bacteriophage or a virus, and I do not suppose that a self-contained organism will be made for centuries. Until that is done the origin of life will remain a subject for speculation." Haldane was being unduly pessimistic about such prospects. In our own lifetime, we are witnessing the successful production of synthetic genomes that are designed for specific purposes by the Nobel laureate Hamilton Smith at the J. Craig Venter Institute[14] in La Jolla, California.

We have also witnessed, in 2000, the sequencing of the human genome as well as those of several other species by Venter's group in a private enterprise as well as the completion of the human genome sequencing with Francis Collins, at the US National Institutes of Health, supported by public funds.

To sum up, four major conditions are of interest:

1. Presence of liquid water
2. Elements needed for metabolism and reproduction
3. Source of energy
4. Evolution of complex organic molecules into primitive half-life

Haldane's article also contained interesting speculations on the nature of viruses and genes, suggesting that they were perhaps parts of organisms that could reproduce inside the special environment of a cell. These concepts were new to those studying viruses and genes at that time. Later, in 1936, Haldane discussed the effect of molecular size on the process of reproduction, and suggested that, if genes were considered to be large molecules, they would not grow and divide like normal cells, but a sister gene would be built near the original, by a process that is similar to crystallization.

In papers in 1944 and 1945, Haldane cautioned that when thinking of the chemical processes involved in the origin of life we must not think in terms of contemporary chemistry only. He observed that in simpler organisms the metabolic processes in a cell are maintained by diffusion, and that various cellular processes became possible only when the changing chemistry had liberated enough energy to maintain an appropriate internal environment.

The idea of alternate ways of looking at the chemical process interested Haldane, and he suggested, in 1954, that metaphosphates might at first have filled the role that was taken over later by adenosine triphosphate (ATP).[15] He suggested later, in 1957, that when ATP appeared later it may have brought about phosphorylations on surfaces of a different nature from the enzymes now used.

Haldane estimated the probability of the appearance of sufficient complexity and activity that could be called "living." In 1952 he estimated that an ocean containing a variety of organic molecules "would by pure chance ... meet specifications calling for between 50 and 200 "bits"[16] in 10^9 years and that a bacteriophage contained about 100 bits.

Brahma

Concerning the probability of the origin of life, Haldane put it in his characteristic manner, invoking the Hindu God of creation, Brahma[17]: "If Brahma, though not eternal, has a time scale of about 10^{15} of our own, his easiest method of creating intelligent beings like ourselves might be to leave a few planets at a suitable distance from their suns for 2×10^9 years, as we light a match with confidence that one of a large number of random events will usually set off a chain of reactions. This is not intended as an argument for Theism, but merely as a plea for precision in the discussion of that hypothesis." On the other hand, Haldane also pointed out the improbability of the appearance of life by mentioning that even one enzyme of the present-day type would need 100 bits, and that 5,000 would be needed for an artificial organism.

Haldane resolved this dilemma by suggesting the independent appearance of two or more systems with a quasi-vital degree of complexity, which can exist in a free state for only a brief period. Accordingly, he was one of the first to refer to the plural "origins" instead of the more usual "origin." He reconciled this possibility with his earlier idea that the existence of molecules of one stereoisomeric type supported a single origin of life, by citing the finding, in 1960, that, in certain nuclear reactions, parity is not conserved, leading to some chiral selectiveness in atoms themselves.

Unity and Diversity of Life

In 1958, in a lecture titled "The Unity and Diversity of life," which he delivered over the All India Radio in India, Haldane reiterated some of his ideas on the origin of life. He emphasized that there is no sharp line between living and nonliving matter. He stated that certain molecules attract others of the same kind to form organized aggregates called crystals. However, whether these very primitive kinds of organization of material processes and objects were able to develop into the more complex organization that is called "life" is not known with any degree of certainty.

Haldane discussed the origins of life several times in his lifetime, contributing new ideas. He participated in a symposium in Tallahassee, Florida, October 27–30, 1963, about a year before his death. He wrote,

> If the minimal organism involves not only the code for its one or more proteins, but also twenty types of soluble RNA, one for each amino acid, and the equivalent of ribosomal RNA, our descendants may be able to make one, but we must give up the idea that such an organism could have been produced in the past, except by a similar pre-existing organism or by an agent, natural or supernatural, at least as intelligent as ourselves, and with a good deal more knowledge.[18]

Haldane's ideas on the origin of life changed with time, as one would expect, with new knowledge and new modes of thinking. Toward the end of his life he considered the possibility that the first organism may have consisted of an RNA molecule that functioned as a single gene. In the following paragraph, from his 1963 paper, he appears to argue against the possibility of half-live systems, which he had suggested earlier in his first statement on this subject in 1929:

> I may be converted in the course of the meeting, but when writing this paper, I am by no means attracted by the theory of a period of many million years of biochemical evolution preceding the origin of life. It seems to me that any half-live systems—for example, catalysts releasing the energy of metastable molecules such as pyrophosphate or sugar—would merely have made conditions less favorable for the first living organisms, by which I mean the first system capable of reproduction. A protein capable of catalyzing such reactions would not multiply in consequence, any more than an enzyme does.[19]

Urzymes

The RNA world hypothesis came under attack recently from a new perspective. The traditional view states that RNA was the original carrier of genetic information, and that RNA derivates called ribozymes translated RNA's genetic code to make proteins that determine the basic functions of life. Charles Carter and colleagues (2013)[20] from the University of North Carolina argued that the process involving RNA and ribozymes and the random interactions that would create the first prototypes of life would have taken far more time than the present age of the universe would allow. He suggested that instead of ribozymes, rudimentary enzymes called *urzymes* decoded RNA to make proteins. Carter's group studied two large groups of enzymes involved in genetic replication and reported that they had a common core, the *urzymes*, which can translate RNA into proteins. This hypothesis has certain advantages, the origins of *urzymes* is extremely old, and they are common across many species and enzymes and consist of basic protein pieces called peptides, which are known to have existed long time ago. On the other hand, it is uncertain whether ribozymes existed that long ago.

Size of the First Organism

In 1963 Haldane addressed the question, How much smaller may the first natural organism have been?

> I suggest that the primitive enzyme was a much shorter peptide of low activity and specificity, incorporating only 100 bits or so. But even this would mean one out of 1.3×10^{30} possibilities. This is an unacceptable, large number. If a new organism were tried out every minute for 10^8 years, we should need 10^{17} simultaneous trials to get the right result by chance. The earth's surface is 5×10^{18} cm^2. There just isn't, in my opinion, room. Sixty bits, or about 15 amino acids, would be more acceptable probabilistically, but less so biochemically. I suggest that the first synthetic organisms may have been something like a tobacco mosaic virus, but including the enzyme or enzymes needed for its own replication. More verifiably, I suggest that the first synthetic organisms may be so constituted. For natural, but not for laboratory life, a semipermeable membrane is needed. This could be constituted from an inactivated enzyme and lipids. I think, however, that the first synthetic organism may be much larger than the first which occurred. It may contain several different enzymes, with a specification of 5000 bits. . . . This should be quite within human possibilities. The question will then arise: How much smaller may the first natural organism have been? If this minimum involves 500 bits, one could conclude either that terrestrial life had had an extraterrestrial origin . . . or a supernatural one (with many religions, but by no means all).[21]

Haldane wrote extensively on the origins of life for over thirty-five years, putting forward novel ideas, and was able to do so more than any other biologist because of his familiarity with what was going on in multiple disciplines, especially in the physical sciences. What made Haldane's theory so different from previous theories of the nineteenth century was that he defined the problem of the origin of life in biochemical terms and gave the concept a concrete form. He presented a testable hypothesis, which was indeed tested and confirmed in experiments by Stanley Miller and Harold Urey in the 1950s.

Haldane considered the possibility of alternate ways of living. For instance, he suggested that life could exist in ammonia and similar nonaqueous environments. Another idea, which interested Haldane, in the context of quantum physics, was Spinoza's view that there may be mind-like correlates in "other modes of being." He also considered that "thought" may exist in the dense matter of a white dwarf star, although we are not accustomed to imagine thought without considering an individual who is capable of a "thought process."

Haldane considered that, because of the likelihood that organisms from a common source were scattered around the universe, an introduction of beings into alien environments would be catastrophic because they are biologically and chemically close enough to be pathogens on one another.

Miller-Urey Experiment

In 1953, almost a quarter-century after the Oparin-Haldane theory was published, experimental evidence was produced at the University of Chicago, which simulated the conditions hypothesized by Haldane and Alexander Oparin on the primitive Earth, which favored chemical reactions that synthesized more complex molecules from simpler organic precursors. This classic experiment was conducted by the graduate student Stanley Miller and his professor, Harold Urey, at the University of Chicago.[22] A mixture of methane, ammonia, water vapor, and hydrogen was circulated through a liquid water solution and continuously sparked by a corona discharge, which represented lightning flashes on the early Earth. After several days of exposure to sparking, the solution changed color. Subsequent analysis indicated that several amino and hydroxy acids, intimately involved in contemporary life, had been produced by this simple procedure. Subsequent experiments have substituted ultraviolet light or heat as the energy source or have altered the initial abundances of gases. In all such experiments amino acids have been formed in large quantities. Much more energy was available in ultraviolet light on the primitive Earth than in lightning discharges. The astrophysicist Carl Sagan and his colleagues also made amino acids by long-wavelength ultraviolet irradiation of a mixture of methane, ammonia, water, and H_2S. It is remarkable that amino acids, particularly biologically abundant amino acids, are made so readily under simulated primitive conditions. However, under oxidizing laboratory conditions, no amino acids are formed, confirming that reducing conditions were necessary for prebiological organic synthesis.

Many of the compounds made in the Miller-Urey experiment are known to exist in outer space. On September 28, 1969, a meteorite fell over Murchison, Australia. Analysis of the meteorite has shown that it is rich with amino acids: over ninety amino acids have been identified so far. Nineteen of these amino acids are found on Earth. The early Earth is believed to be similar to many of the asteroids and comets still roaming the galaxy.

Miller had also performed more experiments, including one with conditions similar to those of volcanic eruptions. This experiment had a nozzle spraying a jet of steam at the spark discharge. In 2008, a group of scientists examined eleven vials left over from Miller's experiments from the 1950s. They found that the volcano-like experiment had produced the most organic molecules, twenty-two amino acids, five amines, and many hydroxylated molecules, which could have been hydroxyl radicals produced by the electrified steam. The group suggested that volcanic island systems became rich in organic molecules in this way.

Jeewanu: Contribution of Krishna Bahadur

Of historic interest, the Indian chemist Krishna Bahadur, from the University of Allahabad, published on organic and inorganic particles that he had synthesized and baptized *Jeewanu*, or "particle of life."[23] Bahadur conceived of the *Jeewanu* as a simple form of the living. *Jeewanu* (ancient Sanskrit for the "particles of life") are synthetic chemical particles that possess cell-like structural and functional properties. They are considered to be primitive cells, or protocells, the simplest life forms. Using photochemical reaction, he produced cell-like spherules (microscopic spherical structures) from a mixture of organic and inorganic compounds. Bahadur named these particles *Jeewanu* because they exhibited the basic properties of a cell, such as the presence of a semipermeable membrane, amino acids, phospholipids, and carbohydrates. Further, like living cells, they had several catalytic activities.

The "Time Window" for the Origin of Life

Recent analyses have indicated that the time frame for the appearance of early life was incredibly short, much shorter than earlier estimates. Extreme shortening of the early period indicates that the early stages of the development of a primitive living system occurred within a very short time period, much shorter than anyone had anticipated.

It is generally agreed that the Earth was formed 4.6 billion years ago. It has been estimated that the bombardment from outer space continued until 3.8 billion years ago. Appearance and persistent of any incipient organic forms had to wait at least until the late bombardment period, which could be considered as the lower limit of the "time window," that is to say, the time during which prebiotic chemical evolution led to the appearance of the first primitive life forms. The upper boundary would be determined by the evidence for the earliest appearance of life on earth. The date of its appearance is being pushed back with the discovery of new information about the date of the earliest fossil life forms. The oldest fossils of bacteria from rocks found in Africa are about 2.7 billion years old. This would indicate a time window of about one billion years. However, scientists at the University of California have found that the earliest fossils found in Australia are aged 3.5 billion years. They contain strings of bacterium-like bodies, similar in appearance to cyanobacteria (blue green algae) of today.[24] Another line of evidence, using isotopes of carbon, also indicates the earliest age to be 3.5 to 3.85 billion years.[25] From these studies, it appears that life on earth emerged extremely quickly, during several hundred million years or even during a few million years. Several authorities have estimated the time for life to evolve to cyanobacteria to be no longer than "10 million years."[26]

Hydrothermal Systems

In contrast to the Oparin-Haldane primordial soup theory, hydrothermal systems,[27] or submarine volcanism, has been mentioned as an alternative source for the synthesis of organic compounds on the prebiotic earth.[28] A major proponent of this hypothesis was John B. Corliss of NASA's Goddard Space Flight Center, who first detected submarine hydrothermal vents at the Galapagos oceanic ridge in 1979.[29] They argued that the environments of hot springs could have provided emerging life with both energy and nutrients. At the same time, such an emerging life is protected against extraterrestrial impacts.

Fluids released from hydrothermal vents on the seafloor were shown to contain carbon dioxide and large amounts of methane.[30] It has been suggested that the mixture of sulphur minerals and hydrogen sulfide under high temperatures and pressures results in free hydrogen and free energy, which can reduce carbon dioxide to methane.[31] It has been hypothesized that in the primitive Earth the earth's mantle was more reducing and the ratio of methane to carbon dioxide in hydrothermal-vent fluids could have been much higher.[32] Even the presence of a few parts per million of methane would have promoted the synthesis of hydrogen cyanide, which is essential for the prebiotic synthesis of amino acids and nucleotide bases.[33] It has been hypothesized that not only organic molecules but even the first living systems appeared not at the Earth's surface but in the ocean depth.[34]

Simultaneously, the microbiologist Carl Woese discovered a most ancient group of living organisms that prefer to live in most extreme conditions of high temperature and pressure as well as anaerobic environments. Woese classified these unusual microorganisms under a new branch of life that he termed archaea. Although lacking the nucleus like bacteria, archaea have distinct characteristics of their own that can be recognized at the molecular level. Woese classified life forms into three domains: archaea, bacteria, and eukarya.[35]

Panspermia

Various theories have been put forward that life on earth has an extraterrestrial origin, either accidentally or deliberately introduced by unknown sources.[36] Some prominent scientists have suggested these possibilities. In 1974 Fred Hoyle and Chandra Wickramasinghe proposed that some interstellar dust was largely organic and that life forms continue to enter the earth's atmosphere, possibly causing new diseases and epidemic outbreaks. In 1973 the Nobel laureate Francis Crick and Leslie Orgel[37] seriously suggested that life may have been deliberately spread by an advanced extraterrestrial civilization, however, Crick [38] later conceded that an early "RNA" world may have originated on earth. There are several objections to theories of panspermia, including the risk posed by the space environment and cosmic radiation to life before it could safely arrive on earth. The main problem with the theory of panspermia is that it still fails to address where life came from. If one believes that life on Earth arose from cells from another planet, then where did those cells come from? If bacteria from another planet gave the genes necessary for evolution to organisms on Earth, then we're still left with the question: how did the alien bacteria develop the genes?

In his classic essay of 1929, Haldane rejected all explanations of the extraterrestrial origin of life on earth.

Notes

1. Charles Darwin, *Correspondence of Charles Darwin*, Darwin Archives, Cambridge University Library, Letter to J. D. Hooker, February 1, 1871.
2. J. B. S. Haldane, "The Origin of Life," in *The Rationalist Annual* (1929); reprinted in J. B. S. Haldane, *Science and Life*, with an introduction by John Maynard Smith (London: Pemberton, 1968), 1–11. Quite independently of Haldane, Alexander Oparin in the Soviet Union proposed a similar hypothesis to explain the origin of life on earth. Oparin's book was published in the Russian language in the Soviet Union in 1924, and a translation appeared in English in late 1929. Haldane's article was written in 1928 and published in *The Rationalist Annual* for 1929.

    Alexander Oparin (1894–1980) was a Soviet biochemist who was noted for his theories about the origin of life, and for his book *The Origin of Life*. In 1924 he put forward a theory suggesting that life on Earth developed through a gradual chemical evolution of carbon-based molecules in the earth's primordial soup. In 1935, along with the academician Alexei Bakh, he founded the Biochemistry Institute of the Soviet Academy of Sciences. In 1939 Oparin became a Corresponding Member of the Academy, and in 1946 a full member. In the 1940s and 1950s he supported the controversial theories of Trofim Lysenko.
3. Haldane was thinking of the work of Baly (1927) of Liverpool, who applied ultraviolet radiation to a solution of carbon dioxide in water, resulting in the synthesis of sugars—organic polymers containing carbon, hydrogen, and oxygen. E. C. C. Baly, W. E. Stephen, and N. R. Hood, "The Photosynthesis of Naturally Occurring Compounds: II. The Photosynthesis of Carbohydrates from Carbonic Acid by Means of Visible Light," *Proceedings of the Royal Society of London. Series A, Containing Papers of a Mathematical and Physical Character* 116 (1927): 212–19.
4. Haldane introduced the term "hot dilute soup" in 1929 to describe the primitive ocean, containing the building blocks of life. It has been widely accepted and used in biology in discussions of the origin of life.
5. Louis Pasteur (1822–1895) was a French microbiologist whose name is closely connected with vaccination, microbial fermentation, and pasteurization. He is remembered for his remarkable breakthroughs in the causes and preventions of infectious diseases. His discoveries have saved millions of lives. He created the first vaccines for rabies and anthrax. His discoveries have supported the "germ theory of disease" and its application to clinical practice. Pasteur is well known to the general public for treating milk and wine to stop bacterial contamination, a process that came to be called *pasteurization*. He is a founder of bacteriology and microbiology. Pasteur made important discoveries in chemistry, in particular the molecular basis for the asymmetry of certain crystals. He was the director of the Pasteur Institute, which was founded in 1887, until his death.
6. Biopoiesis is a process by which living organisms are thought to develop from nonliving matter. According to this theory, conditions were such that, at one time in Earth's history, life was created from nonliving material, probably in the sea, which contained the necessary chemicals. During this process, molecules slowly grouped, then regrouped, forming ever more efficient means for energy transformation and becoming capable of reproduction.
7. Haldane, "The Origin of Life."
8. Sidney W. Fox, ed., *The Origins of Prebiological Systems and of Their Molecular Matrices; Proceedings of a Conference Conducted at Wakulla Springs, Florida, on 27–30 October, 1963, under the Auspices of the Institute for Space Biosciences [of] the Florida State University and the National Aeronautics and Space Administration* (New York: Academic Press, 1965).

    It was at this conference that Haldane and Oparin first met, although they had known of each other for over thirty years, having proposed similar hypotheses on the origin of life. Haldane died on December 1, 1964. It was on his return journey to India after attending this conference that Haldane stopped in London and was operated on for cancer at the University

College Hospital in London. These events were immortalized in his famous poem on cancer that he wrote shortly after his cancer operation:

> I noticed I was passing blood
> (Only a few drops, not a flood),
> So pausing on my homeward way
> From Tallahassee to Bombay
> I asked a doctor, now my friend,
> To peer into my hinder end.

9. Protobionts are systems that are considered to have possibly been the precursors to prokaryotic cells; quite possibly an aggregate of abiotically produced organic molecules surrounded by a membrane or a membrane-like structure.
10. Félix d' Hérelle (1873–1949) was a French-Canadian microbiologist generally known as the discoverer of the bacteriophage, a virus that infects bacteria. He studied medicine in Paris and Leiden, and later studied microbiology at the Pasteur Institute in Paris. In his experiments with a bacterium known to cause enteritis (digestive tract inflammation) in certain insects, he noted certain clear spots (areas free of bacteria) on gelatin cultures of the bacterium under study. Subsequently, while investigating a form of dysentery he happened to mix a filtrate of the clear areas with a culture of dysentery bacteria. The bacteria were quickly and totally destroyed by an unknown agent in the filtrate that Hérelle termed an "invisible microbe"; which was later renamed a *bacteriophage*. He achieved some success initially in using bacteriophages in the treatment of dysentery and other infections, however, the medical use of these agents against such diseases was later replaced by antibiotics and other drugs.
11. H. J. Muller, "The Gene as the Basis of Life," *Proceedings of International Congress on Plant Sciences* 1 (1926): 897–921, and "Bar Duplication," *Science* 83 (1936): 528–30. In the 1926 paper Muller discusses the gene for the origin of all cellular components and activities and the basis for life on Earth through its capacity for replicating its variations. In the 1936 paper he shows how tandem duplication leads to gene number increase and mutational changes in the duplicated genes lead to functional differentiation, both necessary components for the evolution of life from an initial gene-like molecule.
12. J. B. S. Haldane, "Data Needed for a Blueprint of the First Organism," in *The Origins of Prebiological Systems and their Molecular Matrices*, ed. S. W. Fox (New York: Academic Press, 1964), 11.
13. Haldane, "Data Needed for a Blueprint of the First Organism."
14. H. O. Smith et al., "How Many Genes Does a Cell Need?" in *From the Environment to Organisms and Genomes and Back*, ed. K. Zengler (Washington, D.C.: ASM Press, 2008), 279–99.
15. ATP is a nucleoside triphosphate used in cells as a coenzyme. It is often called the molecular unit of currency of intracellular energy transfer, as it transports chemical energy within cells for metabolism. ATP was discovered in 1929 by Yellapragada Subbarao while he was a graduate student at Harvard Medical School, earning his PhD in 1930. However, he never received full credit for his discovery because his PhD supervisor Cyrus Fiske took credit for his work. Subbarao's colleague and fellow student at Harvard, George Hitchings, who shared the 1988 Nobel Prize in Physiology or Medicine with Gertrude Elion, said, "Some of the nucleosides isolated by Subbarao had to be rediscovered years later by other workers because Fiske, apparently out of jealousy, did not let Subbarao's contributions see the light of the day." In his Nobel Prize Lecture, Hitchings wrote, "I was caught up in the Fiske-Subbarao program, and . . . after the discovery of phosphocreatine, this group had detected and isolated adenosine triphosphate."

Later, Subbarao joined Lederle Laboratories, a division of American Cyanamid, (now a division of Wyeth, which is owned by Pfizer), after he was denied a regular faculty position at Harvard. At Lederle, he developed a method to synthesize folic acid, Vitamin B_9, and developed the important anticancer drug methotrexate—one of the very first cancer chemotherapy agents and still in widespread clinical use. He also discovered the drug Hetrazan, which was used by the World Health Organization against filariasis. Under Subbarao's direction, Benjamin Duggar discovered the world's first tetracycline antibiotic, aureomycin, in 1945.

16. A bit is the *basic unit* of information in computing and digital communications. A bit can have only one of two *values*, and may therefore be physically implemented with a two-state device. The most common representation of these values are 0 and 1. The term "bit" is a portmanteau of "binary digit." In *information theory*, one bit is typically defined as the uncertainty of a binary random variable that is 0 or 1 with equal probability, or the information that is gained when the value of such a variable becomes known. The symbol for bit, as a unit of information, is simply *bit*. A group of eight bits is commonly called one *byte*. Claude E. Shannon first used the word "bit" in his seminal 1948 paper "A Mathematical Theory of Communication." He attributed its origin to John W. Tukey, who had written a Bell Labs memo on January 9, 1947, in which he contracted "binary digit" to simply "bit." Interestingly, Vannevar Bush had written in 1936 of "bits of information" that could be stored on the punch cards used in the mechanical computers of that time.

17. Brahma: There is always controversy over who is superior among the Hindu triad—Brahma, Vishnu, and Shiva. Almost all myths, though, agree on Brahma being the creator among the three. Of the other two, Vishnu is the preserver while Shiva is the destroyer. Nevertheless, Shiva, who is usually symbolized by a phallic stone, is also acknowledged as a creator in conjunction with his female consorts, who are accepted as the feminine power complementing his male potency. Brahma is the god of wisdom, and the belief is that the four *Vedas* were delivered from each of his four heads. The *Vedas* are at the apex of all Hindu scriptures. They are texts derived directly from the gods and, as such, are indisputable. Brahma's heaven is believed to contain in a superior degree all the various splendors of the heavens of the other gods.

Earlier myths, though, acknowledge Yama and his sister, Yami, as the creators of the human race. Hindu myths have changed progressively over time. While very early myths often attribute one particular god or goddess with a particular earthly phenomenon such as the creation of humanity, later myths may choose a totally different deity for the same task.

Brahma has red skin and wears white clothes. He rides on a goose. He has four arms. In one arm he carries the *Vedas*, in another a scepter, in a third a *komondul* (a special type of water jug that is stilled used by Hindus during worship), and in a fourth a bow or, variously, a string of beads or a spoon. As mentioned earlier in this text, Brahma has four heads. In this, Shiva who has five, outdoes him.

Brahma generated a female partner, generally acknowledged as Saraswati, out of his own substance. She is his daughter as well as his wife. Of her he gave birth to the human race. Saraswati is the Hindu goddess of wisdom and learning. Her initiation by Brahma at the beginning of all is symbolic of the maximum importance of knowledge and wisdom to all beings. Saraswati was reconciled enough with Brahma to enable him to create the human race in partnership with her. It is also notable that, akin to the fall of Adam from the grace of God and the subsequent loss of Eden, the lust of Brahma signifies the downfall of humanity. Hinduism holds that basic desires hinder total salvation from the cycle of births and rebirths. Total salvation, in Hinduism, is the reconciliation of the individual soul with the Nirguna Brahman.

The Hindu view of the ultimate reality is expressed in the following revelation of the Rig Veda, the oldest Hindu scripture: "Ekam sat vipraha, bahudha vadanti." "Truth is one, the wise call It by various names."

This doctrine recognizes that the ultimate reality possesses infinite potential, power, and intelligence, and therefore cannot be limited by a single name or form. Hindus view the ultimate reality as having two aspects: impersonal and personal. The impersonal aspect of the ultimate reality is called *Nirguna Brahman* in Hindu scriptures. Nirguna Brahman has no attributes and, as such, is not an object of prayer, but of meditation and knowledge. This aspect of the ultimate reality is beyond conception, beyond reasoning, and beyond thought.

18. J. B. S. Haldane, "Data Needed for a Blueprint of the First Organism," in S. W. Fox, *The Origins of Prebiological Systems* (New York: Academic Press, 1965), 12. The conference was held in 1963 in Wakulla Springs, Florida. Haldane died in December 1964, before the book was published.

19. Haldane, "Data Needed for a Blueprint," 15.
20. L. Li, C. Francklyn, and C. W. Carter Jr., "Aminoacylating Urzymes Challenge the RNA World Hypothesis," *Journal of Biological Chemistry* 288 (2013): 26856–63.
21. Haldane, "Data Needed for a Blueprint," 14.
22. In 1953, Stanley L. Miller and his research supervisor, Harold C. Urey, working at the University of Chicago, conducted an experiment that would change the approach of scientific investigation into the origin of life. Miller took molecules that were believed to represent the major components of the early Earth's atmosphere, such as methane (CH_4), ammonia (NH_3), hydrogen (H_2), and water (H_2O). He ran a continuous electric current through the system, to simulate lightning storms believed to be common on the early earth. At the end of one week, Miller observed that as much as 10 to 15 percent of the carbon was now in the form of organic compounds. Two percent of the carbon had formed some of the amino acids that are used to make proteins. Later experiments showed that RNA and DNA bases could be obtained through simulated prebiotic chemistry with a reducing atmosphere. However, skepticism concerning Miller's experiment continued mainly because the early Earth's atmosphere did not contain predominantly reductant molecules, and because this experiment required a tremendous amount of energy that would not have been available in the amounts required.
23. Krishna Bahadur conceived of the *Jeewanu* as a simple form of the living. *Jeewanu* are synthetic chemical particles that possess cell-like structural and functional properties. They are considered to be primitive cells, or protocells, the simplest life forms. Using photochemical reaction, he produced cell-like spherules (microscopic spherical structures) from a mixture of organic and inorganic compounds. Bahadur named these particles *Jeewanu* because they exhibited the basic properties of a cell.
24. J. W. Schopf, *Earth's Earliest Biosphere* (Princeton: Princeton University Press, 1983).
25. M. A. Schidlowski, "A 3800 Million Year Isotopic Record of Life from Carbon in Sedimentary Rocks," *Nature* 333 (1988): 313–18.
26. C. De Duve, "Constraints on the Origin and Evolution of Life," *Proceedings of American Philosophical Society* 142 (1998): 525–32.
27. Iris Fry, *The Emergence of Life on Earth* (Piscataway, NJ: Rutgers University Press, 2000), 117–23.
28. J. B. Corliss, "Hot Springs and the Origin of Life," *Nature* 347 (1990): 624.
29. J. B. Corliss et al., "Submarine Thermal Springs on the Galapagos Rift," *Science* 203 (1979): 1073–83; N. G.
30. Sakai et al., "Venting of Carbon Dioxide-Rich Fluid and Hydrate Formation in Mid-Okinawa Trough Backarc Basin," *Science* 248 (1990): 1093–96.
31. G. Wachterhauser, "Groundwork for an Evolutionary Biochemistry: The Iron-Sulphur World," *Progress in Biophysics and Molecular Biology* 58 (1992): 85–201.
32. J. F. Kasting, "Warming Early Earth and Mars," *Science* 276 (1997): 1213–15.
33. J. F. Kasting and L. L. Brown, "Methane Concentrations in the Earth's Pre-Biotic Atmosphere," *Origins of Life and Evolving Biosphere* 26 (1996): 219–20.
34. J. A. Baross and S. E. Hoffman, "Submarine Hydrothermal Vents and Associated Gradient Environments as Sites for the Origin and Evolution of Life," *Origins of Life and Evolving Biosphere* 15 (1985): 327–45; Holm, "Why Are Hydrothermal Systems Proposed?"
35. C. R. Woese, "Bacterial Evolution," *Microbiology Review* 51 (1987): 221–71; C. R. Woese and G. E. Fox, "Phylogenetic Structure of the Prokaryotic Domain: The Primary Kingdoms," *Proceedings of the National Academy of Sciences of the United States of America* 74, no. 11 (1977): 5088–90.
36. F. Hoyle and C. Wickramasinghe, *Lifecloud* (New York: Harper & Row, 1979).
37. F. H. C. Crick and L. E. Orgel, "Directed Panspermia," *Icarus* 19 (1973): 341–46.
38. Francis Crick, *Life Itself* (New York: Simon and Schuster, 1981).

PART 1930S

9

Human Genetics

> For men are not born equal.
> —J. B. S. Haldane, from *The Inequality of Man and Other Essays* (1932)

Referring to the genetics of humans, Haldane commented, "Man has obvious disadvantages as an object of genetical study."[1] The advantages are the availability of very large populations and data on many serological differences and congenital abnormalities as well as many other genetic markers.

Haldane was an early investigator of the genetic analysis of human pedigrees and populations. His interest in human genetics was an extension of his work on the mathematical theory of natural selection. He investigated various mathematical models of equilibria by balancing natural selection against mutation rates. Extending this principle to human genetics, Haldane considered the example of hemophilia. He deduced that as one-third of the hemophilics in each generation are being eliminated by natural selection, an equal number of new cases are arising in each generation by mutation. That was the beginning of what has now become the extensive field of human mutation rates.

Haldane's contributions to human genetics are entirely theoretical.[2] He developed methods of analysis and tests of significance as the need arose in various situations. However, his ideas have stimulated experimental research of many kinds by other investigators. Much of what is now called "transplantation genetics" began with the discovery of the H-2 locus in the mouse by Peter Gorer, who was pursuing research under Haldane's direction at University College London.

In the 1930s, a small group of individuals including Bernstein,[3] Hogben,[4] Haldane,[5] Fisher,[6] and Penrose[7] started developing statistical methods to understand the transmission of characters in families. That was the formal beginning of the science of human genetics, which later gave rise to medical and clinical genetics. Earlier, Archibald Garrod,[8] physician at Guy's Hospital in London, investigated the recessive inheritance of several human biochemical disorders, such as cystinuria and albinism, however, the significance of Garrod's pioneering work was not fully appreciated for thirty years until Haldane[9] drew attention to it in an interesting paper on chemical individuality in 1937.

From the 1930s onward, Haldane and others steadily developed the methods and applications of human genetics. At first, the methods were mainly statistical and mathematical, but later biochemical and molecular methods contributed to the rapid growth of human genetics and its many extensions such as immunogenetics and biochemical genetics.[10]

For the next thirty years Haldane extended human genetics meticulously, developing methods for the estimation of human mutation rates and mutation impact, linkage and human gene mapping, measurement of the effect of natural selection in human populations, interaction between genetic factors and the environment, estimation of genetic loads and radiation damage, and the impact of infectious disease, among other aspects. Haldane applied Fisher's method of maximum likelihood to estimate the true proportion of recessives in human pedigrees for juvenile amaurotic idiocy. Haldane's influence was far greater than could be judged simply from reading his numerous published papers, because of his readiness to communicate his ideas without reservation to his students and colleagues.

Haldane and Bell[11] pioneered what later came to be known as the "the human genome project." Using statistical methods, they took the first step in 1937 in his estimation of the distance between the loci for the genes for hemophilia and color blindness on the X chromosome map. From these modest beginnings, the human genome project evolved into today's multi-billion-dollar enterprise.

Haldane's analysis was often based on pedigree data collected by others, which occasionally led to errors. One example was his paper on partial sex-linkage, in 1936, which explored the hypothetical possibility of consequences resulting from crossing-over between the human X and Y chromosomes. However, when he found out that it was not supported by analysis of available data later, he abandoned that idea. It was a rare occasion when he went out on a limb, so to speak, but was not supported by further analysis. I recall that during his visit to Glasgow University in 1961, Jim Renwick, who was trained in human genetics under Penrose and Haldane at University College London, asked Haldane in a light-hearted way, "Is there anything new on partial sex-linkage in man?" Haldane took that ribbing with good humor as it was meant to be, and replied, "Look here, I said it was a *search* for partial sex-linkage in the title of my paper!" We all had a good laugh at Haldane's expense, and he joined us happily! That subject had been "dead" for years before Renwick brought it up!

Haldane also pioneered several other areas of human genetics, such as the estimation of genetic damage resulting from radiation in atomic bomb testing, the genetic effects of human inbreeding, and the genetic analysis of complex disease traits. Furthermore, he wrote a great deal on the applications of human genetics for improving the human species, but he was opposed to formal eugenic programs mainly because he did not believe that we had sufficient knowledge of human genetics to benefit from them. Haldane[12] was concerned that such programs could be misused for political and racist purposes as was the case in Nazi Germany. In his plenary address to the 11th International Congress of Genetics, in 1963, Haldane wrote,

> The appalling results of false beliefs on human genetics are exemplified in the recent history of Europe. Perhaps the most important thing which human geneticists can do for society at the moment is to emphasize how little they yet know. This is a thankless task ... the influence of genetical

ideas, such as that of racial purity, has often been disastrous.... It does not much matter if many people know nothing about hybrid maize or progeny tests for milk yield. It matters a great deal if they know nothing about human genetics, because it is a topic of interest to all human beings, and the gaps in their knowledge will be filled by superstition or intellectually dishonest propaganda.[13]

The geneticist Elof Carlson[14] correctly emphasized that the knowledge of population genetics has made abundantly clear the futility of any quick-fix programs that might be erroneously aimed at an elimination of so-called unfit or undesirable individuals in a society from a genetical point of view because of the hidden genetic variability that exists in all societies.

Genetic Loads and the Impact of Mutation

In 1937 Haldane wrote a paper with the deceptively simple title "The Effect of Variation on Fitness,"[15] enunciating a basic principle of population genetics that formed the foundation for assessing the impact of mutation on the population. He showed that the effect of mutation on the fitness of a population is independent of how deleterious the mutant phenotype is but is instead determined almost entirely by the mutation rate. The principle he outlined in that paper became a useful tool later in assessing radiation damage that was produced from the open-air testing of atomic bombs. It was adopted by the BEAR committee (Committee on Biological Effects of Atomic Radiation) of the US National Academy of Sciences in 1956, when the impact of testing nuclear weapons became a subject of great political and social importance. His idea, which was independently used by H. J. Muller and others, provided the first basis for various assessments of the genetic effects of radiation. It also showed that any increase in mutation rate would have an effect on fitness ultimately equal to this increase.

James Crow[16] invented the terminology of genetic loads: "mutation load" for the proportion by which mutation lowers the fitness in an equilibrium population compared with a hypothetical population without mutation, and "segregation load" for the effect of Mendelian segregation in comparison with a nonsegregating equilibrium population. The relative magnitudes of these loads have been a subject of much controversy for several years. Suffice it to say, Haldane's 1937 paper stimulated a great deal of discussion and research. His work not only led to the quantification of various components of natural selection but also opened the door for others to continue.

Disease and Selection

An ingenious idea of Haldane that led to a great deal of epidemiologic research involved the role played by infectious disease in the evolutionary process. The best-known paper of Haldane on this subject, "Disease and Evolution,"[17] was his address to an international conference in Italy in 1949, also titled "Disease and Evolution." Haldane wrote, "the struggle against disease, and particularly infectious disease, has been a very important evolutionary agent, and that some of its results have been rather unlike those of the struggle against natural forces, hunger, and predators, or with members of the same species."[18] Haldane suggested that in every species at least one of the factors that kills it or lowers its fertility must increase in efficiency as the species becomes denser. Otherwise the species would increase without limit. A predator cannot in general be such a factor, because predators are usually larger than their prey and breed more slowly. Lack of food or space can also have a similar effect on density-dependent situations. Competition for food by members of the same species is also a limiting factor. However, Haldane believed that the density-dependent limiting factor is more often a parasite whose incidence is disproportionately raised by overcrowding.

Haldane noted the similarity in geographic distribution of both thalassemia[19] and malaria, especially in the Mediterranean region. The high incidence of thalassemia puzzled everyone. Quite intuitively, Haldane saw a connection between the high incidence of thalassemia and malaria that led to their prevalence in Italy, Greece, and surrounding areas.

Neel and Valentine[20] thought that the higher incidence of thalassemia was due to its higher mutation rate. They suggested a high mutation rate of 1 in 2,500 births. But Haldane[21] offered a different explanation. He wrote, "I believe that the possibility that the heterozygote[22] is fitter than normal must be seriously considered. Such increased fitness is found in the case of several lethal and sub-lethal genes in *Drosophila* and *Zea*."[23] Haldane stated this explanation clearly in his address to the 8th International Congress of Genetics in Stockholm in 1948, which was published in 1949.

Joshua Lederberg,[24] referring only to Haldane's Italian paper, "Disease and Evolution," stated that Haldane never really offered a detailed explanation. Furthermore, Lederberg mistakenly suggested that this phenomenon was known long before Haldane's work. However, Haldane had already explained the mechanism based on the greater resistance of heterozygous carriers (for thalassemia) to malaria, in his earlier paper delivered at the Stockholm Congress in 1948. He was the first one to suggest this explanation, which was supported by the data obtained in Africa by Anthony Allison[25] and others later. Considering that at the time Haldane proposed this hypothesis very little was known about the genetics of thalassemias or other inherited disorders of hemoglobin, Haldane's insight and prophetic vision seem all the more remarkable. David Weatherall[26] provided a fine review of this subject in my book *Infectious Disease and Host-Pathogen Evolution*.

Interestingly, Haldane's prediction proved to be accurate with respect to the prevalence of sickle cell anemia rather than thalassemia. Extensive evidence collected in Africa clearly showed that the carriers (heterozygotes) of the sickle cell trait are protected against the malaria parasite *Plasmodium falciparum*. Evidence for thalassemia is more complicated; although evidence from Sardinia is suggestive, data from other sources is not so clear. This subject has been summarized by David Weatherall.[27]

Altruistic Genes (Kin Selection)

Haldane proposed the existence of genes for altruism in human populations.[28] Darwin saw altruism as a potential problem for his theory of evolution, which was based on the fundamental principles "struggle for existence" and "survival of the fittest." He found it hard to explain how altruism fits into that model of evolution. He was especially concerned about the altruistic behavior of ants and the formation of castes with specialized functions in the ant colonies.

In his classic work *The Causes of Evolution*[29] in 1932 and later in his article "Population Genetics" in 1955,[30] Haldane discussed altruism under the title "Socially Valuable but Individually Disadvantageous Characters." He considered small groups because an altruistic character can only spread through the population "if the genes determining it are borne by a group of related individuals whose chances of leaving offspring are increased by the presence of these genes in an individual member of the group whose own private viability they lower."[31] Haldane cited two cases to illustrate his point. The first one involved broodiness in poultry, which is inherited. In nature, a broody hen is more likely to be caught by a predatory enemy while sitting. On the other hand, a nonbroody hen will not bear offspring, so genes determining this character will be eliminated from the population. With respect to maternal instincts of this kind, Haldane suggested that natural selection would strike a balance in nature. A hen that abandons her eggs or offspring at the slightest danger will not contribute to posterity, but another mother who abandons only under intense pressure will do so, whereas a too-devoted mother will not.

The second case considered by Haldane involved social insects such as bees and ants, where there is no limit to the self-sacrifice and altruistic behavior. In a beehive, the workers and young queens share the same set of genotypes. So any form of behavior in the workers that will benefit the hive will help the survival of the young queens, and will spread throughout the species-change—if they induce unduly altruistic behavior in the queens, they will be eliminated.

Considering small social groups where every individual is a potential parent, Haldane showed that the biological advantages of altruistic conduct outweigh the disadvantages only if a majority of the tribe behaves altruistically. On the other hand, if only a small fraction behaves altruistically, it does not have a significant effect on the viability of the tribe. When the number of individuals in the tribe is small, selection is quite effective. But in large tribes, the initial stages of the evolution of altruism will depend on random survival, not on selection. Haldane pointed out that if any altruistic genes are common at all in the human society, they must have originated when humans were divided into small, inbred, and closed communities.

But Haldane recognized that his discussion did not provide a satisfactory explanation for the spread of congenital altruism in populations. He did not derive a general equation that captured the costs and benefits of altruism in relation to kinship. The two other founders of population genetics, R. A. Fisher and S. Wright, also did not develop a general mathematical model of altruism and kinship. Fisher, like Haldane, seemed to be on the verge of developing a generalized mathematical model of kinship but did not pursue it far enough. In retrospect, it is surprising that none of the founders of population genetics was sufficiently interested to develop a generalized formula linking kinship and altruism. The simplest explanation is that all three were preoccupied with developing their larger, general views on evolution. They were not sufficiently interested to find a solution to what seemed to be a smaller limited problem.

Another reason was mentioned in Lee Dugatkin's book *The Altruism Equation*.[32] None of the three were naturalists. Their primary approach involved mathematics, not animal behavior or natural history. However, both Fisher and Haldane attempted analysis of altruism but did not take it far enough to complete a mathematical formulation. I regret that it did not occur to me to broach that question with Haldane or Wright when I had that opportunity on numerous occasions.

It was left to the naturalist William Hamilton[33] to find a general solution to this interesting problem. Hamilton developed the concept of "inclusive fitness," which is the cumulative result of all interactions, both altruistic and nonaltruistic, among a group of relatives and unrelated individuals. Hamilton proposed that inclusive fitness offers a mechanism for the evolution of altruism. He suggested that this leads natural selection to favor organisms that would behave in ways that maximize their inclusive fitness. Hamilton's rule describes mathematically whether or not a gene for altruistic behavior will spread in a population. Formally, genes should increase in frequency when $rB > C$, where r = the genetic relatedness of the recipient to the actor, often defined as the probability that a gene picked randomly from each at the same locus is identical by descent; B = the additional reproductive benefit gained by the recipient of the altruistic act; and C = the reproductive cost to the individual performing the act.

Nature–Nurture interaction

Among other aspects, Haldane made a systematic analysis of the nature–nurture or the gene–environment interaction systems and their potential applications. For instance, in a 1946 paper titled "The Interaction of Nature and Nurture," Haldane wrote, "A moderate degree of mental dullness may be desideratum for certain types of monotonous but at present necessary work, even if in most or all existing nations there may turn out to be far too many people so qualified,"[34] and later "Meanwhile our efforts should be mainly concentrated on the elimination ... of environments which are unfavorable to all genotypes and of genotypes which are inferior to all environments." Haldane concluded that our efforts should be mainly concentrated on the elimination of environments that are unfavorable to all genotypes, and of genotypes that are inferior in all environments.

Daedalus and Other Applications

Haldane contributed numerous analyses and ideas that have improved the quality of papers in human genetics by others and have greatly advanced the development of this field. Starting with his first book, *Daedalus, or Science and the Future*,[35] his fertile mind had speculated on the genetic future of mankind, predicting widespread use of eugenic selection, in vitro fertilization, and routine production of "test-tube" babies.

In addition to his many publications in technical journals, Haldane expressed numerous original ideas and comments in the popular media that have influenced many readers. Early in the twentieth century, Haldane was one of the few scientists who discussed the ethical impact of science on society. He wrote, "To my mind the greatest danger to which our ethical system is exposed from science is not a debasement of values . . . but the deliberate exploitation of scientific ideas in the interests of unscientific prejudice."[36]

Haldane noted that a great deal of unscientific and immoral information had been advocated in the name of eugenics. But he also affirmed that it had a very great future as an ethical principle. We should seek desirable and inheritable qualities in our spouses. As eugenics is an ethical principle, it should begin at home, like charity, and influence individual behavior before one thinks of it in the context of public policy. Furthermore, it is our duty to envisage, as far as possible, the consequences of our actions. Haldane pointed out that we do not realize how largely a scientifically based code of ethics would depend on statistical data. And new duties would arise from the study of statistical data. It is our duty to acquire the knowledge that will make us competent to moralize our daily activities.

The only branch of science that is concerned with moral conduct as such is anthropology. One branch of anthropology is concerned with human societies and the factors influencing their conduct. It is generally agreed that the magic and religion of primitive peoples are essential parts of their social system, and the missionaries destroy the very foundations of society when they introduce Christianity or Islam. As Haldane put it, the same argument is applied to the Christian society as well; although most of Christian dogma is untrue, the Church makes an essential contribution to the stability of European society. However, while the anthropologist might regard the Church as essential for the stability of society, he would certainly not regard its moral code as correct. This is because the behavior of Christians, like that of other men, has always been a compromise between that dictated by their moral code and their private inclinations. That moral code has never been purely Christian. Other moral codes, based on family pride or patriotism, have also contributed their part.

Haldane noted that we could begin to apply scientific method both to individual morality and to the problem of morality itself, on a quantitative basis, for instance by applying it to social conduct, such as the portion of one's income spent on the pursuit of pleasure. He concluded that it represents "the unification of human effort, the marriage of the mind and the heart, the moralization of science, and the rationalization of ethics."[37]

Notes

1. J. B. S. Haldane, *The Inequality of Man and Other Essays* (London: Chatto & Windus, 1932), 1.
2. Krishna R. Dronamraju, *Foundations of Human Genetics* (Springfield, Ill.: Charles C. Thomas, 1989).
3. F. Bernstein, "Zur grundlegung der vererbung beim menschen," *Zeitschrift Indukt Abstamm Vererbungslehre* 57 (1931): 113–18.

4. L. T. Hogben, "The Genetic Analysis of Familial Traits: 1. Single Gene Substitutions," *Journal of Genetics* 25 (1931): 97–112.
5. J. B. S. Haldane, "A Method for Investigating Recessive Characters in Man," *Journal of Genetics* 25 (1932): 251–55.
6. R. A. Fisher (1918); R. A. Fisher, "The Detection of Linkage with "Dominant" Abnormalities," *Annals of Eugenics* 6 (1935): 187–201.
7. L. S. Penrose, "The Detection of Autosomal Linkage in Data Which Consist of Pairs of Brothers and Sisters of Uncertain Parentage," *Annals of Eugenics* 6 (1935): 133–138.
8. A. E. Garrod, *Inborn Errors of Metabolism* (London: Oxford University Press, 1909).
9. J. B. S. Haldane, "The Effect of Variation on Fitness," *American Naturalist* 71 (1937): 337–49.
10. Krishna R. Dronamraju and C. Francomano, eds., *Victor McKusick and the History of Medical Genetics* (New York: Springer, 2012).
11. J. B. S. Haldane and J. Bell, "The Linkage between the Genes for Colour-Blindness and Haemophilia in Man," *Proceedings of the Royal Society* B123 (1937): 119–150.
12. Personal communication to the author from Haldane.
13. J. B. S. Haldane, "The Implications of Genetics for Human Society," in S. J. Geerts, ed., *Genetics Today*, Proceedings of the XI International Congress of Genetics, The Hague, Vol. 2 (London: Pergamon Press, 1963), cii.
14. E. A. Carlson, *Mendel's Legacy* (New York: Cold Spring Harbor Laboratory Press, 2004), 294.
15. Haldane, "The Effect of Variation on Fitness,"
16. J. F. Crow, "The Possibilities for Measuring Selection Intensities in Man," *Human Biology* 30 (1958): 1–18.
17. J. B. S. Haldane, "Disease and Evolution," *La Ricerca Scientifica, Supplemento* 19 (1949): 2–11.
18. Haldane, "Disease and Evolution," 3.
19. Thalassemias are autosomal recessive inherited disorders that are found in the Mediterranean region. The disorder is caused by the weakening and destruction of red blood cells. Thalassemia is caused by variant or missing genes that affect how the body makes hemoglobin, which carries oxygen. People with thalassemia make less hemoglobin and fewer circulating red blood cells than normal, which results in mild or severe anemia. Thalassemia can cause significant complications, including iron overload, bone deformities, and cardiovascular illness. However this same inherited disease of red blood cells may confer a degree of protection against malarial infection. This selective survival advantage of carriers is known as "heterozygote" advantage. It is responsible for maintaining the disease in populations. This mechanism is similar to that maintaining another disease, the sickle cell disease.
20. J. V. Neel and W. N. Valentine, "Further Studies on the Genetics of Thalassemia," *Genetics* 32 (1947): 38–63.
21. J. B. S. Haldane, "The Rate of Mutation of Human Genes," *Proceedings of the 8th International Congress of Genetics* (Stockholm) 35 (1949): 267–73.
22. A heterozygous individual is one who inherited different forms of a particular gene from each parent. In contrast, a homozygous individual inherits identical forms of a particular gene from each parent.
23. Haldane, "Disease and Evolution," 270.
24. J. Lederberg, "J.B.S. Haldane (1949) on Infectious Disease and Evolution," *Genetics* 153 (1999): 1–3.
25. A. C. Allison, "Protection Afforded by the Sickle-Cell Trait against Subtertian Malarial Infection," *British Medical Journal* I (1954): 290–302.
26. D. J. Weatherall, "J.B.S. Haldane and the Malaria Hypothesis," in Krishna R. Dronamraju, ed., *Infectious Disease and Host-Pathogen Evolution* (Cambridge: Cambridge University Press, 2004), 18–36.
27. Weatherall, "J.B.S. Haldane and the Malaria Hypothesis,"
28. J. B. S. Haldane, *The Causes of Evolution* (London: Longmans, Green, 1932), 119–22; J. B. S. Haldane, "Population Genetics" (1955).
29. Haldane, *Causes of Evolution*.
30. Haldane, "Population Genetics."

31. Haldane, *Causes of Evolution*, 119.
32. Lee Dugatkin, *The Altruism Equation*.
33. W. D. Hamilton, "The Genetical Evolution of Social Behavior I," *Journal of Theoretical Biology* 7 (1964): 1–16.
34. J. B. S. Haldane, "The Interaction of Nature and Nurture," 1946.
35. J. B. S. Haldane, *Daedalus or Science and the Future* (London: Kegan Paul, 1923).
36. Ibid., 1–20.
37. Haldane, *The Inequality of Man and other Essays*, 118.

10

The Marxist Years

> Middle-aged professors and intellectuals were not inclined to lead the life of revolutionaries.
> —J. B. S. Haldane on the Spanish civil war
>
> I require friendship. I want the society of equals who will criticize me, and whom I can criticize. I cannot be friends with a person whose orders I have to obey without criticism before or after, or with one who has to obey my orders in a similar way. And I find friendship with people much richer or poorer than myself very difficult.
> —J. B. S. Haldane[1]

Unlike other British Marxist intellectuals of the 1930s such as Hyman Levy[2] and Lancelot Hogben,[3] Haldane did not grow up in poverty and hardship. His conversion to Marxism was much more complex as he had to travel much farther against all odds to become a Marxist, as compared with others. In a collection of popular essays, *Science Advances*,[4] which was published in 1947, Haldane compared Marx with contemporaries like Faraday, Darwin, and Pasteur, "who are still influencing our lives and thoughts, because their ideas were important not only for their own time, but for many generations to come. These men applied scientific method to new fields. So did Marx."[5] Haldane compared the methods applied by Darwin, Pasteur, and Marx: "Just as Darwin applied scientific method to the problem of man's ancestry, and Pasteur to that of his diseases, Marx applied it to history, politics, and economics."

As a growing child, young Haldane was surrounded by comfort, wealth, social privileges, and a successful family. In 1968 Jack's sister, Nou, recalled in an autobiographical memoir, "It was in many ways a happy household, with devoted parents—perhaps a too devoted and demanding mother—with the best of the period's standards of diet and health, and plenty of books." She continued to describe the ease with which her brother was able to excel at both Eton and Oxford, where he was competing against the best brains of that period:

> He came to terms with Eton (eventually becoming the Captain of the school), learned a lot, including the vast amount of Latin poetry, and enjoyed being a member of Pop, and having the power of the pop cane, but made few friends, other than the Huxleys and Dick Mitchison.[6] . . . When he came up to Oxford however, where he sailed triumphantly through all his exams, he spread

himself in friendships and light and the golden air of the pre-war years for the upper classes.... It had been glorious and romantic.[7]

Besides the Huxleys, their friends included Lewis and John Gielgud, who shared Naomi's early interest in acting.

Jack's father, John Scott Haldane, was a humanitarian who was fully conscious of the fact that his work was saving the lives of thousands of "ordinary" men and women. However, there was no obvious humanitarian creed in young Haldane's early career. He was brought up under the influence of his mother's conservative philosophy of social imperialism. Young Haldane was a staunch supporter of his mother's activities for such organizations as the Victoria League and Children of the Empire, the latter an organization devoted to making people more able to fight for their country and to be useful if they emigrate to the colonies, in other words, teach them to be good citizens of the Empire. Yet, Jack was also aware of the progressive actions of other relatives in the family, namely, his uncle Viscount Haldane, who helped found both the London School of Economics and the Imperial College.

Jack's conservatism was further diluted by his frequent contacts with the miners and other working classes when he accompanied his father into mines and sewers. Their miserable lives in filth and poverty also made a deep impression on young Haldane's mind, and it may have provided some justification later when he turned his attention to socialism and Marxism. However, before his conversion to socialism and Marxism could be completely successful, Haldane had to overcome not only his mother's conservative indoctrination but also his own elite education and family background.

There were at least five other important factors that contributed to Haldane's gradual conversion to socialism and Marxism. The first was his experience in World War I, when he was fighting closely with men of various social classes, mostly working-class men, collaborating in life-and-death situations for mutual survival. It was the first occasion in Haldane's life when he learned to appreciate and respect his comrades in the trenches. From them he learned, as he explained later, "to appreciate sides of human character with which the ordinary intellectual is not brought into contact." Those experiences opened his eyes by dispelling some long-held social prejudices regarding social class and character.[8]

The second factor was his difficulty while studying at Eton, where he was bullied mercilessly, which left permanent scars on his mind and a lifetime resentment against the establishment and authority, any authority.

Third was his disillusionment with the establishment and the governing classes during and after World War I. He realized, like most other fighting men, that the so-called great war was not being fought to achieve any great goals or serve any great values. Yet, it involved a great deal of death and destruction, and large numbers of men were being asked to sacrifice their lives. He was deeply resentful of a government that failed to keep its promise of "homes fit for heroes." He also disapproved of both Liberal Party prime ministers, Asquith and Lloyd George, who plotted with the Tories to oust his uncle Viscount Haldane from his high-level position.

Fourth, some have claimed, as quoted in the *New Statesman Profiles* (1957), that when the *Daily Worker* agreed to air his case for A.R.P. shelters when no other newspaper would give him a chance, Haldane decided to join the Communist Party of Great Britain (CPGB).

Fifth was the impact of the beliefs of his first wife, Charlotte. Charlotte's zeal for socialism was strong, and her desire to visit the Soviet Union became a reality when Haldane's friend Nikolai Vavilov[9] extended an invitation to them to visit Moscow and Leningrad in the summer of 1928. Vavilov had earlier worked under William Bateson at Cambridge University and the John Innes Institution in England. Vavilov was described in glowing terms by a fellow English geneticist, S. C. Harland:

> He could speak at least ten languages ... he was gifted with immense curiosity and a truly gigantic intellect.... Ideas sprang from his mind like a constant succession of balls of fire. He was a man who was entirely selfless in personal life.... Vavilov occupied the posts of President of the Academy of Agricultural Sciences and Director of the Institute of Applied Botany for some twenty years.... During that time he caused to be founded about 400 large research institutes, each staffed by up to 300 research workers. There were institutes for the study of oil plants, fibre plants, rubber plants, desert plants; cotton, tobacco, cereals, tea, citrus fruits and forest trees.... He either led or sent expeditions to all parts of the world in search of every possible variety of all the main economic plants.... By 1930, Vavilov had created a world collection of every known economic plant. I myself saw the great wheat collection of 26,000 varieties growing near Leningrad—a Wellsian dream.[10]

Vavilov arranged for Haldane to lecture in Moscow and Leningrad. Because of Vavilov's importance, they were allowed to make a private visit to Lenin's tomb on Red Square. He threw a colossal party for the Haldanes with champagne and dancing. They made a special visit to the summer palaces of Catherine the Great and Nicholas II. They had the pleasure of seeing two ballets at the Bolshoi Theatre, and two operas.

Charlotte noted in her autobiography that she found the Russians self-critical, as well as critical of foreign ways of living and thinking. Corruption, vagrancy, and maladministration were frequent targets of criticism. However, any criticism of the political regime and of the government was barred. She wrote that by Western standards of comparison the most impressive aspect was the physical vitality and energy of the Russian townspeople and their apparent optimism. The children and young people looked happily toward the future. Almost all the little girls aspired to become doctors, the little boys engineers. Religious worship was not totally prohibited. Some churches were open, and those who wished to attend were financially responsible for their upkeep. However, religious instruction to children was forbidden.

But Charlotte was glad to leave the Soviet Union, especially since they had to cut through a harrowing amount of red tape to acquire their "exit" visas. She commented that the Russians did not regard themselves as Europeans. She noted further that Haldane did not share her lack of enthusiasm. Haldane regarded the high position accorded to scientists in the Soviet society as their most outstanding characteristic.

Gary Werskey and the Visible College

In his book, *The Visible College*,[11] Gary Werskey examined the lives of five leading socialist-scientists—the so-called Red Professors—who rose to prominence during the 1930s in Great Britain: J. B. S. Haldane, Hyman Levy, Joseph Needham, Lancelot Hogben, and J. Desmond Bernal. However, Haldane's background and education clearly set him apart from others of that group. It was but a small step for other intellectuals such as Herman Levy and Lancelot Hogben to embrace Marxism. To achieve that goal, they did not have to travel very far from their poverty-ridden working-class background, whereas for Haldane, who descended from Scottish political and scientific (and economic) aristocracy, it was a major departure from his childhood and family background. To his intellectual colleagues at University College London, it was an amusing sight to see Haldane struggling hard to identify and communicate with the working classes. His background, intellectual style, education, speech, writings, and other aspects of his personality could not be any more different from those of the common man. Yet, he became the chairman of the editorial board of *Daily Worker*,[12] serving in that position for almost ten years (1939–1949). During those years, Haldane contributed numerous articles of excellent quality on popular science. Several of these were later reprinted in the book *What I Require from Life: Writings on Science and Life from J.B.S. Haldane*.[13]

Cambridge Group

Werskey inquired how socialist ideas came to be so rife among the scientific community and in Cambridge University of all places? Scientists are not generally viewed as particularly radical. Science itself is usually presented as unaffected by social or ideological influence, and scientists are expected to be interested solely in the pursuit of scientific truth while remaining "above politics."

The rise of the scientific left was due to two main factors. First was the economic crisis of the 1930s. Scientists who had believed that their work could make the world a better place could only watch in horror as the war destroyed the industrial fruits of their scientific endeavors. In Britain science research funds were cut. In Germany the Nazis appeared to be launching a frontal attack against scientific rationality itself.

The second vital factor was the presence of the growing Communist Party (CP). The CP seized the opportunities the crisis offered and was both able to grow substantially and sink roots deep into the working class movement. The CP built a base in the National Unemployed Workers' Movement. In industry, party members were able to lay the basis for effective rank-and-file organization within the trade unions. The CP also distinguished itself by its struggle against the growing British fascist movement. Linking all this together was the CP's newspaper the *Daily Worker*. In 1932 the paper was selling 20,000 copies on a weekday and 46,000 of the special weekend edition.

The growth of socialist ideas among workers undoubtedly helped the left in the universities. The spark that ignited the scientific left was the growth of fascism. Horrified by Hitler's victory in 1933 and the rising influence of Mosley's British Union of Fascists, many young scientists began to participate in antifascist marches and demonstrations. The scientific left's leaflets and publications highlighted not only how fascism represented a threat to scientific rationality but also the way the Nazis used pseudoscientific theories to back up their racist ideas.

Haldane was a well known and charismatic speaker. Whether it was the Albert Hall or Trafalgar Square, Haldane could pack them in. He turned up to speak at one meeting on the Spanish civil war wearing a beret, having come straight from the Spanish front itself. Even on the obscure subject of "A Dialectical Approach to Biology," Haldane could still draw a substantial crowd.

Air Raid Precautions

It has been said that Haldane's decision to join the Communist Party was simply based on the fact that the Party supported his campaign for better air raid protection (A.R.P.) when no other organization would take him seriously!

The scientific left were able to make use of their scientific knowledge in a daring series of "experiments" with gas and explosives which tested the government's air raid protection procedures and found them sadly wanting. Their work and Haldane's book *A.R.P.* (1938) were to be of great importance in the CP's eventually successful campaign to get proper public provision of air raid shelters.

Haldane's personal observation of air raid attacks in Spain and his demand for proper air raid protection in the event of war, became an important issue that gained some immediacy as World War II loomed. It was to be of great importance in the next major campaign the scientific left became involved in. But equally important was a growing awareness among many young scientists, based on their experience of the antifascist movement, of the power of the collective.

An important focus for the scientific left was building trade union membership among scientists. Most scientists at the beginning of the 1930s still saw themselves primarily as "professionals" and looked to individual advancement rather than collective struggle. Continued cuts in scientific funding began to challenge this complacency.

Russian Influence and the Intellectual Left

Ideological struggle was a vital component of the success of the scientific left. They argued that science is a product of society and that the nature of society affects science too. The beginnings of such an approach had been laid early in the decade in the unlikely surroundings of the Second International Congress of the History of Science and Technology, held in London in 1931. This normally dull event was transformed by the unexpected arrival of a large delegation from the Soviet Union, led by the Bolshevik leader Nikolai Bukharin.[14] It reinforced Haldane's favorable opinion of Soviet science that was formed during his visit to the Soviet Union in 1928.

Bukharin challenged the very notion of what we understand as science. Bukharin and his colleagues put forward a Marxist analysis of science. Boris Hessen, a Russian physicist, demonstrated how Newton's *Principia* was shaped by the social contradictions that followed the English Revolution of 1649. They argued that science is primarily a social activity and one of the major forces for human progress. But its potential for transforming the world is held back under capitalism. The handful of left-wing scientists present listened with delight. They later expanded and developed the Marxist analysis of science, culminating in Bernal's *The Social Function of Science* which appeared in 1939.[15] The success of their endeavor can be judged by the fact that by the end of the decade society's influence on science was accepted not only by the scientific left but also by such classic Liberals as the biologist Julian Huxley.

Many years later, in 1962, the view that science is primarily a social activity was argued successfully by Thomas Kuhn in his book *The Structure of Scientific Revolutions* whose "paradigm concept" had a deep influence on several disciplines in the sciences and the humanities.

The Marxist Philosophy and the Sciences (1938)

In a series of lectures delivered in the University of Birmingham in 1938,[16] Haldane summarized his views of how Marxism may be applied to the scientific problems of his day. Haldane reminded his audience that Socrates described himself as the midwife who helped the unborn thoughts of others into the world and hoped to function in a similar capacity in his own case. Haldane emphasized that the adaptation of the Marxist point of view leads one to look for creative antagonisms in nature and to investigate them but not accept them blindly. He pointed out that the main fallacy of the social application of Darwinism has been the attempt to use it to justify economic and racial inequality. In view of this mistake, Haldane considered it necessary to separate the biological approach to these problems from political and economic considerations.

Haldane expressed optimism that the "Marxist interpretation of our present troubles, economic, political, and ideological, is at any rate something which makes for hope and faith in the future." When he died in 1964, the Soviet Union was still in existence and going strong with its rapid strides in space exploration, creating optimism in its future. However, only twenty years later it collapsed, and today Marxism is rarely mentioned by any one in relation to the Soviet State and science that Haldane admired so much!

He wrote in the Preface, "I do not doubt that I have made mistakes. A Marxist must not be too afraid of making mistakes." What a prophetic statement! Haldane himself may have regretted his close embrace of Marxism and Soviet science when ten years later the suppression of genetic science in the Soviet Union by Trofim Lysenko[17] created a painful dilemma for him. Contrary to the policy adopted by the CPGB, he condemned Lysenko and firmly declared his support for genetics, although the conflict with the Party made him prevaricate at first. However, in 1949, he resigned from the Party itself, severing all his ties to that organization, its members, and its publication the *Daily Worker*. Soon after this debacle, Haldane turned his full attention to

India, and he decided to migrate there permanently seven years later, shortly before his retirement from University College London.

Spain

The next major step in Haldane's political transformation was his experience with the civil war in Spain. In the spring of 1933, Haldane received an invitation to participate in a scientific and literary conference in Madrid, organized by the Cultural and Intellectual Committee of the League of Nations. That was the Haldanes' first visit to Spain. It was the brainchild of the new Republican Government of Spain under the leadership of President Zamora and Prime Minister Azana. France sent a distinguished delegation that included the scientists Madame Marie Curie and her friend Paul Langevin, and the writers Paul Valery and Jules Romains. JBS admired Langevin.[18] Charlotte had wanted to visit Spain for a long time and encouraged JBS to attend the conference. Charlotte enjoyed the social activities enormously, including a bullfight that Haldane at first refused to attend. They enjoyed also visiting the painter Joan Miro in Catalonia.

Here is a strange picture of Haldane, sitting alone in the rain at a bullfight in Madrid. His wife Charlotte wrote,

> Just as the fourth bull entered the ring, a most unusual occurrence for Madrid at this season happened: a sudden, violent thunderstorm broke. hailstones as big as pigeons' eggs came sweeping down. In a twinkling the mass of spectators, protecting their heads with the little leather cushions on which they had been sitting, had emptied themselves into the coveted corridor surrounding the ring. My friends and I dashed out with the crowd, for the thunderstorm had ended the afternoon's show by drenching the sandy arena. When we got under cover we turned round to look for J.B.S. But he was not visible. I returned to the arena. There, in a *barrera* seat, I saw him sitting in solitary state in the pelting rain, his dome protected by his shabby mackintosh, placidly smoking his pipe and waiting for the next fight. Contemplating this unfamiliar vision, stood the bull, whom, in their rush to escape the hail, none of the ring attendants had bothered to remove.[19]

During this period when he was becoming heavily involved in left-wing politics, Haldane was particularly concerned about the emergence of fascism in Germany and Italy. During this trip to Spain he gave his support to the Socialist Party (PSOE) and the Communist Party (PCE) in its struggle with the Falange Española and other extreme right-wing parties.

Haldane's entanglement with Spain was just beginning. In July 1936, storm clouds broke over Europe when the Fascist generals revolted against the Spanish Republican Government. The Italian Fascists and the German Nazis saw in it a great opportunity for intervention and subsequent European domination. The Russians saw an opportunity to try out their aircraft and their pilots. Simultaneously, the International

Brigade got involved and internationally the call went forth: "Aid for Spain! Arms for Spain! Volunteers for Spain!" They were made up of socialist and communist military units volunteering to aid Spain.

Impact of Spanish Civil War

On the outbreak of the Spanish civil war Haldane supported the Popular Front government and was highly critical of the British government's nonintervention policy. Haldane and his wife, Charlotte, both joined the Communist Party and were active in raising men and money for the International Brigades.

Before his first visit to Spain, Haldane had met with the Spanish Ambassador and discussed antigas preparations with him. He visited Spain on three occasions, advising the Spanish government on the precautions to be taken in the event of gas attacks. This involved the best way of protecting civilian inhabitants against the increasing rebel air raids under the command of General Franco. Haldane visited the front and reported extensively to the British public about his observations on the war and war strategy.

Many colleagues of Haldane were not inclined to follow him, either into the Spanish civil war or Marxism. They had had enough fighting and disappointments in World War I. Haldane commented that middle-aged professors and intellectuals were not inclined to lead the life of revolutionaries. They were content with their undisturbed academic life, occasionally referring to Marxism as a nice theory. They followed what Haldane called Housman's[20] profoundly ignoble slogan: "Let us endure an hour and see injustice done."

Of the British writers asked about the war in Spain,[21] 5 (Evelyn Waugh, Eleanor Smith, and Edmund Blunden, among them) favored the Nationalists, 16 were neutral (including T.S. Eliot, Charles Morgan, Ezra Pound, Alec Waugh, Sean O'Faolain, H. G. Wells, and Vita Sackville-West), and 106 were for the Republic, many of them passionately. As for Spain, there is no doubt where the poets of the Spanish language—those who are now remembered—stood: García Lorca, Antonio Machado, Rafael Alberti, Miguel Hernández, and Pablo Neruda, among others.

It has been suggested that the nonintervention of the British Government in the face of the rising right-wing dictatorships in Europe alone would have driven Haldane into the arms of the Communist Party. From the mid-1930s onward, a number of other factors propelled him in that direction. In addition to his daily contacts with the problems of the refugees, Haldane suffered a personal indignity when the BBC imposed a ban on his radio broadcast on the causes of war. At first he was invited to present his account to counterbalance the speeches of the three other speakers, Sir Norman Angell, Den Ange, and Lord Beaverbrook. However, when the authorities saw the text, their nerve failed them and they had second thoughts. Today his main point would seem totally innocuous, simply stating that capitalism is one of the main causes of war. Haldane compared the University of Oxford, which had conferred an honorary degree on Sir Basil Zaharoff, the arms king, with the vice-chancellor of Leeds university, who had recently censured a lecturer for making an antiwar speech.

Haldane concluded that, if we really wanted peace, "we must examine all the causes of war, economic and technical, as well as psychological and political. We must be prepared to associate with all sorts of people, from Bishops to Bolsheviks, who share our view." He offered to tone down the socialism in his text, but the BBC still refused to let him speak. So the broadcast was not delivered, however the *Daily Herald* published the text on November 3, 1934. And the complete broadcast was published as a book after the war.[22]

Other events were sweeping the international scene, and they played an important part in Haldane's political conversion. Shortly afterward, Italy, under the fascist government of Mussolini, invaded Abyssinia and, in 1936, Germany reoccupied the Rhineland.

Although Haldane spoke and wrote about the Spanish civil war quite extensively, his impact on it was not significant. On the other hand, the war had a great impact on him, which was very profound and long lasting. His visits to Spain never lasted more than a few months. Yet, his experiences convinced him that only the Communists would stand up to the fascist dictatorships. His visits to the frontline were not unmixed with romance and excitement. Like Hemingway, Haldane too felt the need to recapture his youth. However, science was never too far from his mind. He retained his scientific objectivity, but he was not above enjoying the limelight and glory of the International Brigade. One of its members remembered meeting Haldane at a Christmas Party that was arranged before they took off for another battle: "He struck me as a big shaggy bear enjoying a picnic."

One must not get the impression that Haldane abandoned his scientific research. During the mid-1930s, Haldane published numerous scientific papers in professional journals, a few hundred popular articles in the newspapers, and at least five books, including a storybook for children, *My Friend Mr. Leakey*.[23] It was also during those years that Haldane discovered an important aspect of the impact of mutation on human populations (later called "genetic loads") and laid the foundations of human and medical genetics including the first human gene map, an important forerunner of what is now called the "human genome project."

A.R.P.

Haldane's participation in the Spanish civil war transformed him into a scientist-spokesman who could speak authoritatively on all matters related to war. His main concern was that the British Government was not prepared to defend the country against aerial bombing. Haldane believed that it was naive not to have defense-preparedness in view of the volatile situation. He traveled widely, lecturing on air-raid-precaution (A.R.P.). His book *A.R.P.* was published in 1938 by Victor Gollancz.[24]

An incident in aerial bombing shook him severely. He was sitting on a park bench in Madrid during an air raid. He did not seek shelter because he believed that he should be "as brave as the other citizens of Madrid, women and children as well as men" (as he stated in a radio broadcast). Whatever the reasons were, Haldane remained sitting on the bench during the air raid. A bomb dropped nearby, killing an elderly lady who was sitting on the same bench instantly. It upset him more than anything he had ever encountered in World War I.

When he returned home from Spain, Haldane proceeded directly to his laboratory at University College London (UCL). The door was locked, even though he instructed the steward never to lock his laboratory. To a man of Haldane's temperament, the solution was obvious, by putting his shoulder to the door and bursting it open. He spent that night in his huge armchair, working as he had planned.

Aid Spain

Haldane traveled a great deal, lecturing and appealing for funds in aid of Spain. In addition to his academic activities he joined the *Daily Worker* as its science correspondent and later served as the chairman of its editorial board for several years. These activities brought him into contact with large numbers of working-class people, which he enjoyed. He preferred staying with a worker's family than in a hotel. His feelings were reciprocated by many who enjoyed meeting the great man. One exception was the group organized by Sir Oswald Mosley's[25] British Union of Fascists who attempted to disrupt Haldane's meetings. Haldane met the opposition head-on, and was advised, on one occasion, by his friends to leave the lecture hall by a back entrance. But he refused to do so, preferring to face his opposition at the main entrance. He proudly displayed the bruises for several days afterward!

Haldane's experiences in Spain enabled him to campaign for efficient air raid precautions. The government still appeared to be dragging its feet even though the threat from German bombers was imminent and in fact became a serious threat only a year later. There were many converts in the general population who had no left-wing connections of any kind. Part of his effort was writing his book on A.R.P. Gas warfare was still regarded as a major threat by many. Haldane suggested that in the event of a gas attack babies can be safely shut up in airtight boxes. When one mother became anxious, he shut himself in a box for an hour, saying that he could have stayed for two hours with no significant ill-effects.

Another hazard was the danger from high explosives. Haldane kept on pestering the British authorities on the need to build deep shelters. In 1939 his pleas were rejected by Sir John Anderson, who was to become the minister of home security very soon. Haldane pointed out that the above-ground shelters would be futile against an all-out attack by the German air force, which, according to his estimates, would kill at least 50,000 immediately.

Haldane's increased visibility in the war effort made him an attractive candidate, for some, to run for a seat in the Parliament. He had rejected a similar offer when he was much younger. But by 1939 he was a mature scientist who was far better educated and informed than any member of the parliament. However, whether Haldane was suited for a life in politics is a different matter. He was not a patient man, nor did he excel in tact and diplomacy. He would be the last man one would think of in relation to political intrigue and infighting. Furthermore, as a socialist and a Marxist he would be an ideal target for the Conservatives. Perhaps he was aware of these difficulties himself because he declined to run for the Parliament on more than one occasion. To one local Labour Party official who invited him to run for office, Haldane wrote, "I am not, and never have been a member of the Labour or of any other political party and it would not look too good if I joined up *ad hoc*.... I do not think I should make a good M.P. and, indeed, I hope to get back to scientific work completely.... I actually believe that my scientific work is of some value and that I should make a mess of a political career."[26]

Earlier, in 1928, in *Possible Worlds*,[27] Haldane wrote about science and politics:

> And until politics are a branch of science we shall do well to regard political and social reforms as experiments rather than shortcuts to the millennium. The time-scale of evolution so far has been the geological time-scale, in which we expect no substantial change in less than a hundred thousand years. We may be thankful if we have speeded up this progress a thousand-fold and at the end of a long life can leave the world noticeably better than we found it.

Education of Politicians

Gary Wersky's[28] collective biography of the British left-wing intellectuals of the 1930s discussed at length the leading roles of Haldane, Needham, Bernal, Levy, and Hogben in the social and political emancipation of the British left. Among these, Werskey characterized Haldane as the best-known and most vocal figure in the Cambridge counterculture.

In the 1920s Haldane wanted to be known as a "political" animal, but only on his own terms. In *Possible Worlds*, he declared, "I have not to take many paces outside my laboratory to see the need for political and social reform. As a skilled manual worker and trade unionist, I have a strong idea where I should find my political affinities." However, he also declared that a talented scientist in a laboratory is of more use to his fellows in the laboratory than out of it. He was opposed to any involvement in the "pettiness" of party politics. Haldane advocated the introduction of "scientific attitude" into politics. He argued that while the nation's "material basis" was scientific, its intellectual framework was "prescientific." And unless this disharmony were corrected, the civilization would undergo the fate of past cultures. The ruling classes, according to Haldane (in 1928), were "grossly ignorant of the mental attitude which has led to scientific discoveries." Among Haldane's recommendations were the abolition of hereditary wealth and the provision of "absolutely free and equal schooling." If these measures fail to be enacted, he predicted that "we shall probably go the way of the dodo and the Kiwi." He remarked that *until* educational inequalities were removed "my political views are likely to remain on the left." Sadly, he came to the conclusion that no political party was fully equipped to understand science and technology adequately to make any difference. He believed that a knowledge of science should be spread among socialists, and a knowledge of socialism should be spread among scientists.

He considered the education of politicians. In *The Inequality of Man*,[29] Haldane pointed out that the usual course of study for would-be politicians is history. But he added, "I think that the study of history is somewhat fallacious owing to the enormous changes which have taken place." Writing in 1932, Haldane added, "up till fifty years ago every State was based on the presupposition that most of the population would have to spend the greater part of their time in hard physical work. That is no longer the case. It seems to me that facts such as that make the lessons of history a little dubious in their application to modern problems."

Haldane's words were prophetic as far as the future course of science was concerned. He wrote, "It is quite possible, I think, that as the ideals of pure science become more and more remote from those of the general public, science will tend to degenerate more and more into medical and engineering technology, just as art may degenerate into illustration and religion into ritual when they lose the vital spark."[30]

Parents

By now, Haldane's political estrangement from his parents was complete. In fact his parents had already concluded earlier that Jack had permanently severed his ancestral and political heritage. It was especially painful for his mother, who had spared no effort in molding young Jack's political ideology. However, the scientific link with his father continued, and further strengthened when Jack too, like his father, was elected to the Royal Society in 1932. However, JBS did not share his father's enthusiasm for vitalism. John Scott Haldane adopted an antimechanist approach to biology; he saw his work as a vindication of his belief that teleology was an essential concept in biology. His views became widely known with his first book *Mechanism, Life, and Personality* in 1913. Haldane treated the organism as fundamental to biology: "We perceive the organism as a self-regulating entity," he argued, and "every effort to analyze it into components that can be reduced to a mechanical explanation violates this central experience."[31]

Socialist Scientists

Among the intellectual-scientific left of the 1930s, there were four others besides Haldane who provided exceptional leadership: Lancelot Hogben, Hyman Levy, J. Desmond Bernal, and Joseph Needham.

Lancelot Hogben (1895–1975)

Haldane's contemporaries Hyman Levy and Lancelot Hogben grew up in an environment of poverty. They also lacked the intellectual lineage and the distinguished ancestry that Haldane was so proud of. Hogben grew up in a household of small parental income and several siblings. In his autobiography, Hogben wrote, "Unlike many writers of memoirs, I can boast of no distinguished forebears. I came of poor but intellectually dishonest parents and was spared the experience of material poverty because they, with their offspring, lived as pensioners of my maternal grandparents, and in the same establishment, for the first eleven years of my life."

His father was a Plymouth Brethren evangelist, which led to young Hogben's isolation from his contemporaries. He was born on December 9, 1895, two months ahead of schedule. His mother, a fundamentalist who was inspired by the circumstances of his birth, vowed that her son would become a medical missionary. He was named Lancelot, after a Methodist missionary, the Reverend Lancelot Railton. His mother's "unwavering concern" for his spiritual welfare limited his friendships. She discouraged him from accepting any invitation from any household that did not conform to her evangelical standards.

Unlike Haldane's elite education at Eton and Oxford, Lancelot entered for the final B.Sc. course in zoology as an evening student at Birkbeck College in the fall of 1912, then foremost among the night schools of London University.

Hogben noted one admirable quality of Haldane that I have mentioned earlier. In his autobiography Hogben wrote, "no man I have known has at all times been more willing to admit that he could have been wrong and no one less reluctant to invite the opportunity of discussing the possibility."[32]

Hyman Levy (1889–1975)

Hyman Levy was born on March 7, 1889, into a poor Scottish working-class family in Edinburgh. His father, a self-employed picture-framer, was an exile from tsarist Russia. Young Hyman was no stranger to threats of violence and anti-Semitism from an early age. While these early experiences undoubtedly shaped his later politics, the real influence came from his mother, who regularly attended the speeches delivered at "The Mound"—Edinburgh's version of the Hyde Park Corner—listening to orators like Keir Hardie. Levy was educated at George Heriot's School and studied mathematics and physics at the University of Edinburgh. He enjoyed a successful academic career as a professor of mathematics at Imperial College of Science and Technology in London, and later the dean of the Royal College of Science.

Levy was in the Labour Party from 1920 to 1931, and then in 1931 he joined the CPGB. Despite his theoretical allegiance to the principles of communism, Levy strongly protested the way the Russian Communists treated Jews. As a direct result of this controversy, he was expelled from the Communist Party in 1958. Levy's political life was derived from his active role in the National Union of Scientific Workers (NUSW). He was a member of its Executive as well as the chairman of the important Propaganda Committee. He was also a member of the National Education Subcommittee of the Labour Party and urged the inclusion of science in its committees, but his advice was not accepted.

J. D. Bernal

In their youth, J. D. Bernal[33] and Joseph Needham shared a Cambridge background and a similar philosophical approach toward science, its history, and its applications. Later, their interests diverged—Bernal toward Russia and Needham toward China. Bernal was born on May 10, 1901, in Nenagh County, Tipperary, Ireland. Bernal received his early education in England, but, like Haldane, he was not happy at school. In 1919 he went to Emmanuel College, Cambridge University, with a scholarship. He studied both mathematics and science, qualifying for a BA degree in 1922, which he followed by another year of natural sciences. While at Cambridge he became known as "Sage," a nickname given to him about 1920 by a young woman because he seemed to possess unlimited knowledge. He is best known for pioneering X-ray crystallography in molecular biology. One of his students was Max Perutz,[34] who arrived as a student from Vienna in 1936 and started the work on hemoglobin that would occupy him most of his career.

Bernal's books include his 1929 work *The World, the Flesh and the Devil*[35] (called "the most brilliant attempt at scientific prediction ever made" by Arthur C. Clarke), and the aforementioned *The Social Function of Science*, in which he argued that science was not an individual pursuit of abstract knowledge and that the support of research and development should be dramatically increased.

Bernal arrived in Cambridge as a Catholic, but soon joined the CPGB in 1923. He became a prominent intellectual in political life, particularly in the 1930s. In 1931 he was deeply influenced by the famous meeting on the history of science, where he met the Soviet intellectuals Nikolai Bukharin and Boris Hessen, who gave an influential Marxist account of the work of Isaac Newton. However, in 1949 Bernal's endorsement of Soviet agriculture under Lysenko, which was authorized by Stalin as official Soviet orthodoxy, damaged the British scientific left. Haldane himself, who was in fact one of the founders of population genetics, exercised greater caution and condemned Lysenko Joseph Needham.

Joseph Needham[36] was born in London on December 9, 1900. He was the only child of a Harley Street physician father and an alcoholic mother who was a composer of salon music and popular songs. Young Needham grew up in a difficult household of contrast and conflict, but finally decided to study medicine under his father's influence. However, he came under the influence of Professor Frederick Gowland Hopkins at Cambridge University and switched to biochemistry. Here he came under the direction of Haldane, who was second-in-command of the biochemical laboratory.

Although Needham's career as a biochemist and an academic was well established, it developed in unanticipated directions during and after World War II, when he directed almost all his research into the history of science and civilization in China until his death in 1995.

Needham's hero was H. G. Wells from a very early age. He was inspired by Wells's scientific adventures and utopian novels. Needham's socialism was closely entwined with his scientific interests and religious convictions. He considered science and socialism to be two sides of the same process.

But entertaining these ideals was not followed by practical action by these intellectual leaders. Haldane, Needham, and Bernal were slow to wake up to fulfill their roles to support workers' rights. They were not in the forefront of taking part in such trade union activities as strikes and the demand for workers' rights. One exception was Haldane. In his youth Haldane did take a modest part when the Oxford Tram workers went on strike to demand higher wages!

The most significant aspect of the collective biography of these individuals has been the interaction between their socialism and their scientific profession. These dominant personalities of British science were not "above" politics. They were, in fact, structurally and ideologically committed to strengthening capitalism at home and abroad. This was manifested in four different ways: (1) overt propaganda on behalf of science-based monopolies, (2) advising the state on how to increase the productive and military resources of the nation, (3) promoting scientific theories and practices that support class domination and bourgeois ideology, and (4) reinforcing elite control within the science community. The combined effect of these policies is to predefine the boundaries of socialist attitudes narrowly within the scientific community.[37]

Worldview

Each of these scientists had eye-opening episodes or events that transformed their worldview. For Haldane, it was his visit to the Soviet Union in 1928. He was greatly impressed by the favorable treatment accorded to their scientists. Charlotte wrote, "Lenin, who had spent most of his life in Europe, returned home convinced that only large-scale electrification, and applied scientific techniques on the vastest possible plan, could secure a sound economic basis for the Revolution. So it came about that, at the time of our visit in 1928, the scientists and the factory workers were the most favoured classes."

There was a consensus among the socialist scientists that science was the lynchpin and arbiter of their politics, and they held the elitist view that the needs of their profession were more compelling than those of the rest of the community.

Haldane's Interactions

Because of his sensational bestseller, *Daedalus, or Science and the Future*, which appeared in 1923, and collections of excellent popular essays that were included in the book *Possible Worlds* and other publications, Haldane was firmly established by 1930 as the intellectual hero of the younger generation. A member of that generation, Isaiah Berlin,[38] a young Haldane admirer from the 1920s, commented that Haldane was "one of our major intellectual emancipators." Thus, Haldane was ahead of the rest of the socialist scientists of the 1930s as a scientific popularizer.

Each of these leaders addressed a specific type of audience. Levy chose to popularize science in more explicitly political contexts, within the labor movement's ideology. He was a brilliant and witty speaker, who became a leading spokesman for the Left Book Club. Unfortunately, while his lectures were most lucid and entertaining, his books were often tough to read, leaving the reader less than satisfied with unfulfilled expectations.

Hogben's highly successful and popular books, such as *Mathematics for the Million*,[39] soon eclipsed Levy's books, as they were especially directed at educating the lay men and women to understand the practical problems of the day, such as public health and malnutrition, in an easily understandable language.

Bernal and Needham preferred to address more restricted groups such as a summer school or a specific group of scientific workers. Haldane preferred very large audiences. His famous family name, flamboyant style, scientific reputation, and, above all, great ability to simplify complex ideas suited his purpose. His talks were lucid, informative, and entertaining. Both in speech and popular writing, Haldane had no peers.

His immense knowledge helped him to cover a great range of subjects. Being a polymath, he was especially talented in cross-fertilizing ideas among multiple disciplines not only in science but also in politics, economics, military affairs, ethics, philosophy, and many more. For Haldane as well as the others, new opportunities opened up when new magazines appeared both in Britain and the United States, such as *Fact, Science and Society, Modern Quarterly*, and the *Marxian Quarterly*, among others. Haldane, Hogben, Bernal, Needham, and Levy dominated the field of scientific socialism and functioned as its chief spokespersons during the 1930s, first in Britain and later in the entire English-speaking world.

Haldane and Hogben

Haldane and Hogben came from very different social backgrounds, which had a significant impact on their approach to science and its popularization. Haldane's knowledge as a polymath was much more evident in his popular essays. Both were biologists, but the range of topics covered in both Haldane's scientific work and his writings is not to be found in Hogben's scientific contributions or his popular essays. However, Haldane and Hogben shared sufficient biological interest to collaborate with Julian Huxley and F. A. E. Crew in establishing the Society for Experimental Biology (SEB) in 1932.[40] Hogben also became a founder-member of the fledgling National Union of Scientific Workers. He was a conscientious objector in World War I and was sent to prison, whereas Haldane not only fought in that war but also was proud of his exploits and bravery.

Haldane and Levy

As children, Haldane and Levy grew up in far different worlds. Levy grew up in a Edinburgh slum, whereas Haldane grew up in the comfortable home of his physiologist-father in Oxford. Yet, they became comrades in the CPGB later. A consequence of the interrelationship between Levy's political and professional activities was his increasingly dialectical view of science. On the other hand, Haldane and others were more inclined to isolate science from its social context. This dichotomy was reflected in their writings.

Haldane displayed an ambivalent attitude regarding the likelihood of the Labour Party achieving a socialist society and the inclusion of scientific research in its official program. In *Daedalus*, he considered whether Labour might provide better support for biological research than capitalism, but later despaired at the degree of scientific illiteracy among its leaders. On the other hand, he also wondered whether capitalism has a vested interest in protecting the scientific worker though it may not always give him a living wage.

Notes

1. J. B. S. Haldane, *Keeping Cool and Other Essays* (London: Chatto & Windus, 1938), 277.
2. Hyman Levy (1889–1975) was a Scottish mathematician, political activist, and Emeritus Professor at Imperial College, London. Levy was in the British Labour Party from 1920 to

1931. He joined the British Communist Party in 1931. However, he became disappointed by the way the Russian Communists treated Jews, and published on the topic, leading to his expulsion from the party in 1958.
3. Lancelot Hogben (1895–1975) was a British experimental zoologist and medical statistician. He was one of the founders of the Society for Experimental Biology (SEB). He is best known for popular books on science, especially *Mathematics for the Million* and *Science for the Citizen*. In later life, he preferred to describe himself as "a scientific humanist." He was a vigorous opponent of the British eugenics movement.
4. J. B. S. Haldane, *Science Advances* (London: Allen & Unwin, 1947).
5. J. B. S. Haldane, "Marx," in *Science Advances* (London: Unwin Brothers Limited, 1947), 14.
6. Naomi Mitchison, "Beginnings," in *Haldane and Modern Biology*, ed. Krishna R. Dronamraju (Baltimore: Johns Hopkins University Press, 1968), 299–301.
7. Richard Gilbert Mitchison (1894–1970) married Naomi. He was a Labour Member of the Parliament for several years and was later made a Labour peer.
8. J. B. S. Haldane, personal communication to the author, 1962.
9. Nikolai Vavilov (1887–1943) was a prominent Russian botanist and geneticist, best known for having identified the centers of origin of cultivated plants. Today, the N. I. Vavilov Institute of Plant Industry in St. Petersburg maintains one of the world's largest collections of plant genetic material. Vavilov was the head of the institute from 1921 to 1940. In 1968 the institute was renamed after him in time for its seventy-fifth anniversary.
10. From Charlotte Haldane's autobiography *Truth Will Out* (London: Vanguard Press, 1950), 40–41.
11. G. Werskey, *The Visible College: The Collective Biography of British Scientific Socialists of the 1930's* (New York: Holt, Rinehart and Winston, 1979).
12. The *Daily Worker* was the organ of the Central Committee of the Communist Party of Great Britain. The first edition was produced on January 1, 1930. On October 1, 1935, the first eight-page *Daily Worker* was produced.
13. Krishna R. Dronamraju, ed., *What I Require from Life: Writings on Science and Life from J.B.S. Haldane*, Foreword by Sir Arthur C. Clarke (Oxford: Oxford University Press, 2009), 3–128.
14. Nikolai Bukharin (1888–1938) was a Russian politician.
15. J. D. Bernal, *The Social Function of Science* (London: Routledge, 1939).
16. J. B. S. Haldane, *The Marxist Philosophy and the Sciences* (London: Allen & Unwin, 1938).
17. Trofim Lysenko (1898–1976) was a Soviet pseudo-scientist who was responsible for the suppression of Mendelian genetics in the Soviet Union from the late 1940s to the 1970s.
18. Paul Langevin (1872–1946) was a prominent French physicist who developed the Langevin dynamics and the Langevin equation. He was president of the Human Rights League from 1944 to 1946. Previously a doctoral student of Pierre Curie and later a lover of Marie Curie, Langevin is also famous for his two US patents with Constantin Chilowski in 1916 and 1917 involving ultrasonic submarine detection.
19. Charlotte Haldane's autobiography *Truth Will Out* (London: Vanguard Press, 1950), 40–41.
20. Alfred Edward Housman (1859–1936), usually known as A. E. Housman, was an English classical scholar and poet, best known to the general public for his cycle of poems *A Shropshire Lad*. Lyrical and almost epigrammatic in form, the poems were mostly written before 1900. Their wistful evocation of doomed youth in the English countryside, in spare language and distinctive imagery, appealed strongly to late Victorian and Edwardian taste, and to many early twentieth-century English composers (beginning with Arthur Somervell) both before and after World War I. Through its song-setting, the poetry became closely associated with that era, and with Shropshire itself. Housman was counted one of the foremost classicists of his age, and has been ranked as one of the greatest scholars of all time.
21. Personal communication from Naomi Mitchison.
22. J. B. S. Haldane, *A Banned Broadcast and Other Essays* (London: Chatto & Windus, 1946).
23. J. B. S. Haldane, *My Friend, Mr. Leakey* (London: Cresset Press, 1937).
24. J. B. S. Haldane, *A.R.P.* (London: Victor Gollancz, 1938).
25. Sir Oswald Mosely (1896–1980) was an English politician and Member of the Parliament, and founder of the British Union of Fascists (BUF). Mosley opposed free trade and associated

closely with Nazi Germany. He was interned in 1940 and the BUF was proscribed. He was released in 1943, and politically disillusioned in Britain he moved abroad in 1951, spending most of the remainder of his life in France.

26. J. B. S. Haldane Archives, University College London.
27. J. B. S. Haldane, *Possible Worlds and Other Essays* (London: Chatto & Windus, 1927), 199.
28. Werskey, *The Visible College*.
29. J. B. S. Haldane, *The Inequality of Man and other Essays* (London: Chatto & Windus, 1932), 224.
30. Haldane, *The Inequality of Man*, 237.
31. J. S. Haldane, *Mechanism, Life, and Personality: An Examination of the Mechanistic Theory of Life and Mind* (London: J. Murray, 1914).
32. Lancelot Hogben, *Lancelot Hogben: Scientific Humanist; An Unauthorised Autobiography*, ed. Adrian Hogben and Anne Hogben (London: Merlin Press, 1998), 124.
33. John Desmond Bernal (1901–1971) was one of the United Kingdom's most well-known and controversial scientists. Bernal was a pioneer in developing X-ray crystallography and its applications in molecular biology. He became known as "Sage," a nickname given to him about 1920 by a young woman working in C. K. Ogden's book shop in Cambridge. In 1927 he was appointed as the first lecturer in structural crystallography at Cambridge, becoming the assistant director of the Cavendish Laboratory in 1934. There he started applying his crystallographic techniques to organic molecules. He also worked on the structure of liquid water, showing the boomerang shape of its molecule (1933). Dorothy Hodgkin, Max Perutz, and Aaron Klug started their research careers under Bernal. In 1939 Bernal argued in *The Social Function of Science* that science was not an individual pursuit of abstract knowledge and that the support of research and development should be dramatically increased. The originator of the Science Citation Index (SCI), Eugene Garfield, stated that Bernal's idea of a centralized reprint center was in his thoughts when he first proposed the as yet nonexistent SCI in *Science* in 1955.
34. Max Perutz (1914–2002) was an Austrian-born British molecular biologist, who shared the 1962 Nobel Prize for Chemistry with John Kendrew, for their studies of the structures of hemoglobin and globular proteins. He also won the Royal Medal of the Royal Society in 1971 and the Copley Medal in 1979. At Cambridge he founded and chaired (1962–1979) the Medical Research Council's Laboratory of Molecular Biology, fourteen of whose scientists have won Nobel Prizes. Perutz's contributions to molecular biology in Cambridge are documented in *The History of the University of Cambridge: Volume 4 (1870 to 1990)* (Cambridge: Cambridge University Press, 1992). He began his career in crystallographic research under J. D. Bernal.
35. J. D. Bernal, *The World, The Flesh and The Devil* (London: Jonathan Cape, 1929).
36. Joseph Needham (1900–1995) was a biochemist who received his early training under F. G. Hopkins and J. B. S. Haldane at Cambridge University, specializing in embryology and morphogenesis. Later, he spent many years researching science and civilization in ancient China, publishing several monographs on that subject, which ultimately reached twenty-four volumes. Politically, he was sympathetic to socialism and British Labour Party and was included in the *Visible College* by Gary Werskey, *The Visible College: The Collective Biography of British Scientific Socialists of the 1930s* (New York: Holt, Rinehart and Winston Ltd, 1979), 376.
37. Werskey, *The Visible College*.
38. Isaiah Berlin (1909–1997), was a Russian-born British social and political theorist and philosopher. He excelled as an essayist, conversationalist, and raconteur and as a brilliant lecturer who improvised, rapidly and spontaneously, richly allusive and coherently structured material. He translated works by Ivan Turgenev from Russian into English and, during the war, worked for the British Diplomatic Service. In its obituary of the scholar, *The Independent* stated that "Isaiah Berlin was often described, especially in his old age, by means of superlatives: the world's greatest talker, the century's most inspired reader, one of the finest minds of our time."
39. L. Hogben, *Mathematics for the Million* (New York: Norton, 1937).

40. The Society for Experimental Biology was established in 1923 at Birkbeck College in London to "promote the art and science of experimental biology in all its branches." The original founders of the Society were Julian Huxley, J. B. S. Haldane, F. A. E. Crew, and Lancelot Hogben. (In *Lancelot Hogben: Scientific Humanist*, ed. Adrian and Anne Hogben [London: The Merlin Press Ltd, 1998], 79).

The Society has an international membership of approximately two thousand biological researchers, teachers, and students. The main activities of the Society are the organization and sponsorship of scientific meetings, the publication of relevant research, and the promotion of experimental biology through its education, public affairs, and career development programs.

PART 1940S

11

Lysenko Controversy

Disappointing Role of Haldane

> Now I am not going for one moment to suggest that there is not a very grave danger for science in so close an association with the State.
>
> —J. B. S. Haldane in *The Inequality of Man and Other Essays* (1932), 136

This chapter outlines the series of events that led to the response of the scientific left in Great Britain, especially Haldane, to the so-called Lysenko controversy, which resulted in the suppression of the science of genetics in the Soviet Union during the 1940s and later. But Haldane had already predicted that possibility in 1932. First, we must examine Haldane's political evolution, starting from his youth, when he was deeply under the influence of his mother's conservative beliefs, to his ultimate role as a leading Marxist of the scientific left in Great Britain in the 1940s. Haldane's silence, even when his friend, the eminent scientist N. I. Vavilov, was persecuted and killed, is inexplicable. I describe here the events that led to his controversial role in the so-called Lysenko controversy. These events were also indirectly responsible, among other factors, for his decision to spend his last years in India.

During the 1930s, largely influenced by his first wife, Charlotte, Haldane's mild interest in socialism bloomed into an active interest in full-fledged Marxism. It was also one of the two major crises of his life, the other being the divorce scandal that involved his wife Charlotte, in which he was cited as a co-respondent and which resulted in his dismissal from Cambridge University.

Haldane's Political Evolution

Haldane's political evolution underwent several radical transformations before he could embrace Marxism. As a young boy, he was much influenced by his mother's conservative politics. Descended from a long lineage of leisured country gentlemen and military men, Louisa Kathleen Haldane was an empire loyalist—an uncompromising Tory and an enthusiastic supporter of Joseph Chamberlain's brand of social imperialism.

Haldane's years at Eton saw his transformation from a bright boy into a brilliant young man. He was never really a part of the Eton crowd. At first he was bullied mercilessly. Of those years, his sister, Naomi, wrote, "He came from his first half desperately miserable and longing not to go back. He told me some of the tortures and I made up my mind (I was approaching eight) that I must convince my parents that they should take him away. I tried, but it was no use; this was good for him."[1] She added, "Later, he came to terms with Eton, learned a lot, including a vast amount of Latin poetry, and enjoyed being a member of Pop, having the power of the pop cane, but he made few friends, other than the Huxleys and Dick Mitchison."[2]

Contrary to his mother's expectation, the Eton years had in general led to his disaffection from the mainstream values of upper-class life and a greater appreciation of the liberal and intellectual side of the Haldane line of the family. Later, at Oxford, it seemed only natural that he joined the University Liberal Club and the local Cooperative Society.

According to his sister, JBS sailed triumphantly through all his exams at Oxford, and spread himself in friendships and light and the golden air of the prewar years for the upper classes. His friendships were very wide, with young men of completely varied interests. But many of them were killed in the war. Naomi wrote that Jack never made friends again in the same way. His friends were men and women in his own discipline, work friends. The play world was gone forever.[3]

Naomi wrote that their mother used to read Montefiore's *Bible for Home Reading* but excluded the salacious parts, which were an expected pleasure later. She wrote further, "We had a set of strict ethical principals which were slightly harder live up to because there was no supernatural sanction behind them. We had no religious conflicts with our parents, but in time we began to wriggle in our ethical bonds, but lying, for instance, was apt to make us both rather uncomfortable."[4]

Mother's Political Interests

As a young boy, Jack was deeply influenced by his mother's politics and philosophy. Louisa was an Empire loyalist and a founder of the Imperialists' Club of Oxford. The club promoted, among other things, empire goods. Within his own home, JBS faced two conflicting philosophies. His mother was an uncompromising Tory and an empire loyalist, descended from a long line of loyal military officers who formed the backbone of the empire. She supported the social imperialism of Joseph Chamberlain, who became gradually more conservative, beginning to the left of the Liberal Party and ending to the right of the Conservatives.

By the time of his eleventh birthday, Jack was already echoing his mother's ideas in politics; he wrote in his diary for September 18, 1903, "My mater and I are very sorry to hear of Mr. Chamberlain's resignation, but hope he will get into office again soon." Louisa recruited her children in various activities designed to serve the cause of empire one way or another. On one occasion, she took him to an assemblage of the Children of the Empire in a village to teach them to be good citizens of the empire. JBS learned early that the empire is not a lot of little countries but one big one. Mrs. Haldane combined an aristocratic humanitarianism with an extreme contempt for the urban proletariat and tradesmen.

Compromise

On the other hand, the Haldane side of the family was noted for their Liberal outlook. JBS learned of Liberal ideals through his father's investigations into mining diseases and disasters, which taught him early the poor living conditions of the miners and the dangers they faced in their work to earn their daily bread. His father involved JBS in physiological experiments from an early age and taught him the risks faced by the scientific workers. Such experiences made a deep impression on young JBS, who came to regard the welfare of such workers as one of the many responsibilities that had to be borne by their intellectual and social betters. That sentiment continued to stay with JBS as he later, in his undergraduate years, became a supporter of a eugenics movement that had as one of its goals curbing the reproduction of the supposedly unintelligent lower classes. One can see here a *compromise* between mother's Toryism and father's Liberal heritage. It reflects also his mother's sense of social responsibility in a new context. The same theme continued when he fought, as an officer of the Black Watch, in the trenches of World War I with soldiers from the lower classes, and still later when he moved to India, realizing his lifelong dream to provide science education for the less privileged, as he had previously done for the working classes through his *Daily Worker* articles.

Jack soon realized that both politically and personally his mother and Uncle Richard (Lord Haldane)[5] hated each other. In her autobiography, Louisa made it clear that although she liked her mother-in-law, her insistence that Louisa should "appreciate" her older son (Richard) made things difficult between them. Louisa wrote, "his omniscience, his self-satisfaction and his sneers at the ideas and loyalties of people who disagreed with him, were enough to account for my dislike. Nothing, however, but realization of the harm that he and his party were doing to the country would have accounted for and justified the strength of my antagonism."[6] However, Uncle Richard maintained a cordial relationship with his brother (her husband) as well as being a caring uncle to his nephew Jack. He was a frequent visitor to their Oxford home (Cherwell). Louisa, in her autobiography, recalled fondly that Jack was given a chemical set by Uncle Richard. She wrote, "It was a wonderful box which contained enough ingredients to make a long series of explosions and 'stinks' which kept him and Austin Lane Poole (now President of St. John's) happy and wonderfully out of mischief during my short season in bed."[7]

Soon, however, young Jack was drawn more and more into the scientific activities of his father, who treated him more like an adult colleague. Jack was deeply conscious of a long line of successful male ancestors, both in science and politics—role models to emulate. Jack's intellectual gifts enabled him to learn almost any subject faster than any of his contemporaries. Intellectual excitement of science discovery, collaborating with father in experiments, and the influence of the careers of his politician-uncle Lord Haldane and of his great-uncle the physiologist Burdon Sanderson[8] are among the important factors in young Jack's life that veered him away from his mother's circle of influence.

Uncle Richard rescued his nephew Jack during the war. Jack's sister Naomi wrote (in "Beginnings") that most of the officers of the Third Battalion of the Black Watch were killed, but Jack survived because his father asked him to help in his work at headquarters, countering the gas attacks at Ypres. This was facilitated by Uncle Richard, who was a powerful politician and a member of the government. Jack returned to battle later but was hit by a shell and was picked up by the Prince of Wales and sent to recuperate, which saved his life once again.

Haldane's Early Years in Science and Socialism: Transformation

Haldane was an exception among these exceptional Cambridge men. He was the only scientist in that crowd who did not possess an academic qualification in science. He was also different in the sense that he became a controversial public figure quite early in his life. Besides his brilliance, his family background and distinguished Scottish lineage were contributing factors. He quickly became known for his futuristic predictions about "test-tube babies" and genetic manipulation. He was also unique in his involvement in a sensational divorce and his successful challenge to be reinstated as the Dunn Reader in Biochemistry at Cambridge University. Haldane made significant contributions to three different fields of science simultaneously during his Cambridge period: physiology, genetics, and biochemistry.

Haldane became a cultural hero to the educated younger generation. A member of that generation, Isaiah Berlin[9] later recalled that Haldane was, along with Aldous Huxley, "one of our major intellectual emancipators." Haldane's sensational book, *Daedalus*, was a bestseller. There were leading articles in the newspapers commenting on his book. Bertrand Russell and Robert Graves responded with lengthy articles. By the mid-1920s, Haldane was quickly attaining the status of a successful public figure.

Figure 11.1 Haldane making a political speech in Trafalgar Square in the 1940s.

Needham and Bernal were overawed by Haldane's larger-than-life public persona. Although later they became famous figures in their own right, theirs was of a different kind, both quantitatively and qualitatively.

At this stage, two publications enhanced his reputation as a brilliant intellect even further. He was portrayed as a major protagonist in two novels by successful writers. Aldous Huxley's Shearwater in *Antic Hay*[10] was the physiologist who was deeply engrossed in physiological experiments while his wife was serially bedding his friends and associates. Another character, Mr. Codling in Ronald Fraser's *The Flying Draper*,[11] was portrayed as a corrupter of youth who, after writing *The History and Probable Future of Morality*, was chased by jealous rivals from his university chair. The resemblance to JBS was unmistakable. All these publications and events enhanced his reputation as a daring intellect. But intellect aside, what about JBS's politics?

In the pre-Depression era, the colleges of Cambridge were not known to be seedbeds of radical politics. There was little of what today would be called left-wing activity on the campus. Indeed, political activity of any sort was hard to find. The local branch of the National Union of Scientific Workers (NUSW) was no more than a debating society. An appeal to deunionize the NUSW in 1927 was supported by many senior dons, including F. Gowland Hopkins. Haldane and Needham were inactive members of the Labour Party. Bernal and Needham were not even members of the NUSW. Haldane was a member but not an active one, which is surprising because of the support he received from the Union when he appealed his dismissal from Cambridge University.

In the 1920s, Haldane wanted to be seen as a budding man of political interests. But he was reluctant to leave the laboratory, where he was still trying to build his career as a scientist. However, we can see his leanings, though briefly, in two of his major popular books of that period: *Daedalus, or Science and the Future* and *Possible Worlds and Other Essays*.[12] Both were addressed to the general reader with a smattering of scientific knowledge.

In *Daedalus*, Haldane expressed support for the Labour Party, saying, that party "alone among political organizations includes the fostering of research in its official programme. Indeed as far as biological research is concerned labour may prove a better master than capitalism, and there can be little doubt that it would be equally friendly to physical and chemical research."[13] Earlier, he commented that capitalism, though it may not always provide the scientific worker a living age, would always protect him, "as being one of the geese which produce golden eggs for its table." He also made a passing reference to scientific research in Russia, which he regarded as first-rate. This last comment, in particular, was prophetic as indicative of his great admiration for Russian science, which occupied his attention for several decades, until his death in 1964. When Sputnik was launched in 1957, Haldane felt vindicated in his lifelong faith in Soviet science.

In *Daedalus*, Haldane further referred to "Marxian socialism" as one of the results of practical importance arising from materialism. Once again, his reference to Marx and socialism, though brief, was clearly indicative of an idea that occupied his attention for much of his life. In *Possible Worlds*, a collection of essays written in the years following *Daedalus*, Haldane returned to the topic of Marxism.

It became obvious throughout the 1920s that Haldane was reluctant to involve himself in active party politics. He did not wish to embroil himself in what he regarded as the "pettiness" of party politics. He was also not prepared to consider that the knowledge of human biology had advanced far enough to impact on the activities of human society. On the other hand, he suggested that the scientific attitude and rigor could benefit the way human activities including politics are conducted.

Political Transformation

Unlike some of his colleagues, Haldane would have to repudiate a thoroughly bourgeois background first before he could affirm communism. His severance of the bonds that attached him to the political and social conventions of his class was an elaborate and tortuous process, yet JBS landed on his feet squarely when so many others failed that test. And it was not without much satisfaction that Haldane was able to thumb his nose against the establishment, which was in its own unique way an answer to all the indignities he suffered at Eton and Cambridge, among other places and incidents.

The social and political influences that shaped Haldane's early life set him apart from others. The conventional elite education he received was in contrast to the uniqueness of his home. The education he received at Eton and Oxford prepared him well for the needs of an English gentleman. Haldane's background in the Church of England and his elite education assured him a coveted position in the civil service or an equally comfortable position in business or industry. In his essay "On Being Finite," Haldane expressed some satisfaction that he was not one of those who decided to join the British civil service and serve the empire.

> Except for a few troubled spots like Aden, the British Empire is dead, and its servants who survive it are disillusioned old men and women. Those who worked for the World Revolution, whether primarily against capitalism or colonialism, were farther sighted, and have been more fortunate, including, in my opinion, those who died fighting. But the world revolution, like a swift stream, has its eddies, including recrudescences of various kinds of tyranny.[14]

Haldane wrote, "If Marx was right, and I happen to agree with him, inventors are doing more for world revolution than many politicians. Changes in productive 'forces' generate a social instability which brings about change in productive relations, and hence in social structure. Those inventors who are Marxists are quite aware of this."[15] Concerning his own work, Haldane hoped that it would outlive him, not merely for a human lifetime, but for very much longer.

Various political and social reasons have been credited for the leftward drift of intellectuals in the 1930s. Economic hardship and depression in the United States and the rapid rise of fascism in Europe are among these. For those searching for a better alternative, the Soviet Union offered an attractive prospect. For Haldane, who had long been an admirer of Soviet science, the choice readily offered itself. But that was still not quite enough to cross the line and join the Communist Party of Great Britain (CPGB). When he was desperately seeking an outlet to voice his concerns and ideas on air raid precautions in 1936, the Communist Party's daily newspaper, *The Daily Worker*,[16] was the only medium that offered him the much-needed publicity. Typically for Haldane, he never did anything halfway; soon he joined its editorial board and served as its chairman for ten long years. He believed in wholehearted commitment!

Trofim Lysenko

First, some historical background about Lysenko and the controversy that bears his name. Trofim Denisovich Lysenko[17] (1898–1976) was born in a peasant family in Ukraine and studied at the Kiev Agricultural Institute. He was a Soviet agronomist who rejected the science of Mendelian genetics and supported the hybridization theories of the Russian horticulturist Ivan Michurin.[18] He adapted them to a powerful political and pseudoscientific movement termed Lysenkoism. No one views Lysenko's theories of agriculture as a scientifically valid program today. Lysenko's scientific interests fall under four categories: vernalization, developmental physiology of plants, plant breeding, and genetics.

Vernalization

In 1927, working at an agricultural experiment station in Azerbaijan, Lysenko embarked on the research that would lead to his 1928 paper on vernalization, which drew wide attention due to its practical consequences for Soviet agriculture. Severe cold and lack of winter snow had destroyed many early winter-wheat seedlings. By treating wheat seeds with moisture as well as cold, Lysenko claimed to have induced them to bear a crop when planted in spring.

The term "vernalization" is a translation of "Jarovization," a word Lysenko coined to describe a chilling process he used to make the seeds of winter cereals behave like spring cereals and the alleged inheritance of this trait in a Lamarckian manner (i.e., inheritance of acquired characters). Vernalization is an established fact. On the other hand, the genetic induction of winter cereals into spring cereals through a few years of partial vernalization, as Lysenko claimed, is *not* an established fact. Some of the results obtained may seem Lamarckian, but are in fact due to Lysenko's erroneous methods and the heterogeneous nature of his materials, which helped natural selection to operate. In order to prove that vernalization had, in fact, occurred, Lysenko should have repeated the experiments without his previous errors and using controls.

Lysenko also claimed that he was able to convert spring wheats into winter wheats, which are far more frost-resistant, by sowing in autumn for two to four years without vernalization. Lysenko wrote, "When experiments were started to convert hard wheat into winter wheat it was found that after two, three, or four years of autumn planting (required to turn a spring into a winter crop) *durum* becomes *vulgare*, that is to say, one species becomes converted into another. *Durum* wheat with 28 chromosomes is converted into several varieties of soft 42-chromosome wheat, nor do we, in this case, find any transitional forms.... The conversion of one species into another takes place by a leap."[19]

Lysenko's claim of converting one species into another is a remarkable feat that would have surprised Darwin himself! Julian Huxley[20] offered a simpler explanation: The seed was of mixed type and the winter conditions killed off all but the winter-resisting types—the result of straight selection.

Lysenko entered politics and was for a time a vice-president of the Supreme Soviet. He was twice awarded the Stalin Prize, and the Order of Lenin. In 1938 he was elected president of the Lenin Academy of Agricultural Sciences, a position previously held by his opponent Vavilov. In 1940 he began attacking the statistical methods used in genetics, and genetics itself. Lysenko claimed that heredity can be altered by grafting and claimed that the Mendelian results can be explained by a process of "assimilation" of external conditions by the organism over a series of generations.

His unorthodox experimental research earned the support of the Soviet leader, Joseph Stalin, who saw in Lysenko's theories a justification of the Soviet State, where the environment triumphs over genetics, and the disparagement of Western Mendelian genetics. In 1940 Lysenko was appointed director of the Institute of Genetics within the USSR's Academy of Sciences, and his anti-Mendelian doctrines were further secured in Soviet science and education by the exercise of political influence and power. Scientific dissent from Lysenko's theories of environmentally acquired inheritance was formally outlawed in 1948, and for the next several years, opponents were purged from long-held positions, and many were imprisoned. Many geneticists of international reputation, such as G. D. Karpetchenko, A. S. Serebrovsky, and N. I. Vavilov, disappeared during the 1940s. Until 1964, Lysenko's ideas reigned supreme in Soviet agriculture.

N. I. Vavilov

The great Russian botanist and agronomist N.I. Vavilov[21] had traveled widely and organized a series of expeditions to regions where various crop-plants were presumed to have originated. As a young man, he studied with William Bateson in England. It is a sign of his scientific distinction that Vavilov was elected a foreign member of the Royal Society of London.

In the summer of 1928, Haldane and his first wife, Charlotte, visited Soviet Union as guests of Vavilov. Charlotte recorded that Vavilov had a vivid regard for Haldane, whose pioneering work on the mathematical theory of natural selection had just begun. In her autobiography, *Truth Will Out*,[22] Charlotte revealed that they were neither Communists nor even dialectical materialists. Vavilov arranged lectures by Haldane to geneticists in Moscow and Leningrad. At that time, Vavilov was at the height of his scientific and political popularity in the Soviet Union. The Haldanes visited Vavilov's famous Institute and plant collections at Leningrad and his experimental station at Djetskoie Selo., where they were shown the summer palaces of Catherine the Great and Nicholas II. Among others, they met Vavilov's younger colleague the brilliant cytologist G. D. Karpetchenko,[23] who was to succeed Vavilov but who later became a victim, along with Vavilov, of the persecution directed by Lysenko.

At the conclusion of their visit, Charlotte wrote,

> I had no regrets at leaving the Soviet Union in 1928, but, on the contrary, a feeling of relief. The experience of having to receive an exit visa had not been pleasant. I was very conscious of the difference in atmosphere, as soon as I set foot aboard the little German steamer that brought us to Stettin. It was also a physical relief to encounter again clean beds and lavatories, the amenities of one's own Western "bourgeois" world, which are always taken for granted.[24]

Left-Wing Intellectuals

Haldane's conversion to Marxism and socialism was not so unique as it might seem. In the 1930s, it was almost de rigueur for many intellectuals, especially scientists, to embrace Marxism or socialism, or even join the Communist Party, as Haldane and J. D. Bernal in the UK and Herman J. Muller in the United States had done. They were also united in their opposition to the rise of Nazi Germany during the 1930s. Others who belonged to this group include the Nobel laureate P. M. S. (Patrick) Blackett, Joseph Needham, Lancelot Hogben, Julian Huxley, and N. W. Pirie, among others, who led the way to a social emancipation based on a scientific society. They argued in numerous articles, books, and speeches that a new social policy based on science must be developed and that a scientifically educated public is more likely to support such a policy.

In 1960, C. P. Snow[25] estimated that a poll of the two hundred brightest physicists under the age of forty in 1936 would have revealed that "about five would have been communists, ten fellow-travellers, fifty somewhere near the Blackett position (noncommunist, but activist and fairly far left), a hundred passively sympathetic to the Left. The rest would have been politically null, with perhaps five (or possibly six) oddities on the Right. Of those two hundred, many have since occupied positions of eminence. It is interesting that none of them has drastically altered his political judgment. There has just been a slight stagger, a pace and a half to the Right, no more. The scientists who got under the shadow of the Communist party have come out but have stayed (like Haldane) on the extreme Left, The changes have not been any more dramatic than that. There have not been any renunciations or swings to religious faith, such as a number of writers of the same age, once left wing, have gone in for. The scientists' radicalism had deeper roots."[26]

In his book, *The Visible College*,[27] a collective biography of British scientific socialists of the 1930s, Gary Werskey presents a detailed and sympathetic account of the social movements of some left-wing scientists. The scientists covered by Werskey are J. D. Bernal, J. B. S. Haldane, Lancelot Hogben, Hyman Levy, and Joseph Needham. Others, who are briefly mentioned, include P.M.S. Blackett, N. W. Pirie, W. A. Wooster, and C. H. Waddington. The five primary scientists were selected because they shared certain common qualities: their commitment to socialism predated the so-called Red

Decade of the 1930s; their participation at various times in party politics, whether on the left wing of the Labour Party or in the CPGB; and finally, through their many books and articles, their authorship of a large and distinctive body of socialist thought.

Science and Politics

Haldane and others, who shared a coherent subculture of science and politics, have been collectively characterized as "Cambridge men." Their self-absorption and communality, sharing a tribal social structure and values that isolated them from the nonscientific world outside, helped to give their society a cohesive structure. This counterculture was tailor-made for a person of Haldane's intellect and stature. He excelled in the sciences but was also ready to quote from the classics at the slightest opportunity. In his *Memories*,[28] Julian Huxley commented on Haldane's propensity to show off his classical scholarship.

Impact on British Left-Wing Scientists

The impact of Lysenkoism on Soviet science and society deeply shocked British left-wing scientists and intellectuals in general. It was hard to believe that the Soviets changed their earlier policy to respect the opinion of international science. Yet, that was exactly what was happening in the case of genetics. Scientific authority became subservient to political authority. Party authority and Marxist orthodoxy rather than scientific evidence ruled the roost as far as science policy was concerned. As early as 1937 it became clear that further attacks would be made on Soviet biology. Ironically, the foremost victims outside the Soviet Union were the most ardent supporters of Soviet science and leading advocates of classical genetics, including Haldane, Hogben, Bernal, and Needham. Their strategic response was to offer alternative scientific explanations for Lysenko's findings rather than attack Lysenko himself. Thus Haldane suggested that Lysenko's admirable and exceptional results were not caused by environmentally induced alterations in the genetic material but were probably due to virus infections that commonly afflicted the plant types that were being used in Lysenko's experiments.

In the early years of the Lysenko controversy, Needham and Haldane argued that the debate would enrich Soviet science and Haldane suggested that Lysenkoism would eventually advance biological theory and practice in the Soviet Union. In retrospect, such comments, which were made before the excesses of Lysenkoism and Vavilov's death became widely known, seem so naïvely optimistic. In opposition to the Aryan superiority and all forms of scientific racism that were being advocated by the Nazis, the Bolsheviks abolished the testing of intelligence. They closed the Medical Genetics Institute in 1936 and its director, Dr. S. G. Levit, was arrested and shot. Back home in Britain, ironically, Haldane was simultaneously fighting the neo-Lamarckian views of E. W. MacBride (1866–1940), an embryologist and professor at Imperial College in London. Shocked as they were, the inner circle of the socialist-scientist group did not change their views on the interrelationships between socialism, science, and public participation.

The Lysenko controversy became complicated partly because Lysenko's work was published in the Russian language, which none of his critics in the Western world were able to understand. For this reason, while some of his colleagues, such as R. A. Fisher and C. D. Darlington, were rushing to judgment, Haldane stated that he would like to read the translated works before reaching a decision about Lysenko's research. This was seen as prevarication on Haldane's part, because he was caught in a dilemma, partly due to his own decisions and to distant events that were beyond his control.

It was also Haldane, with remarkable prescience, who correctly foresaw the disastrous consequences of the impact of Lysenko on Soviet science. In *The Inequality of Man and Other Essays*, which was published in 1932, Haldane wrote,

> Now I am not going for one moment to suggest that there is not a very grave danger for science in so close an association with the State. It may possibly be that as a result of that association science in Russia will undergo somewhat the same fate as overtook Christianity after its association with the State in the time of Constantine. It is possible that it may lead to dogmatism in science and to the suppression of opinions which run counter to official theories, but it has not yet done so.[29]

Yet, barely a year later Lysenko was confirming Haldane's worst fears, which came true because of the close association and support enjoyed by Lysenko with the Soviet premier Joseph Stalin.

Haldane's assessment of Vavilov

As a prominent geneticist and a Marxist as well as a great admirer of Soviet science, Haldane was caught in a difficult dilemma. In his book *Science Advances*,[30] which contained several popular essays that were first published in the *Daily Worker*, Haldane included an essay on Vavilov. As it had been widely reported in the Western press that Vavilov had just been murdered in Siberia, Haldane's article appeared to be an obituary of some sort.

Haldane commented that, from the viewpoint of scientific theory, Vavilov's most important work was on the geographical distribution of varieties of crop plants. Haldane acknowledged that the work of Vavilov's institute was cut down to some extent before the war, but he attributed that partly to Lysenko's invention of vernalization, not to any political harassment. He had also hoped that the work of such scientific institutes would be resumed after the war. Haldane thought that Vavilov's successors should be able to continue his valuable work to improve the crop plants of their country (and the rest of the world) and contribute to our knowledge of variation, and to throw new light on the origin of agriculture, and by implication, our civilization.

Coming at a time when deep emotions were aroused by the imprisonment and death of Vavilov in Siberia, and the suppression of all genetic research in the Soviet Union, Haldane's mild statement was viewed as a prevarication by his colleagues. Instead of a strong condemnation of Lysenko, which was expected of him, Haldane went out of his way to recognize Lysenko's contribution to Soviet agriculture, while bestowing faint praise on Vavilov.

In the same collection, Haldane, in "Genetics in the Soviet Union," made a valiant effort to point out what was right about Lysenko's science. For instance, he wrote that Lysenko's attacks on genetics made that science stronger, leading to the elimination of wasteful research.

In defense of Lysenko, Haldane wrote:

> But scientific pioneers are not infallible. Pasteur did more for the theory and practice of fermentation than any other man. Yet he made some big mistakes. Having discovered that the usual agents of fermentation, such as yeasts and bacteria, were alive, he denied the possibility of fermentation by nonliving substances. Yet today thousands of different enzymes are known, about twenty have even been crystallized by Sumner, Northrop, and others, mostly in the U.S.A. In the same way Lysenko, who is right in pointing out that the majority of characters showing Mendelian inheritance are of little economic importance, is quite wrong in supposing that none of them are.[31]

However, Haldane carefully avoided condemning the excesses committed by the Soviet authorities in suppressing genetics, which included the destruction of text books and the imprisonment and deaths of Soviet geneticists. The manner in which he downplayed the official harassment of Soviet geneticists while praising Lysenko proved to be the last straw, which led to Haldane's total estrangement from his genetic colleagues in Great Britain. Although he resigned from the CPGB in 1949, Haldane's continuing support of Soviet science increased his alienation from many of his scientific colleagues in Western countries.

Haldane's political alienation in Great Britain was closely followed by his increasing interest in India as a possible future home. He had long admired India's cultural and philosophical foundations and developed an increasing admiration for Prime Minister Nehru's[32] political neutrality as well as his intellectual standing.

At this juncture, Haldane was only too happy to accept an invitation to participate in the Indian Science Congress in 1952. After that visit, he started corresponding with P. C. Mahalanobis,[33] director of the Indian Statistical Institute (ISI) in Calcutta, about his move to India, which occurred in 1957.

In his article "Lysenko and Genetics,"[34] Haldane began by examining Lysenko's ideas to separate those which he agreed with from others where he thought Lysenko was mistaken. Haldane agreed with Lysenko that so-called pure lines of stocks could still be subjected to a selection process, resulting in other approximately pure lines. Haldane wrote that "Lysenko is quite right in stressing the importance of selecting 'elite strains of seed' from so-called pure lines." He also thought that Lysenko was essentially correct in his estimation of the breadth of the zone of isolation needed for different crops, although this has nothing to do with Mendelian genetics.

In an undated letter to Mikulas Teich, quoted in Teich's "Haldane and Lysenko Revisited," Haldane accepted Lysenko's claim that he was able to produce autumn wheats from spring wheats and conversely, but he rejected the general claim that "acquired characters are inherited."[35]

But, Lysenko questioned the biological significance of Mendelian ratios, which he characterized as merely statistical artifacts. Haldane thought Lysenko was clearly mistaken in this regard, but he hoped that the translation of Lysenko's paper from Russian to English may have introduced some errors. However, he agreed with Lysenko on a different aspect, that is, selection in the first hybrid generation is important if the hybrids are not between pure lines. Haldane emphasized that so-called pure lines are ideals that are rarely realized and many agricultural varieties are far from pure lines. Haldane thought Lysenko was badly mistaken in believing that you could improve a breed of animals by improving its food.

Another major point of difference concerned the role of grafting. It was known before Lysenko that virus diseases in such plants as tomatoes can be transmitted, among other methods, by grafting. Later research has shown that besides disease-producing viruses, it is possible to transmit viruses that have no obvious effect on a plant but immunize it to one or more of the disease-causing agents. Lysenko claimed that certain transmissible agents could alter the shape of the fruit. Haldane's general comment about various claims made by Lysenko is that so little is known about the range of effects produced by grafting and other methods of vegetative propagation that Lysenko's claims cannot be ruled out without a thorough investigation. However, he disagreed with Lysenko when he claimed that he could induce a permanent color change in a white-fruited cherry by grafting.

Haldane also pointed out that Lysenko was quite mistaken when he claimed that hereditary properties can be transmitted from one breed to another without the transmission of chromosomes. He wrote further, "In the same way Lysenko was wrong if he referred to the theories of current genetics, such as the three-to-one ratio and the like, as 'fantasies'. They are not fantasies, but approximations." He pointed out that although the Copernican and Newtonian systems were also approximations, they were great advances on the systems of Ptolemy and other earlier astronomers. He concluded that "posterity will rank Mendel with Copernicus or Kepler, though hardly with Newton" Quoted in Teich 2007.

In the same article Haldane acknowledged Russian criticism of his mathematical theory of evolution because it did not include "real biological interrelations." Haldane readily agreed that a mathematical consideration of even the simplest evolutionary problems is difficult.

J. D. Bernal (1901–1971) and Joseph Needham (1900–1975)

Both the crystallographer J. Desmond Bernal and the biochemist and historian of science Joseph Needham came from Catholic families, Bernal came from Ireland and Needham from England. Both also came from troubled families, but both were encouraged to study science by a favorite parent—the father's influence in the case of Bernal and the mother's help in the case of Needham. Both attended public schools,

Bernal at Bedford School and Needham at Oundle in Northamptonshire. Both studied at Cambridge University, and both experienced a need to situate their science and politics within religious frameworks. Both opposed the military-type discipline glorifying the empire that was expected of them by the public school curriculum; this was indicative of their true sentiments, which became more obvious in later years. Young Needham admired H. G. Wells, for his science fiction and utopian novels as well as his socialist beliefs, despite his parents' disapproval. Bernal, on the other hand, was more outspoken about his origins. An Irish Catholic at an English public school, he was deeply conscious of the suppression of his country by Britain. Bernal came to believe more and more that violence was the only means that could lead to liberation of his country.

Needham's socialism and Bernal's nationalism were closely related to their religious beliefs and scientific ambitions. Needham saw both socialism and science as virtually synonymous with reference to certain historical epochs. Bernal, on the other hand, saw early in his career that scientific progress should go hand-in-hand with Ireland's economic and political independence. All three would have to be a part of the same process. However, Bernal was committed to revolutionary socialism after an exhaustive study of the Marxist writings, which led to his conviction that Roman Catholicism was both antiscientific and reactionary.

Needham's socialist beliefs and his scientific career, which were closely entwined, were undergoing rapid changes. Although it was a great disappointment to his parents, he abandoned his initial interest in a medical career and was drawn more and more toward biochemistry, under the influence of the great Cambridge University biochemist Frederick Gowland Hopkins.[36] Compared with Bernal, however, his political transformation to socialism was slow and gradual. Haldane and Bernal, on the other hand, had already begun their political estrangement from the privileged worlds that had provided their education and social prominence. In reality, Haldane's seniority (he was eight years older than Needham and nine years older than Bernal) provided him the much-needed time for his political transformation from his bourgeois roots, which was not required for Needham and Bernal.

Response of the Scientific Left to the Lysenko Controversy: Disappointing Role of Haldane

Predictably, each of the members of the "Visible College" reacted in a different manner to the Lysenko crisis. Haldane's position has already been discussed at length. Of the four leaders of the scientific left of the 1930s in Great Britain, Haldane's position was unique and most difficult as well as controversial. Among the four individuals discussed by Werskey, Levy was a mathematician, Bernal was a crystallographer, and Needham was a biochemist, whereas Haldane was the only geneticist. Consequently, he was expected to take a stand on Lysenko's pronouncements, publications, and his very prominent role in the persecution of geneticists in the Soviet Union, especially their leader, Vavilov, who was not only an outstanding geneticist but also a personal friend of Haldane. Fellow scientists, friends, and the public looked to Haldane for a fitting response. Would Haldane be forthcoming with an appropriate response, an absolutely clear condemnation of Lysenko and his disastrous impact on Soviet science? The silence was deafening.

Julian Huxley's Evaluation

Haldane wrote an article in the *Modern Quarterly*, "In Defence of Genetics."[37] In the words of Haldane's close friend, Julian Huxley, who wrote more in sadness than anything else, "Professor Haldane does not deal with what I consider the main issue, namely the official banning of Mendelian genetics on the basis of a scientific party line. He is only concerned with an appraisal of the views of Lysenko and his followers (which, he points out, "has been made much more difficult by ill-informed criticism of [Mendelian] genetics by supporters of Lysenko in this country [Britain])."

Huxley, quoting Haldane, commented, "He [Haldane] says that Lysenko's speech made him realize for the first time 'the idealistic character of Mendel's formulation of his results,' because Mendel had spoken of the transmission in heredity of 'differentiating characters' (as opposed to genetic units)." Huxley continued, "However, whether on this point Mendel was or was not under the influence of Thomist philosophy, as Haldane suggests, is irrelevant to the situation today (as Haldane himself later implies). Geneticists quite early realized the illegitimacy of speaking of the inheritance of *characters* (except as an occasional form of convenient short-hand) and began talking in terms of the inheritance of *hereditary factors*—which were later styled *genes*."[38]

Huxley noted with approval that Haldane took his stand unequivocally in favor of neo-Mendelism and neo-Darwinism. Haldane wrote, "I am a Mendelist-Morganist, although Mendel used an idealist terminology, and Morgan wrote of the mechanism of heredity. But Morgan and his colleagues made the very great advance of showing that heredity has a material, not a metaphysical basis."[39]

Lysenko doubted the existence of genes. Haldane responded,

> We do not cease to believe in atoms because they can be split. Nor need we cease to believe in genes because they can be changed. On the contrary, if they were unchangeable, I, as a Marxist, could not believe in them.
>
> In a recent discussion in London, some Marxists went so far as to deny that there was a material basis of inheritance. There is good reason to doubt that any parts of a cell are *only* the material basis of heredity. Genes certainly play an active part in a cell's ordinary life. But a Marxist can no more deny a material basis for heredity than for sensation or thought.[40]

Huxley commented, "I am extremely glad that Haldane has made this statement. But it is worth pointing out that many of the speakers on Lysenko's side in the discussion *did* deny the existence of genes, or their significance as a material basis for heredity; and it is worth wondering what would have happened if Haldane had made these same statements in Moscow during the session of the Academy of Agricultural Science."[41]

Huxley also noted that Haldane affirmed the Mendelian interpretation of the value of wide crosses and was skeptical of the claims of Lysenkoists that "acquired characters are inherited."

Haldane noted, "The results of experimental work are not available until they are published in such a form that they can be repeated." This is an unobtrusive and indirect way of drawing attention to the unscientific procedure of the Michurinites in not publishing their results in an adequate way. Haldane concluded, "Marxist circles have made wholly unjustifiable attacks" on the profession of genetics.[42]

Bernal

If Haldane appeared to be evasive in condemning Lysenko and his theories, J. D. Bernal's position presented an even more extreme situation. Julian Huxley wrote, "When he [Bernal] is not dragging in subjects which have nothing to do with the genetical issues, he is concerned with the Marxist thesis ... that new social conditions are bound to give rise to new kinds of scientific theories, and indeed to new kinds of scientific theories."[43] On this basis, Bernal defended the views and actions of the Michurinites, and justified the drawing of a distinction between bourgeois and Soviet science.

The distinction between Haldane and Bernal was partly based on the fact that Bernal, as a physicist, owed no loyalty to genetics, whereas Haldane, as a founder of population genetics, was keenly aware that his professional integrity was at stake. Huxley thought that Bernal had unfairly evaded the major issues—the legitimacy of officially condemning a whole branch of science as false, antiscientific, and so forth, and the scientific validity of the results obtained and the methods employed by the Michurinites. Huxley wrote,

> If Professor Bernal were a geneticist instead of a physicist, he would realize that Lysenko's theories are, scientifically speaking, largely nonsense—meaning that they do not make scientific sense.... Far from being in a state of "most unsatisfactory confusion" as Bernal asserts, classical genetics is in an extremely healthy and vigorous condition, and shows every sign of being able to cope satisfactorily with the numerous new developments to which it is giving rise.[44]

Mathematician G. H. Hardy

Cambridge men set their own standards. Any consideration that one's pure research may have some practical application (Heavens no!) is not only out of the question but outrageous! The best example of this kind is the mathematician G. H. Hardy, who addressed this problem with great emphasis and clarity in *A Mathematician's Apology*:

> I shall ask, then, why is it really worthwhile to make a serious study of mathematics? What is the proper justification of a mathematician's life? And

my answers will be, for the most part, such as are to be expected from a mathematician: I think it is worthwhile, that there is ample justification. But I should say at once that my defence of mathematics will be a defence of myself, and that my apology is bound to be to some extent egotistical. I should not think it worthwhile to apologize for my subject if I regarded myself as one of its failures.

Some egotism of this sort is inevitable, and I do not feel that it really needs justification. Good work is not done by "humble" men.... A man who is always asking "Is what I do worthwhile?" And "Am I the right person to do it?" will always be ineffective himself and a discouragement to others. He must shut his eyes a little and think a little more of his subject and himself than they deserve.[45]

Haldane and Hardy shared more than a Cambridge memory. In 1908, while Haldane was still a student at Eton, Hardy published an important paper on the constancy of relative gene frequencies in populations. This work, which later came to be known as the Hardy-Weinberg equilibrium, turned out to be the fundamental equation of population genetics.

There is an important difference between Haldane and Hardy with respect to genetics. Hardy's contribution, though extremely important in the history of genetics, was an isolated instance in his mathematical career. Haldane, on the other hand, was one of the founders of the extensive theory of population genetics to which he made important contributions throughout his life, as well as to several other branches of genetics.

Notes

1. Naomi Mitchison, "Beginnings," in Krishna R. Dronamraju, ed., *Haldane and Modern Biology*, ed. (Baltimore, MD: Johns Hopkins University Press, 1968), 299–305; 301.
2. Richard Gilbert Mitchison, Baron Mitchison (1894–1970), Labour MP for Kettering from 1945 to 1964. He married the writer Naomi Haldane (daughter of physiologist John Scott Haldane and sister of J. B. S. Haldane) in Oxford in 1916. They had six children, including the immunologist Professor Avrion Mitchison of University College London, to whom this book is dedicated.
3. Mitchison, "Beginnings."
4. Mitchison, "Beginnings," 304.
5. Richard Burdon Haldane, 1st Viscount Haldane of Cloan (1856–1928), uncle of J. B. S. Haldane, Scottish lawyer, philosopher, and statesman who instituted important military reforms while serving as British secretary of state for war (1905–1912). He was educated at the universities of Göttingen and Edinburgh. He sat in the House of Commons from 1885 until his elevation to the peerage in 1911. In 1912 Haldane became lord chancellor in H. H. Asquith's Liberal government. In May 1915, however, when Asquith formed a wartime coalition ministry, he excluded Haldane, who was unjustly accused of being pro-German because of his German education. By the end of the war his political orientation had moved to the left, and he served once again as lord chancellor in Ramsay MacDonald's first Labour Party government. Lord Haldane was keenly interested in educational reforms. He was associated with the Fabian socialists Sidney and Beatrice Webb in founding the London School of Economics in 1895. He discussed the philosophical consequences of Albert Einstein's theories of physics in *The Reign of Relativity*. His *Autobiography* was published posthumously in 1929.

6. Louia K. Haldane, *Friends and Kindred* (London: Faber and Faber, 1961), autobiography, 150.
7. Haldane, *Friends and Kindred*, autobiography, 219.
8. John Burdon-Sanderson (1828–1905), great-uncle of J. B. S. Haldane and physiologist, received his medical education at the University of Edinburgh and at Paris. In 1871 he reported that *Penicillium* inhibited the growth of bacteria, an observation that places him among the forerunners of Alexander Fleming. When the Waynflete Chair in Physiology was established at Oxford University in 1882, he was chosen to be its first occupant. In 1895 Sanderson was appointed Regius Professor of Medicine at Oxford, resigning the post in 1904. He was Croonian Lecturer to the Royal Society in 1867, and was awarded the Royal Medal.
9. Isaiah Berlin (1909–1997), was a Russian-born British-Jewish social and political theorist, philosopher, and historian. He excelled as an essayist, conversationalist, and raconteur; and as a brilliant lecturer. From 1957 to 1967 he was Chichele Professor of Social and Political Theory at the University of Oxford. He was president of the Aristotelian Society from 1963 to 1964. In 1966 he played a crucial role in founding Wolfson College, Oxford, and became its first president. Berlin is popularly known for his essay "Two Concepts of Liberty," delivered in 1958 as his inaugural lecture as Chichele Professor of Social and Political Theory at Oxford. Berlin is widely quoted for his statement that J. B. S. Haldane was the intellectual emancipator of his generation.
10. *Antic Hay* is a comic novel of ideas by Aldous Huxley, published in 1923. The story depicts the self-absorbed cultural elite in the turbulent times following the end of World War I. The book follows the lives of a diverse cast of characters in artistic and intellectual circles. It indicates Huxley's ability to dramatize intellectual debates in fiction. The book was condemned for its immorality because of its open debate on sex. It was banned for a while in Australia and burned in Cairo. The manuscripts for the novel are part of the collection of the University of Houston Library. The character Shearwater in *Antic Hay* is a comic version of JBS Haldane, who was supposed to be an absent-minded biologist who was engrossed in his experiments while his friends took his wife to bed. Haldane was not married at that time.
11. Ronald Fraser, *The Flying Draper* (London: The Book Depository, 2006; first published in 1924).
12. J. B. S. Haldane, *Daedalus, or Science and the Future* (London: Kegan Paul, 1923); and *Possible Worlds and Other Essays* (London: Chatto & Windus, 1927), Haldane's second book, and first of several books of collected essays, includes his famous essay "On Being the Right Size."
13. Haldane, *Daedalus*, 7.
14. J. B. S. Haldane, "On Being Finite," *Rationalist Annual* (1965), 200. Haldane's last contribution to the *Rationalist Annual*, written by him in 1964 and posthumously published in 1965, after his death in India on December 1, 1964.
15. J. B. S. Haldane, *On Being the Right Size* (Oxford: Oxford University Press, 1985), 114.
16. The newspaper *Morningstar* was founded in 1930 as the *Daily Worker*, organ of the Communist Party of Great Britain (CPGB). Since 1945 the paper has been owned by the Peoples Press Printing Society. It was renamed the *Morning Star* in 1966. During the period in World War II when the Soviet Union was in alliance with Germany the *Daily Worker* ceased to attack Nazi Germany and advocated policies that some perceived as seeking to undermine the war effort. For this reason in January 1941 the newspaper was suppressed by the wartime coalition's (Labour) home secretary, Herbert Morrison. The ban was lifted in September 1942 following a campaign supported by Hewlett Johnson, the Dean of Canterbury, and J. B. S. Haldane.
17. Trofim Lysenko (1898–1976) was a controversial Ukrainian agricultural scientist, who claimed that he could convert spring wheat into winter wheat by pretreating the seeds before planting and demonstrating its inheritance in future crops. His claims were viewed with skepticism in Western countries, and he was called a "pseudoscientist." More serious was the support he received from Stalin and the consequent suppression of Mendelian genetics in the Soviet Union as well as the persecution of many geneticists, including their leader, Vavilov.
18. Ivan Vladimirovich Michurin (1855–1935) was a Russian practitioner of selection for fruit trees. Michurin made a major contribution in the development of genetics, especially in the field of pomology. In his cytogenetic laboratory, he researched cell structure and

experimented with artificial polyploidy. Michurin studied the aspects of heredity in connection with the natural course of ontogenesis and external influence, creating a whole new concept of predominance. He proved that predominance depends on heredity, ontogenesis, and phylogenesis of the initial cell structure and also on individual features of hybrids and conditions of cultivation. In his works, Michurin assumed a possibility of changing genotype under external influence.

Michurin was one of the founding fathers of scientific agricultural selection. He worked on hybridization of plants of similar and different origins, cultivating methods in connection with the natural course of ontogenesis, directing the process of predominance, evaluation and selection of seedlines, and acceleration of the process of selection with the help of physical and chemical factors.

Vernalization is the practice of exposing plants (or seeds) to low temperatures to stimulate flowering or to enhance seed production. By satisfying the cold requirement of many temperate-zone plants, flowering can be induced to occur earlier than normal. This process has been used to eliminate the normal two-year growth cycle required of winter wheat.

19. JBS Haldane Archives, University College, London, and Julian Huxley's book *Heredity East and West* (New York: Henry Schuman, 1949), 223, also see J. B. S. Haldane's article, "In Defence of Fenetics," *Modern Quarterly* (NS) 4 (1949): 194–204; 194. This paper is hard to obtain but a summary is presented in Huxley's book *Heredity East and West*, 223.

20. J. S. Huxley, *Heredity East and West: Lysenko and World Science* (New York: Schuman, 1949). Julian Huxley and Eric Ashby visited the Soviet Union in 1946 and attended a lecture by Lysenko, but Huxley wrote it was impossible to have a meaningful dialogue because the language and terminology used by Lysenko and his supporters was very different from what Western scientists normally use in scientific discussions.

21. Nikolai Ivanovich Vavilov (1887–1943) was a prominent Russian geneticist, best known for identifying the centers of origin of cultivated plants. He studied under William Bateson at Cambridge University and was a friend of Haldane. He devoted his life to the study and improvement of wheat, corn, and other cereal crops that sustain the global population. Initially, Vavilov occupied important positions in Soviet agriculture but lost his positions when he was persecuted by Lysenko, and eventually died in a prison camp in Siberia.

22. *Truth Will Out* is an autobiography written by Haldane's first wife, Charlotte. At the invitation of N. I. Vavilov, the Haldanes visited Russia in 1928. They received a cordial welcome, but JBS and Charlotte differed in their impressions of Russia. While Haldane admired the Soviet system, especially its preferential treatment of scientists, Charlotte did not like the restriction of civil liberties and was concerned that they were being watched all the time. She was only too glad to leave. (See C. Haldane, *Truth Will Out* [London: Vanguard Press, 1950], 339.)

23. G. D. Karpetchenko (1899–1941) was a Russian biologist and a colleague of N. I. Vavilov. He worked at the Institute of Applied Botany near Leningrad, but collaborated with geneticists in other countries. He also traveled abroad to the John Innes Horticultural Institution (JIHI) in London. He was arrested by the NKVD under the false grounds of belonging to an alleged "anti-Soviet group" led by Vavilov. He was sentenced to death and executed on July 28, 1941. He worked on cytology and created several hybrids. Among his contributions is his seminal work on allopolyploids, culminating in his creation of a fertile offspring of radishes and cabbages, the first instance of a new species obtained through polyploid speciation during experimental cross-breeding.

24. C. Haldane, *Truth Will Out*, 124.

25. Charles Percy Snow (1905–1980) was a British novelist, scientist, and government administrator. Snow earned a doctorate in physics at the University of Cambridge, where, at the age of twenty-five, he became a fellow of Christ's College. After working at Cambridge in molecular physics for some twenty years, he became a university administrator, and, with the outbreak of World War II, he became a scientific adviser to the British government. In 1950 he married the British novelist Pamela Hansford Johnson. In the 1930s Snow began the eleven-volume novel sequence collectively called "Strangers and Brothers" (published 1940–1970), about the academic, public, and private life of an Englishman named Lewis Eliot and the

corrupting influence of power. As both a literary man and a scientist, Snow was particularly well equipped to write a book about science and literature; *The Two Cultures and the Scientific Revolution* (1959) and its sequel, *Second Look* (1964), constitute Snow's most widely known—and widely attacked—position. He argued that practitioners of either of the two disciplines know little, if anything, about the other and that communication is difficult, if not impossible, between them. Snow thus called attention to a breach in two of the major branches of Western culture, a breach long noted but rarely enunciated by a figure respected in both fields. Snow acknowledged the emergence of a third "culture" as well, the social sciences and arts concerned with "how human beings are living or have lived."

26. See C. P. Snow, *The Rede Lecture* (Cambridge: Cambridge University Press, 1959).
27. Gary Werskey, *The Visible College: The Collective Biography of British Scientific Socialists of the 1930s* (New York: Holt, Reinhart, and Winston, 1979), 376. Historical and political evolution of leading figures of the British scientific socialist movement in the 1930s. The author delves into the underlying causes that gave birth to this movement and evaluates its impact on Great Britain and the rest of the world.
28. J. S. Huxley, *Memories* (New York: Allen & Unwin, 1970), and *Memories II* (New York: Allen & Unwin, 1973).
29. J. B. S. Haldane, *The Inequality of Man and Other Essays* (London: Chatto & Windus, 1932).
30. J. B. S. Haldane, *Science Advances* (London: Allen & Unwin, 1947).
31. Haldane, *Science Advances*.
32. Jawaharlal Nehru (1889–1964), friend of Haldane, freedom fighter and national leader of India, and first prime minister of India 1947–1964, followed a neutral international policy during the Cold War years, which attracted Haldane to India.
33. Prasanta Chandra Mahalanobis (1893–1972), director of the Indian Statistical Institute (ISI), Calcutta (now Kolkatta), offered appointments to both J. B. S. Haldane and Helen Spurway, which facilitated their move to India.
34. J. B. S. Haldane, "Lysenko and Genetics," *Science and Society* 4 (1940): 433–37.
35. Mikulas Teich, "Haldane and Lysenko Revisited," *Journal of History of Biology* 40 (2007): 557–63.
36. Frederick Gowland Hopkins (1861–1947), biochemist, Nobel laureate in Physiology or Medicine 1929, and professor of biochemistry at Cambridge University, appointed Haldane to a readership in biochemistry (1923–1933). From 1930 to 1935, Hopkins served as president of the Royal Society of London and in 1933 served as president of the British Association for the Advancement of Science.
37. Haldane, "In Defence of Genetics."
38. J. S. Huxley, *Heredity East and West* (New York: Henry Schuman, 1949), 223–24.
39. Haldane, "In Defence of Genetics."
40. Haldane, "In Defence of Genetics."
41. Huxley, *Heredity East and West*, 224–25.
42. Haldane, "In Defence of Genetics."
43. Huxley, *Heredity East and West*, 227.
44. Huxley, *Heredity East and West*, 225–26.
45. G. H. Hardy, *A Mathematician's Apology* (Cambridge: Cambridge University Press, 1940), 65–66.

12

Helen

Helen Spurway (1915–1977)[1] is aptly described in a chapter on Haldane by Professor Ioan James in his book *Remarkable Biologists*: "A woman of irrepressible energy, she was able to expound her views at great length, at top pitch and with a ferocity that was not easily quenched."[2]

Helen first came to Haldane's attention in 1933 as an eighteen-year-old undergraduate at University College London, where he was a forty-one-year-old professor of genetics and later of biometry. Already famous for his popular articles and public pronouncements in the press, Haldane, who had just been elected to the Royal Society, was a shining star (if not the star) at University College London. Spurway, the young upstart, was known to proclaim loudly to one and all that she had two goals: get her PhD in genetics from University College and marry Professor Haldane! She achieved both.

Their marriage plans were interrupted by World War II, but they got married as soon as the war was over, in 1945. Spurway played many roles: intellectual companion, wife, trusted devil's advocate. She critiqued and questioned Haldane's ideas mercilessly and was the only individual who could do so without raising his ire. During Haldane's lectures she frequently sat at the back of the lecture hall. It was not unusual to hear Spurway's strident voice interrupting him and questioning his statements. To the great amusement of everyone, what often followed was a private dialogue between them for several minutes with no apparent recognition of the rest of the audience.

They enjoyed a close, loving relationship for thirty years until Haldane's death in 1964. JBS and Spurway were widely known as a scientist-couple similar to their friends the French biochemists Rene and Sabine Wurmser[3] and the geneticists Boris and Harriet Ephrussi.[4] The Haldanes were often seen at international conferences during postwar years, making friends with other scientists and their families. Their intimate working relationship could hardly escape the attention of the most casual observer. Haldane was not an easy man to live with. He lived by certain high ethical and moral principles and expected the same of others. His life revolved around science, and he had few other interests. Spurway met these challenges quite successfully for almost thirty years, which speaks highly of her own intellectual standards and abilities.

Spurway also acted as a mediator between Haldane and the rest of the world. She was his eyes and ears. Those who found Haldane unapproachable had at least another recourse—they asked for Spurway's help. Of course, as we shall see later, she had her own problems as well.

Physiological Experiments

Spurway was a courageous woman who participated in Haldane's physiological experiments[5] even though they were often painful and ended in convulsions. However, her math performance slightly improved under experimental conditions while breathing toxic gases, whereas other participants—all men, though intellectually distinguished, failed even the most elementary tests!

Spurway was a zoologist whose initial interests and PhD thesis were in the genetics of *Drosophila* (fruit flies). She reported the discovery of an interesting mutant known as *grandchildless*,[6] which means the flies carrying that gene produced sterile offspring, hence no grandchildren. She conducted an impressive series of breeding experiments with European newts, and in collaboration with Professor H. G. Callan at St. Andrews University in Scotland, studied the chromosomes of the parents and their hybrids. She was one of the pioneers who studied the genetics of interspecific hybrids, throwing light on the role of genetic mechanisms in the speciation and evolution of European newts.[7] Recently digitized Wellcome Trust Archives in London contain the draft of a letter that Haldane wrote to Professor Callan in St. Andrews. Haldane asked Callan to appoint Spurway to a teaching position in his department if he should predecease her. He was forthright about her faults, "notably a refusal to publish," but praises her "capacity for working a 12 hour day." However, if they should both perish, then Callan was urged to "hurry up and grab the newts," a valuable research material.

Animal Behavior

Spurway was especially interested in animal behavior, partly because of her close association with Haldane and her friendship with the founder of ethology, Konrad Lorenz, in the postwar years. By the late 1940s, Lorenz, who shared a Nobel Prize later, was already famous for his brilliant research in animal behavior, especially in birds and mammals. His two early books, *King Solomon's Ring* (1949) and *Man Meets Dog* (1950), quickly established his reputation as a founder of modern ethology, which had been initiated by Julian Huxley through his classic study of courtship behavior in the great crested grebe long before. Spurway fell in love with Lorenz when he was visiting London in 1949.[8]

In an interesting paper on the dog-cat relationship,[9] Spurway explained that their relationship represents a commensalism by which both species satisfy drives otherwise starved of sign stimuli. She commented that cats seem to provide the stimuli releasing the chasing instincts of dogs, and dogs provide the stimuli releasing the flight instinct of cats. Haldane and Spurway collaborated in several projects in animal behavior. Spurway's zoological interests may have influenced Haldane to consider these topics in his research.[10] Some of these investigations were concerned with the physiology of breathing in such animals as newts and lung-breathing fishes. In other papers, problems of instinct and imprinting in animals received special attention.

Haldane and Spurway made an extensive analysis of Karl von Frisch's observations of communication in the European honeybee *Apis mellifera* and summed up their findings:

> The dance conveys about 5 cybernetic units of information concerning direction, of which the average recipient receives at least 2.5 . . . between 100 and 3000 metres the number of turns made in a given time fall off linearly with the logarithm of the distance. At greater distances they fall off more slowly. The number of abdominal waggles made per straight run increases by 1 per 75 metres between 100 and 700 metres. It is suggested that this is the principal means by which distance is communicated.[11]

Other studies involved the possible inheritance of birdsong, food gathering in birds, nest-building activity of wasps, and the rhythm of breathing in newts and fish.

Haldane often referred to Spurway's ideas and publications in his own lectures and writings with great care. It has long been tradition of Hinduism to regard animals as one's own kin, but in the Christian world it was Charles Darwin who placed it at the center of his theory of evolution. It was typical of Helen Spurway that she simplified Darwinism in the Indian context by saying that "Darwin converted Europe to Hinduism."[12]

Another aspect of Hinduism that influenced Spurway's ideas was what she called the concept of Vaishnava and Saiva biology. The Vaishnava sect of Hinduism worships Vishnu, the Hindu God who is largely concerned with the preservation and maintenance of life, whereas Saivaites worship the God Shiva (or Siva), who maintains the cyclic process of destruction of cosmic order and its regeneration. When we speak of natural selection we are being Saivaite in our mode of thinking. And when we refer to adaptations, and so forth we are referring to Vaishnavite mode of thinking. The two coexist and their delicate balance maintains all we know about the universe.[13]

Virgin Births?

Shortly before the Haldanes moved to India in 1957, Spurway became the subject of much unwanted publicity and notoriety. It started quite innocuously as an extension of one of her research projects. In guppies,[14] Spurway has been studying genetics and parthenogenesis (from Greek *parthenos* = virgin + *genesis* = birth). As a possible extension of her research, in 1956, Spurway proposed in an article that virgin births, although extremely rare, might occur in human populations. The scientific basis of her claim was her research with guppies, where parthenogenesis had been known to occur. It is the process by which a female can give birth to offspring without fertilization, that is, without involving a male partner. All the offspring would be female. Parthenogenesis is known to occur in many species of animals but has never been proved to occur in the humans.

Spurway further proposed that a woman who claimed a pregnancy by parthenogenesis could be tested by a skin grafting operation if the child were born alive. Normally, no skin graft from one human being to another (except between identical twins) "takes" permanently. A normal child's cells are different from the mother's, because they have some of the father's antigens. A successful graft from child to mother would show that the child had received no antigens from any other source.

Spurway's claim received sensational press coverage in several countries. The medical journal *Lancet* commented, "If it does occur at all, it is extremely rare, . . . A rare event which is hard to prove is likely never to be reported at all if it is also . . . 'known' to be impossible. . . . Possibly some of the unmarried mothers whose obstinacy is condemned in old books . . . may have been telling the truth."[15]

Time magazine covered the report under the title "Medicine: Parthenogenesis":

> Britain's interest in a topic long pigeonholed by science was spurred by a report that Eugenist Helen Spurway gave at University College in London. Among humans, she declared, virgin birth could not happen in the case of a hermaphrodite, who would not be self-fertile. However, parthenogenesis might occur. This is the process by which an ovum begins to divide spontaneously, without having been fertilized by a sperm—perhaps after it has made up for the missing male chromosomes by a form of doubling. It is almost certain that the offspring of parthenogenesis would be a female, since the ovum contains only female chromosomes.[16]

The *South African Medical Journal* discussed the subject in an editorial on August 25, 1956, "The Virgin Births Newspaper Feature": "Dr. Helen Spurway, lecturer in Biometry and Eugenics in London, set the cat among the pigeons by some remarks about possibilities in mammals, and even man, in a lecture she recently gave entitled 'Virgin Births' concerning her observations on the guppy (*Lebistes reticulates*), a small live-bearing fish." The progeny of such procreation can never show genetic features which are not found in the mother."

The British tabloid *Sunday Pictorial* went one step further. It initiated an inquiry, searching for mothers who believed that they had produced a parthenogenetic infant. After screening several false claims, one mother–daughter pair was found to fit the bill. However, a skin graft from daughter to mother was shed in approximately four weeks, so the mother's claim could not be upheld.

In later years in India, Spurway abandoned her research on parthenogenesis but continued her work on the genetics of guppies. She also chose several other organisms for her research. One particular study involved the tussore silk moth *Antheraea mylitta*.[17]

Spurway's Social Role

As the wife and companion of a famous scientist, Spurway attended numerous State banquets and receptions over the years. This experience brought her into contact with many eminent and successful leaders in science, politics, business, and the arts all over the world. She carried out this role with great intelligence and dignity, though she often considered them a waste of her time just as Haldane did. Her personal lifestyle and tastes paralleled Haldane's. Their household was devoid of the usual decorations and feminine touches one normally expects in a home. Instead, one was apt to find tons of books and papers everywhere in the living room, dining room, bedrooms, and even in the bathrooms. Spurway wore little makeup or jewelry and dressed in a spartan manner. Her conversion to Hinduism was so complete that she regarded herself as a Hindu widow after Haldane's death. Accordingly, she wore the traditional widow's white and gray clothing. Her decision to continue living as Haldane's widow was, I am sure, mainly due to the intense feeling and closeness she felt with his spirit.

Spurway conscientiously attempted to maintain a household suitable for Haldane's work and instructed the domestic staff accordingly. Not touching his books and papers in the living rooms no matter how much dust they gathered was one of her frequent admonitions to their servants. Haldane and Spurway adopted a lifestyle that, for all its simplicity, suited their work and intellectual interests very well. They were, in their own special way, highly compatible. Indeed, it is hard to imagine how else they could have lived had they not found each other.

Spurway's Problems

During her lifetime, Spurway had occasional disagreements with colleagues and strangers, many of them involving what others may consider to be trivial matters. What made it more tragic was that Haldane always felt duty-bound to defend her no matter how silly the situation was. That was indeed the situation when she blamed their colleague at University College London, Hans Kalmus, in a trivial matter and Haldane compounded the problem by taking her side. He too blamed Kalmus beyond all reason, although Kalmus proved to be a loyal colleague of several years who took part voluntarily in painful physiological experiments. Similar situations arose with respect to other individuals, an anthropologist at University College London, a student at the Indian Statistical Institute in Calcutta, and others.

Many considered it unfortunate that although they admired Haldane greatly, they avoided the Haldanes' company because of Spurway. When conflicts arose between Haldane and the director of the Indian Statistical Institute, Prasanta Mahalanobis, Spurway exacerbated that conflict by urging Haldane to resign. Other colleagues thought that the differences with the director were not so great as to warrant a resignation. It has been said that Haldane suffered deep psychological scars when he was severely harassed in his Eton years, which made him intolerant of authority, any authority no matter how slight it was. And Spurway was not helpful in those situations.

I had known Spurway when she collaborated with me in research on the color preference behavior of butterflies and the nest building activities of wasps, and I enjoyed her critical commentary and brilliant suggestions. As a young student of Haldane, when I was writing my first scientific papers, Spurway was most generous in her advice and support. Although her own writings were far from lucid, she had the unusual ability of being able to instruct younger scientists, probe their young minds, and stimulate research.

I also recall Spurway walking around the suburbs of Calcutta looking for caterpillar food or a wasp's nest and reporting back to Haldane on any interesting observations she may have made on local animal and plant species. There is also the endearing picture of Spurway acting as Haldane's constant companion and a generous hostess to all of us. All our research group, and any visitors who dropped by, were invited to join the Haldanes for lunch or dinner several times a week, which gave us an excellent opportunity to converse with both Haldane and Spurway informally. It was a pleasure to listen to Haldane's views on a great number of scientific topics and occasionally his recitations of classical poetry in Latin and Greek. Our visitors included some of the famous scientists of the day, such as Julian Huxley, T. Dobzhansky, Ernst Mayr, Harlow Shapley, Harold Jeffreys, George Gamow, Joshua Lederberg, Antoine Lacassagne, and Jacques Monod.

Dress and Looks

Spurway attracted immediate attention wherever she went. During the years I knew her, she wore masculine clothes and a very short haircut. She wore no makeup and walked in sandals. But her most unique characteristic was her voice. It defied description—high pitched, but neither masculine nor feminine. As one of my colleagues remarked, it was unforgettable! She made a distinct mark at once on strangers. In contrast to Haldane, who was most careful and tactful in choosing his words, Spurway was outspoken, often laced her conversations with profanity and rudeness, partly to express frustration in certain situations, or to show her contempt for some situation or someone. More interestingly, she mixed four letter words in her classroom lectures; one typical context, referring to *Drosophila* flies, which were used for teaching genetics, she said, "the fruitflies have started fucking!" Needless to say, she made few friends, in contrast to Haldane who knew how to make friends when he wanted to. It was not unusual for Haldane to make amends when she offended some one.

Police Dog Incident

One evening Spurway and a visiting American research assistant, William Clarke, were walking back to the lab at University College after dining at a nearby pub, the Marlborough Arms, that was often frequented by the Haldanes. On the way they had to pass two policemen and their police dog.[18] The dog's tail was stretched across the path, a tempting target for Spurway who had been in a rebellious mood now for several days. According to her companion, William Clarke, she stepped on the tail gently, and moved on. The dog was not harmed in any way, and in fact stood up and wagged the tail afterward. However, there were other versions of what exactly happened. One former colleague of Spurway informed me (present author) that she kicked the dog gently in passing. Others remarked that she was quite inebriated. Whichever version was nearer truth, it is also clear that the policeman acted in haste, mistaking them for two students on the lark, and arresting them without verifying their credentials. That officer was not aware, and did not bother to find out, that she was Mrs. Haldane, wife of one of the most famous scientists at University College London, and herself a member of the faculty at the same institution. Clearly a "himalayan miscalculation"[19] was committed by the policeman with some help from Spurway herself! The policemen apparently questioned Spurway, and she was only too happy to express her opinion of policemen in general in no uncertain terms. Spurway and Clarke were both arrested instantly, and Spurway was charged with assaulting a police officer and public intoxication as well as disorderly conduct. Her companion was charged with obstructing the police.

The entire unfortunate episode was greatly embarrassing to the government and to the administration of University College London. There was certainly no intention to arrest a university lecturer and a visiting American scientist. The authorities would be only too happy to forget the whole incident in exchange for a small fine. That was, of course, far from what Spurway had in mind. Considering the political climate at that time coupled with Haldane's public announcements that he was considering moving to India, that particular incident seemed to have been specially arranged by providence and appropriately tailored to suit their particular circumstances. Far from settling the case quietly, the Haldanes were in complete agreement to publicize what they regarded as the criminal nature of the government and the shoddy work of the police. It provided JBS with an excellent opportunity to denounce the government in no uncertain terms, in an atmosphere that was far more favorable to his cause than anything he could have hoped for. It reminds one of the famous pronouncement of T. H. Huxley, who reacted by saying, "God has delivered him into my hands," upon hearing Bishop Wilberforce's ignorant attack on Darwinism, during their famous debate at Oxford on June 30, 1860.

Over the last fifty years since I first met the Haldanes, I have heard many amusing stories about them, and here is one more about Spurway. This is a favorite Haldane/Spurway story of Sir Arthur C. Clarke. Shortly after her arrest in relation to the police dog incident, Spurway was taken out to a lunch by a *Daily Express* reporter, possibly hoping for an interview. At one point, she hastened to tell the reporter, "Look here, I must tell you something, I only go to bed with Fellows of the Royal Society!"

Notes

1. Helen Spurway was Haldane's second wife. Their plans for a married life were interrupted by World War II, but they got married in 1945 as soon as the war was over. They were a devoted couple, close colleagues and teachers who worked together and were always seen together. The name "Spur-way" means "bridle path." Helen told me, in 1960, that it is not a common name; however, today you can find several Spurways on the Internet. When I first met Helen, her mother was living in a retirement home in Bournemouth.

    In our informal gatherings with colleagues and students, Helen considered it her duty to bring up topics for discussion that we would normally hesitate to mention, such as obtaining funds to attend scientific meetings or to purchase reprints of papers, or how to find additional

office space in the building, and so forth. She saw herself as a mediator between Haldane and the rest of the world. She was, in fact, helpful most of the time.
2. Ioan James, *Remarkable Biologists* (Cambridge: Cambridge University Press, 2009), 140–45.
3. Rene Wurmser, (1890–1993) and Sabine Filitte-Wurmser were biochemists from the University of Paris and close friends of Haldane. Rene Wurmser wrote in my edited volume, *Haldane and Modern Biology* (1968): "I first met J.B.S. Haldane at the Congress of General Physiology in Stockholm in 1926. Haldane presented some experiments on tetany by overbreathing, making the demonstration on his own person. The scene was a bit painful and no doubt surprising for those unaware of Haldane's practice of running personal risks, a habit which his father had inculcated to him during his childhood." Rene and Sabine visited us in 1963 in Bhubaneswar, where the Haldanes were living then. R. Wurmser, "Haldane as I Knew Him," in *Haldane and Modern Biology*, ed. K. R. Dronamraju (Baltimore: Johns Hopkins University Press, 1968), 313–17.
4. Boris Ephrussi (1901–1979) and Harriet Ephrussi-Taylor (1918–1968) worked in genetics and developmental biology at the University of Paris, at Gif-sur-Yvette. I met them first in 1961 with the Haldanes at the Marine Biological Institute at Roscoff in Brittany. I recall that Harriet congratulated Haldane on receiving the Feltrinelli Prize from the Academy of Sciences in Rome, except that she noted that the name "Haldane" was changed to "Haldano" in the Italian press. Later that summer we visited their laboratory at Gif-sur-Yvette. We had a most pleasant time. Boris showed us slides of hybrid cell fusion, which had just been discovered by his colleague Barski and later played an important part in the development of somatic cell genetics. Harriet gave us a private seminar on bacterial genetics. And they took us for a most sumptuous lunch at a lovely restaurant in the country. After Haldane's death, I ran into Boris again in San Francisco in 1967, when we were both attending the American Tissue Culture Association meeting. Sadly, Harriet was only fifty years old when she died of breast cancer in 1968.
5. Haldane followed in his father's footsteps in using his own body for testing various concepts and ideas in physiology. During World War II, he conducted several experiments in diving and inhalation of gases such as carbon dioxide, nitrogen, helium, and hydrogen. These were often painful, ending in convulsions. Helen Spurway was one of the volunteers who participated in these experiments with much dedication and patience.
6. Helen Spurway discovered the mutant *grandchildless* in the fruit fly. H. Spurway, "Genetics and Cytology of *Drosophila subobscura*: IV. An Extreme Example of Delay in Gene Action Causing Sterility," *Journal of Genetics* 49 (1948): 126–40.
7. H. Spurway, "Genetics of specific and subspecific differences in European newts," *Symposia of the Society for Experimental Biology* 7 (1953): 200–237.
8. H. Spurway, personal communication to the author, 1959.
9. H. Spurway, "The Escape Drive in Domestic Cats and the Dog and Cat Relationship," *Behaviour* 2 (1953): 81–84.
10. H. Spurway, personal communication to the author, 1959.
11. J. B. S. Haldane and H. Spurway, "A Statistical Analysis of Communication in *Apis mellifera*, and a Comparison with Communication in Other Animals," *Insectes Sociaux* 1 (1954): 247–83.
12. "Darwin converted Europe to Hinduism"—Helen Spurway commented that in accepting the essential aspects of Darwinism, Europeans are also accepting some essential tenets of Hinduism, because certain principles such as kinship with other species, which are a central part of Darwinism, are closer to Hindu beliefs than Christianity. Haldane pointed out that Christian theologians drew a sharp distinction between humans and other species.
13. H. Spurway, "Vaisnava and Saiva Biology," *Zool. Jahrb. Syst.* 88 (1960): 107–16.
14. H. Spurway, "Hermaphroditism with Self-Fertilization, and the Monthly Extrusion of Unfertilized Eggs, in the Viviparous Fish *Lebistes reticulatus*," *Nature* 180 (1957): 1248–51.
15. J. W. Nicholas and H. Spurway, "Parthenogenesis in Human Beings," *The Lancet* 268, no. 6932 (July 1956): 47–48.
16. "Medicine: The Problem of Pain," *Time*, November 2, 1955.

17. H. Spurway and K. R. Dronamraju, "The Biology of the Two Commercial Qualities of Cocoons Spun by *Antheraea mylitta* (Drury) with a Note on the Cocoons of the Related *A. assama* (Westwood)," *Genetica Agraria* 45 (1959): 175.
18. Spurway herself narrated the incident with the police dog and her arrest to me. She said her brief stay at Holloway prison was not uncomfortable and she was treated courteously.
19. "Himalayan miscalculation" was first used by Mahatma Gandhi in his "Autobiography" to refer to a major error committed by him in his peaceful struggle for India's independence. M. K. Gandhi, *An Autobiography: The Story of My Experiments with Truth* (Boston: Beacon Press Paperback, 1957), 469.

13

Popularizing Science

Scientific popularization is largely a twentieth-century phenomenon. However, two notable science fiction works of the nineteenth century, Jules Verne's *Twenty Thousand Leagues under the Sea* and Mary Shelley's *Frankenstein*, stimulated popular interest and imagination in science. One must also consider Charles Darwin's books *Voyage of the Beagle, The Descent of Man, Insectivorous Plants,* and *The Origin of Species* and others, as well as Alfred Russell Wallace's books on natural history and travels as popular and semipopular works related to science.

Early in the twentieth century, popular interest in science was further stimulated by the fictional works of H. G. Wells, such as *The Time Machine, War of the Worlds, The Invisible Man,* and *The Island of Dr. Moreau*. However, the popularization of Wells was mainly concerned with fiction based on science, not scientific popularization.

Science popularization in the early twentieth century was pioneered by James Jeans and Arthur Eddington, whose writings were mainly in cosmology, astronomy, and astrophysics. Science popularization of a much broader genre started in the 1920s with Julian Huxley and J. B. S. Haldane, whose professional background was in the biological sciences. Lancelot Hogben was another. Of these, Haldane covered a far greater range of sciences than any other writer. He tackled numerous topics in biology, geology, ecology, animal behavior, astronomy, cosmology, physics, chemistry, paleontology, space exploration, physiology, and mathematics. He also wrote on such nonscientific topics as theology, military affairs, politics, and government. No matter what he wrote, there was always a scientific connection in Haldane's writings. Haldane was prolific in his writings, especially the art form of the short popular essay, which may very well have started with him.

Starting, in 1923, with his first popular book, *Daedalus or Science and the Future*,[1] Haldane paid special attention to the ethical issues arising from the social applications of science. Indeed, the physicist-author Freeman Dyson credited Haldane and Einstein as the pioneers in the early twentieth century who emphasized that the progress of science is destined to bring enormous confusion and misery to mankind unless it is accompanied by progress in ethics. In his essay, "*Daedalus* after Seventy Years," Dyson wrote, "The ethical standards of scientists must change as the scope of the good and evil caused by science has changed. In the long run, as Haldane and Einstein said, ethical progress is the only cure for the damage done by scientific progress."[2]

Haldane's essays are characterized by an enormous depth and breadth of knowledge that is rarely seen in popular writing. Another well-known writer of science (and science fiction), Sir Arthur C. Clarke, once wrote, "J.B.S. Haldane was perhaps the most brilliant science popularizer of his generation. Starting in 1923 with *Daedalus; or Science and the Future*, he must have delighted and instructed millions of readers. Unlike his equally famous contemporaries, Jeans and Eddington, he covered a vast range of subjects.... He also wrote a workmanlike Novella, 'The Goldmakers', and a charming tale for children, *My Friend Mr. Leakey*."[3]

Haldane's popular essays were mostly written in his "spare" time. He used to carry a writing pad and pen in his briefcase at all times. Whenever he had a few minutes' leisure he started writing an article on some scientific topic. Occasionally, it was a scientific paper, but more often it was a popular article that he would send to a newspaper. Haldane's popular essays originally appeared in such publications as the *St. Louis Post-Dispatch*, the *Saturday Evening Post*, the *Atlantic Monthly*, and the *Manchester Guardian*, among others.

Haldane's success as a popular writer owed much to his first wife, Charlotte Franken. Charlotte was working as a journalist for the *Daily Express* in London when she came across an abridged version of Haldane's sensational book *Daedalus, or Science and the Future* in the *Century Magazine* of New York. As she put it, "I found an article that set my imagination aflame." She learned that the author was someone called J. B. S. Haldane, a biologist who specialized in making experiments on himself with some substance called "acid sodium phosphate." She wrote, "His imagination seemed to equal his physical courage." She was captivated by the "humorous audacity" of the author who was making startling predictions about the biological future of the human race, including "a fantastic but matter-of-fact account of the growing of a human foetus in the laboratory."[4] After meeting Haldane, she soon discovered that he was endowed with not only a phenomenal intellect, but also a colossal memory and admirable gifts of verbal fluency. The short version of *Daedalus* had shown that his written style also was brilliant. She learned that, unlike most of his scientific colleagues, Haldane had an unusual enthusiasm for the popularizing of science, and "a firm conviction that this was both desirable and necessary." She found out that he greatly admired two great pioneers of popular scientific journalism, Thomas Henry Huxley and Ray Lankester. Haldane was more than willing to accept Charlotte's suggestion that he should practice it as a hobby in his spare time and get paid for it. She became his secretary and agent. As she had anticipated, there was a large market, especially in the United States.

Among many outstanding articles by Haldane that appeared in the popular press, I summarize a few in what follows that illustrate the quality and content as well as the writing style of his essays.

"On Being the Right Size"

One of the most brilliant essays on popular science written by Haldane was included in his early book of collected essays: *Possible Worlds and Other Essays*. It is titled "On Being the Right Size"[5] and is noted for its originality, lucid presentation, and an interdisciplinary approach to a complex problem. He presented a masterly summary of the physiological and morphological conditions that determine the size of an animal species. He lamented the lack of interest among zoologists in the relative sizes of various animals and the underlying mechanisms. He wrote, "In a large text-book of zoology before me I find no indication that the eagle is larger than the sparrow, or the hippopotamus bigger than the hare ... yet it is easy to show a hare could not be as large as a hippopotamus, or a whale as small as a herring. For every type of animal there is a most convenient size, and a large change in size invariably carries with it a change of form" (20). Several years before Haldane, D'Arcy Thompson, in his On Growth and Form,[6] wrote lucidly about the importance of size; however, Haldane's approach was more comprehensive, involving multiple disciplines, and has had a significant impact on other biologists and writers.

This essay represents the finest popular exposition of complex scientific interactions on several different levels. Body size is related to several factors, such as the animal's locomotion, habitat, and body weight, as well as such physiological functions as respiration and digestion. With reference to man's size, Haldane pointed out that any increase in size should be matched by a proportionate increase of the bone size to carry the extra weight, otherwise the bones would be crushed. Each species has special requirements and adaptations that need to be changed if the size is to increase.

The relative proportions of the limbs and body sizes in different animals such as the gazelle, hippopotamus, and a giraffe are mechanically sound. Gravity, which is a major nuisance for large animals, poses no problem to smaller animals. Thus, a mouse can survive the fall from a great height onto soft ground whereas larger animals would die instantly. Gravity poses no danger to insects. However, water and wetness can be serious hazards to them because surface tension of the water grips the insects until they drown. Their long proboscises enable them to drink water from a distance.

Tall animals require higher blood pressure and tougher blood vessels. Haldane pointed out that a small animal, such as a rotifer, if enlarged a thousandfold would require a thousand times as much food and oxygen per day and would excrete a thousand times as much waste products (23). However, the surface area of the skin (through which oxygen enters) may be increased by some special device such as folding, thus increasing it in proportion to the animal's bulk. Haldane wrote: "Comparative anatomy is largely the story of the struggle to increase surface in proportion to volume" (23).

With respect to flying, Haldane wrote, "An angel whose muscles developed no more power weight for weight than those of an eagle or a pigeon would require a breast projecting for about four feet to house the muscles engaged in working its wings, while to economize in weight, its legs would have to be reduced to mere stilts" (25).

Other observations of Haldane include the need for an animal to keep warm. The loss of heat from the body surface of an animal must be balanced by a food supply proportional to its surface area and not to its weight. One of the advantages of size is that it enables one to keep warm. All warm-blooded animals at rest lose the same amount of heat from a unit area of skin. Haldane wrote,

> Five thousand mice weigh as much as a man. Their combined surface and food or oxygen consumption are about seventeen times a man's. In fact a mouse eats about one-quarter its own weight of food every day, which is mainly used in keeping it warm. For the same reason small animals cannot

live in cold countries. In the arctic regions there are no reptiles or amphibians, and no small mammals.... The small birds fly away in the winter, while the insects die, though their eggs can survive six months or more of frost. The most successful mammals are bears, seals, and walruses. (25–6).

Haldane's central idea was to show that there is an optimum size for every organ and every organism. No single organ can be enlarged in one direction without many changes in the other organs and systems, which would be so complex that the species would be endangered. If a flea were as large as a man, it could not jump a 1,000 feet in the air because, in addition to the resistance of the air, the jump would require an expenditure of energy proportional to the jumper's weight. However, if the jumping muscles form a constant fraction of the animal's body, the energy developed per ounce of muscle is independent of the size (27).

In his later years, Haldane encouraged a research project by one of his associates in India, T. A. Davis,[7] to measure the pressure in the root cell sap of tall palm trees in India. Root pressure is osmotic pressure within the cells of a root system that causes sap to rise through a plant stem to the leaves. Root pressure occurs in the xylem of some vascular plants when the soil moisture level is high either at night or when transpiration is low during the day. When transpiration is high, xylem sap is usually under tension, rather than under pressure, due to transpirational pull. At night in some plants, root pressure causes exudation of drops of xylem sap from the tips or edges of leaves. Davis measured root pressure by attaching a pressure gauge to the cut root. A plant needs both root pressure and transpirational pull.

"How to Write a Popular Scientific Article"

Haldane once wrote an article in the *Daily Worker* advising his younger readers on how to write a popular scientific article.[8] He warns at the outset that the task is not easy, as one has to set fairly high standards. He suggests that the article should be aimed to interest, or even excite the readers but not give them complete information: "You must therefore know a very great deal more about your subject than you put on paper. Out of this you must choose the items which will make a coherent story.... It may take you twelve hours' reading to produce an intellectually honest article of a thousand words. In fact you will have to educate yourself and your public" (4).

When the article is completed, he recommends giving it to a fairly ignorant friend or put it away, to be reexamined after six months. He suggests other improvements, such as using a period instead of a comma or a semicolon as often as possible. Other comments of Haldane may seem fairly obvious but are seldom observed by the novice writer. For instance, a slow series of steps leading to a conclusion are preferable to sudden jumps in ideas and conclusions, and the order of the phrases in a sentence should correspond with the temporal or causal order of the facts.

For whom are you writing? Your subject matter should be matched to your audience. For instance, an article on the history of eighteenth-century physics would not be suitable for a daily newspaper. Other important points: "you are not trying to show off; nor are you aiming at such accuracy that your readers will be able to carry out some operation" (4).

Haldane's popular science writing stemmed from his firm belief that scientific development has been generally beneficial to society and that future social progress will be closely tied to the success of science and its applications. He suggested further that a popular article on science should include news of some recent advances from technical journals as well as speculations regarding future developments. Furthermore, a scientific article can be made more appealing to the public by including one or two humorous anecdotes about the subject.

"Darwinism and Its Perversions"

Haldane began this article with the following sentence: "Most Marxists are Darwinists ... Nevertheless Darwinism has been used to defend highly anti-democratic ideas. The fact is, I think, that Darwin went badly wrong, not in his account of how evolution happened, but in his comments on the process."[9] The theory of natural selection can be restated in modern terminology. If a number of animals or plants in a population carry a gene that makes them fitter than the rest of the population, in the sense that on the average they leave more descendants behind them, that gene will tend to spread through the population. An animal that looks after its young is fitter, in the Darwinian sense, than one that does not, because more of them survive to maturity. However, modern work has confirmed Darwin's idea that natural selection is the main driving force of evolution. Haldane wrote, "Unfortunately Darwin did not stop here. He wrote of natural selection favoring the good and rejecting the bad." Darwin even ventured further: "And as natural selection acts solely by and for the good of each being, all corporeal and mental endowments will tend to progress towards perfection." Of course Darwin realized that most lines of descent in the past had ended in extinction. He thought, however, that this was mainly due to competition by "betters" or more perfect species. Haldane explained that the survival of the fittest does not necessarily make a species fitter in any intelligible sense of the word. Fossil records indicate that many species have progressively increased in size but that increase in size has often been the prelude to extinction. The larger species died out, while smaller ones lived on.

Haldane affirmed that natural selection is the main agent of evolution. It certainly prevents animals from losing useful organs and instincts. But it is a blind force, not necessarily beneficial, in the long run. Many species have become extinct as the result of natural selection, which led them down blind alleys.

"The Starling"

Haldane wrote a great number of popular essays on science and nature, especially common animals including birds. One essay on the starling was included in a book of collected essays titled *Everything Has a History* (1951).[10] The following is a summary of that essay.

Migrant birds return to England to spend the summers. Some have flown a short distance, but others, such as the swallows, have flown all the way from South Africa. Other birds that have wintered in Britain are going northward and eastward. One of these species is the starling. It is much more social than most British birds. Many of them are found in huge communal roosts with up to a hundred thousand members. They may fly for twenty miles each day to feed. The robin, on the other hand, spends much of its time in quite a small area, and quarrels violently with other robins except its mate. In starlings, the males far outnumber the females, and a female starling may have two husbands.

The starlings that spend the summer, and breed, in England, spend their lives in England, however, those that arrive in October and leave in March breed in Europe. The Scottish winter visitors go to Norway, and the English winter visitors go as far east as Sweden and Russia. Another interesting difference: The more sedentary race have lighter-colored beaks, especially in December and January, while those going farther have dark-colored beaks. The reason for the differences in behavior is related to the size of their sex organs. During the winter, the continental birds have very small ovaries and testicles, and show no sexual activity. On the other hand, in the British race, they increase in size in autumn, followed by a great deal of love-making. The continental starlings fly back to their breeding places, where their sexual organs mature in spring. The British birds not only never leave Britain but also never move permanently away from their homes. Haldane commented that it looks as if the emotions of birds toward their homes and their mates were similar. The two races appear to differ genetically, and do not breed together.

Other differences were noted between the two groups of starlings. The continental starlings do not roost in towns, whereas the British starlings do so. Some of their roosts in London are well-known tourist sites, such as the Marble Arch, St. Martin's, and Trafalgar Square. The British birds are less social than the continental ones. There is no conflict between the two races. There are slight differences in the feathers and in the eye colors of the two races. Haldane speculated that their separation did not go back to even ten thousand years, because England was cooler then, which would not be suitable for winter refuge. Furthermore, it would be at least another twenty thousand years or more before the two races of starlings evolve into a pair of distinct species. This is an interesting situation in evolution and speciation. Haldane concluded, "I wish we could resurrect Darwin, if only for five minutes, to tell him about it."

On Research and Writing

In contrast to the theoretical nature of much of his own research, Haldane often discussed in his popular essays many practical and technological applications. It would be incorrect to say that Haldane was a total stranger to experimental research. He conducted Mendelian breeding experiments in his early research and later conducted painful physiological experiments on himself, his wife, and his students. However, it would be correct to say that his most important scientific work was his mathematical theory of natural selection. I have heard of amusing stories about Haldane's clumsy attempts to sort a few *Drosophila* flies under the microscope while visiting laboratories of other scientists (for instance, Jack Schultz's lab at Caltech). These are likely to be true, as his clumsiness was also evident in his physical movements, which was perhaps the main reason for his falls and bone fractures during his later years in India.

Because of his lack of expertise or experience in active experimental research, Haldane tended to overappreciate all experimental work—perhaps to an extent that is surprising to find in a scientist of his eminence. At the same time, he continued to produce extremely clever theories and results using little or no equipment. His ideas and theories have led to much experimental work by others in diverse fields of science.

Haldane outlined his view of basic research in an article that appeared in the *New York Times Magazine*:

> We can do something to redress the balance today by encouraging research, particularly on apparently useless topics such as the biology of slime moulds, the psychology of birds, and the sociology of primitive human communities. It is the study of exceptional substances and processes—such as the alkali metals and halogens, which are not found uncombined in nature, electric conduction in metals, and radioactivity—which have led to our mastery of physics and chemistry. We do not seem to have learned the lesson, however, and still spend vast sums on research on agricultural plants, cancer, madness, and so on. Such work will not give us the fundamental information which we need, nor the attitude to life and death which makes happiness possible even today.[11]

Haldane deplored the lag in development between the physical and chemical sciences on one hand and the biological, psychological, and sociological sciences on the other. He believed that the socially and individually harmful effects of science mostly arose from the backwardness of biological and social sciences, which might even kill off humankind. In an essay written on the occasion of the New York World Fair of 1964, he wrote that happiness is a byproduct, and "if you aim at it directly you won't achieve it." Haldane's prescription for his own happiness was to indulge in research of a very basic nature using the simplest possible methods but obtaining very profound results. He was very good at this kind of approach, which earned him the title "England's most clever and eccentric biologist" by the Nobel laureate Jim Watson in his famous book *The Double Helix*.[12]

Haldane firmly believed that basic research and popular science writing should go hand-in-hand because an understanding of science and its applications holds the answers to most, if not all, social evils. For him, science was a part of everyday life. He seldom thought of it as a career in the usual sense and downplayed the importance of academic degrees. In fact, he once wrote an article in the *Hindu* newspaper (from Chennai, India) titled "The New Caste System," in which he ridiculed and deplored the excessive emphasis placed on academic degrees while ignoring real scholarship and knowledge. When he moved to India, he was faced with a situation entirely different from what he was used to England, because at the time of his arrival science instruction in India was regarded mostly as imparting knowledge. Haldane's view was just the opposite. He wrote:

> But the most important part of science, in my opinion, is not knowledge but method. Scientific method cannot be explained, but only demonstrated. Research is rather like poetical composition. There are rules for both, but you will not become a poet at all, let alone a great poet, by adhering to the rules. Nor will you make great scientific discoveries by following "scientific method" as laid down by writers on that subject. Both the great poets and the great scientists adhere to the normal canons as a general rule, but do not hesitate to violate them from time to time. Their violations may sometimes become the canons of art or science in future ages.[13]

"On Expecting the Unexpected"

Haldane was a member of the Rationalist Press Association, contributing an essay each year to its annual publication. One of these was titled "On Expecting the Unexpected," which was published in 1960, after he moved to India. It was included in the collection: *Science and Life*.[14] The essay was devoted to the rare capacity of expecting the unexpected, which plays a vital role in the advancement of science. The fundamental character of Haldane's discussion gives it a timeless quality. According to Haldane, the popular picture of scientists looking down a microscope or looking at a test tube is quite misleading, because the great advances in science came about as a result of the original thinking of a few men and women. Einstein comes to mind. The public is generally ignorant of how such important steps originate in practice. It is also true, as Haldane pointed out, that the usual form of scientific publications does not explain how an idea originated. Haldane was interested in inculcating the scientific spirit that is necessary for carrying on innovative research. He was especially concerned with the sources of original ideas, as these provide the raw material for achieving successful scientific research.

Haldane recognized two primary sources of original ideas.

1. One source of original ideas is theoretical research, and this was the case with both Einstein and Haldane. Haldane wrote, "If Newton's theory of gravitation is true, the equator should be farther from the earth's centre than are its poles.... If light consists of electromagnetic oscillations, radio communication should be possible. If infectious diseases are caused by small living beings, it should be possible to poison these beings with some of the substances already known to be poisonous to larger plants and animals." Haldane emphasized that once a theory becomes mathematical, it becomes possible to design experiments and observations that would otherwise not be possible with any degree of accuracy.
2. The second source of ideas is our own environment, which can be a rich source for those who possess the unusual capacity to observe objects and phenomena that are not noticed by most individuals. During the process of classifying a group of individuals, for instance, such exceptions may be found. Haldane pointed out that

a few individuals such as Rutherford and Darwin had had the exceptional quality of expecting the unexpected. He stated that he found two pupils in India who possessed this quality. Haldane's approach was in contrast with that of J. D. Bernal, who stated in his *World without War* that one of the objects of education should be to train people in the capacities of creative thought. Haldane responded that he "cannot train people in the capacity to make observations of this sort," but he can detect it and "bring it out" by explaining to them that their observations are interesting. He concluded that such capacities can be either discouraged or encouraged but cannot be taught under most existing educational systems. In the past, educational systems for the most part were dominated by the clergy, who were dogmatic in their approach to science in general.

Dissemination of Scientific Knowledge

Haldane wrote extensively on the problems of dissemination of scientific knowledge. In a popular article devoted to that subject, Haldane listed four stages: obtaining data, classification, analysis, and publication. Referring to science-fiction, Haldane stated that it does advance science but only very rarely. Occasionally, he used to read science fiction as a break from his mathematical labors. He enjoyed the writings of H. G. Wells, Olaf Stapledon, and Arthur C. Clarke, among others. In the 1950s, when many scientists of his stature preferred not to be associated with science fiction, Haldane did not hesitate to speculate freely on the future possibilities of space travel but also wrote science fiction himself. In 1937 he wrote a storybook for children, *My Friend Mr. Leakey*, that combined some ingenious science fiction and fantasy with an Indian character called Chandrajotish. India was never very far from his mind.

Haldane emphasized that science cannot be adequately disseminated through textbooks and lectures. He urged large-scale dissemination of scientific knowledge to millions of people, and he himself contributed a great deal to achieving that goal. Haldane's writings were full of passion and force and were calculated to arouse public interest to follow his point of view.

Astronomy

Throughout his life, Haldane wrote many articles on astronomical subjects, simplifying complex concepts and ideas for the lay reader. One of his bedside books was C. W. Allen's *Astrophysical Quantities*. In India, he was an avid star-watcher. His popular articles in the press often referred to the Sanskrit and Greek names of stars and planets. The following are examples of his popular writings on this subject.

"Is There Life on the Planets?"

In this essay, writing from London, Haldane stated that five planets could be seen with the naked eye at the time of his writing (in 1939): "Venus is an evening star, setting about an hour and a half after the sun. Jupiter is the brightest star in the sky after Venus has set, and Saturn is to be seen to the east of Jupiter. If you go to work in the morning before daybreak, you may see Mercury and Mars in the East, rising before the Sun."[15]

Haldane's essay contained its share of speculations. He wrote, "It seemed natural to speculate that they [planets] were inhabited. But before it was possible to say whether life as we know it could exist on the planets, a lot more information was needed. And our knowledge of the planets has not increased very greatly in the last fifty years, although we have found out vastly more about the distant stars and nebulae."

Furthermore, about Mars he wrote,

> we can see the solid surface of Mars, whereas in the case of Jupiter, and probably Venus and Saturn, we can only observe the tops of clouds, which may consist of drops of liquid, or of solid dust. We can follow seasonal changes on Mars. During the winter each pole develops a white cap which doubtless consists of frost. This frost may be frozen water. But it may be solid carbon dioxide, which is used in the refrigerating industry under the name of dry ice, and is only solid at temperatures far below the freezing point of water. There are also colour changes elsewhere which may be due to vegetation.

Regarding the possibility of life on the planets, Haldane wrote, "if there is life on Mars it is probably more like that of the bacteria which live without oxygen in black mud than to those of familiar animals and plants. So perhaps we had better make our own planet fit for rational beings before we colonize others." We know a great deal more today.

"Simplifying Astronomy"

Twenty years later, he wrote another essay, "Simplifying Astronomy," while living in India. Here is an excerpt[16]:

> At present [March 1959] Canopus, Agastya, or Alpha Carinae, is conspicuous in the south about sunset. This is a very distant star, but produces so much light that it is the second brightest of all stars in the sky, even though if the sun were removed to the same distance we should not be able to see it with the naked eye.... About an hour and forty minutes after Canopus two fairly bright stars reach the southern median. These are Gamma Velorum and Zeta Puppis almost due north of it. These are the two hottest stars which appear bright enough to be easily recognized. More accurately Gamma Velorum is a double star consisting of two components moving round their common centre of gravity of which the hotter and brighter is what is called a Wolf-Rayet star (after the names of two astronomers).
>
> Wolf-Rayet stars show a continuous spectrum, corresponding to a surface temperature of about 80,000 degrees, and also bright lines. In fact their

spectrum is superficially like that of a gas discharge lamp with a fluorescent background.

Haldane pointed out some of the stars in Trisanku, or the Southern Cross, have very hot surfaces, perhaps because they are younger stars. He suggested various observations and studies one could make without using complex expensive apparatus. Curiously, he mentioned a moral reason for not using expensive research equipment.

"Some Autumn Stars"

Another excellent article was titled "Some Autumn Stars" (169–73). Most people are familiar with the sun, moon, and planets, but they know very little about the background stars like the sun. Haldane stated that it is hard to find books that give up-to-date information about specific stars (true even today!), so he decided to write a series of popular articles about some of the more conspicuous stars that are seen at various times of the year in India. If a star is bright enough to have a name, it is always near enough to measure the distance. The unit of distance is a parsec.[17]

> This is 30 million kilometers, and it takes light three and a quarter years to travel so far.... We can also measure the heat reaching us from the brighter stars.... In the case of double stars revolving round a common centre of gravity we can also calculate the mass. Finally we can measure the rate at which a star is approaching or receding, from its spectrum, and if it is near enough we can often measure how fast it is moving sideways. (170)

Among the stars visible in India in September evenings, two bright stars stand out because they are not far apart in the south, and set before midnight. These are the planet Saturn and the brightest star in the constellation Scorpius, which is called Antares. About Antares, Haldane wrote:

> Antares is a red double star, or more accurately, a pair of stars of which one at least is about 20 times as massive as the sun, and of about 400 times its radius. However, its temperature is only about 3,000 degrees, or half that of the sun. But it is so large that it puts out about 2,000 times as much light per second as the sun. So though it is about 120 parsecs away, it is the fifteenth brightest star in the sky. (171)

I have quoted this paragraph from Haldane's essay to show how Haldane was able to condense so much data in a small space and present it so lucidly that what would be ordinarily daunting to read in a scientific paper or a textbook becomes intelligible to any lay reader. Haldane was aware that many lay readers have a fear of numbers, so he uses the technique of analogy to introduce a number by comparing with an object that is already well known. He writes: "its temperature is only about 3,000 degrees, or half that of the sun." Most readers would not have any difficulty imagining a star that puts out half as much heat as the sun. At the same time, he introduces the term "parsec" and the kind of data that interest scientists.

The next very bright star to be seen northward in the Milky Way, mentioned by Haldane, is Altair, in the constellation of the Eagle. It is interesting to note that Haldane referred to Altair as "one of our nearest neighbours, just under five parsecs away," thereby drawing attention to "astronomical" distances in space. One needs a whole new concept of distance in space.

Continuing with Altair, we learn that "it is about twenty times as bright as the sun, not quite twice as heavy, and about one and a half times as hot." Both the sun and Altair belong to what is called the main sequence of stars. Haldane suggested that all stars might start as members of the main sequence, in which there is a simple relation between mass and light output. Altair is a much more representative star than most of the bright ones.

Moving further north in the Milky Way, one comes across Deneb, the brightest star in the constellation Swan. Its distance is uncertain but probably about 500 parsecs away. It is by far the brightest of all the stars except Regel in Orion, which is easily visible with the naked eye. It is only about twice as hot as the sun although it has 4000 times its area. Such stars are very rare. Very close to Deneb is the star Vega, in the constellation Lyre, which is a main sequence star about eight parsecs away. The Earth's axis recently pointed to Polaris, and it pointed to Vega 13,000 years ago, and will do so again 13,000 years hence.

If a star is moving toward the earth the light characteristic of any particular change in the electrons surrounding an atom shifts toward the violet, it is moving away, toward the red. Such shifts occur in the spectrum of Polaris with a period of about thirty years.

Though not much hotter than the sun, Polaris is ten times heavier, so Haldane suggested that the star revolving round it must be a bit further from it than Saturn is from the sun. Sirius, the brightest star, has a faint companion that goes round it every five years, and was first detected spectroscopically, and only later seen with a telescope. Haldane hoped that the companion of Polaris would be detected in the future, and that has been accomplished since he wrote the article in 1959.

"The Pleiades and Orion"

In one of his essays in astronomy, Haldane discussed the Pleiades or Krittika, which is one of the most striking of the star groups to be seen in late autumn in the Indian sky.

"The Pleiades are a natural group, that is to say they are not merely in the same direction but close together. So all the stars in the group must agree in four measurable characters. Because their distances are nearly the same, they agree in certain respects. The distance is, in fact, about 130 parsecs, or 420 light-years, that is to say their light that reaches us today started off when the Mughal emperor Babur[18] invaded India."[19]

All these stars are moving away from us at about the same speed of 20 kilometers per second. There are about 120 stars that fall into this class. The stars in such a cluster must be moving relative to each other, or they would fall together. Occasionally, if one moves too fast, it will be able to escape from the gravitational pull of the others.

The cluster is dominated by hot blue stars, displaying a wide range of luminosities, which have formed within the last 100 million years. Dust that forms a faint reflection of nebulosity around the brightest stars was thought at first to be left over from the formation of the cluster (hence the alternate name Maia Nebula after the star Maia), but is now known to be an unrelated dust cloud in the interstellar medium that the stars are currently passing through. It has been estimated that the cluster will survive for about another 250 million years, after which it will disperse due to gravitational interactions with its galactic neighborhood.

Aldebran or Rohini, rises soon after the Pleiades. It is a double star about 21 parsecs away, red in color, cooler than the sun and much more luminous. Orion, or Kalpurush, is not a "natural" system like the Pleiades. All the bright stars in it are distant, but some are very far away. It also includes some clusters. Three stars forming the belt are at a distance of about 200 parsecs.

The belt of Orion consists of the three bright stars Alnitak, Alnilam, and Mintaka. Alnitak is approximately 800 light-years away from earth and is about 100,000 more luminous than the sun. Alnilam is approximately 1,340 light years away from earth and is 375,000 times more luminous than sun. Mintaka is 915 light years away and is 90,000 times more luminous than Sun. Mintaka is a double star. Both stars orbit around each other every 5.73 days. All three are extremely hot stars emitting about ten thousand times as much light as the sun from only about a hundred times its area. The other brightest stars in Orion include Regel, the brightest at the southern end, which is a quadruple star, and the second brightest, Betelgeuse at the northern end, is so huge that if it replaced our own sun it would reach almost halfway to the earth.

"The Origin of Life"

Although Haldane's essays were generally considered "popular" and written in a mostly nontechnical language, they often contained novel ideas and concepts that would be of great interest to professional scientists. A fine example of this kind of writing is Haldane's essay "The Origin of Life," which was published not in any scientific journal but in the *Rationalist Annual* in 1929. It is discussed in another chapter (see Chapter 8, "Origin of Life"). The Rationalist Association is a charity registered in the UK that promotes reason and evidence in the understanding of life, and a humanist outlook. Its forerunner, the Rationalist Press Association, was founded in 1899, as a free-thinking secular publisher. Haldane was a lifelong member of the Rationalist Press Association and contributed an article annually to the *Rationalist Annual*. Several of his finest essays were published in that journal. In 1968, they were published together in a collected volume.[20]

Movies for Toads

Haldane wrote that primitive men took it for granted that animals can think and they have souls. Christian philosophers have always denied that other species are capable of reasoning or had any rights. Darwin accepted the view that the higher animals possessed most of the human faculties. It is very difficult to be sure whether an animal is thinking independently or only picking the clues given by the experimenter. Some horses in Germany were supposed to tap out answers with their hooves for certain elaborate sums, but the psychologists who examined them carefully concluded that they were watching their teacher's facial expressions and breathing carefully and reacting accordingly.

Haldane narrated the experiments on hens by Dr. Honigman, a refugee who came to London during World War II. He put hens in a cage with a narrow gap in the floor. Under this gap there was a board that moved on rollers and carried rows of wheat grains. Every second grain was glued down so that the hen could not remove it. Some hens learned very quickly to peck at alternate grains. But they failed completely when only every third wheat grain was free. Some other birds were able to count up to five because they were able to distinguish between four and five eggs.

Some animals recognize moving pictures because they will only eat moving objects. A toad can flick out tongue with very great speed and accuracy and bring back a small insect or other food into its open mouth. If two toads are competing for the same food, and one of them is successful, the disappointed toad may flick its tongue at the eye of its rival. Honigman made films of toads catching flies and showed them to other hungry toads. They are seen flicking at the successful toads in the film.

Honigman continued his studies, showing films of moving worms to dragonfly larvae, which live under water and shoot out their jaws to capture their prey. It is interesting that moving pictures in these experiments captured the attention of only very small animals with very simple minds such as toads and dragon fly larvae. Larger animals such as dogs and cats can recognize that the moving picture is not real, although some dogs may watch films showing vigorous action for short durations. Haldane made the point that one is likely to get good results in research if one is really fond of a particular species as Honigman was about toads.

"What I Require from Life"

An earlier essay, which was written while he was in London, discussed his requirements in life[21]: first, work "which is hard but interesting," and a decent wage; second, freedom to pursue his activities; third, health, "fit for work and enjoyment in the intervals"; fourth, friendship, "of my colleagues and comrades in scientific and political work." He wrote, "I cannot be friends with a person whose orders I have to obey without criticism before or after, or with one who has to obey my orders in a similar way. And I find friendship with people much richer or poorer than myself very difficult."

Besides these basic human needs, Haldane also required adventure, which he sought in his physiological experiments. Other things he desired (but did not demand) included a private room with some books, good tobacco, a motor car, and a daily bath. He considered himself exceptionally lucky, as his requirements were mostly fulfilled. He wrote that he would like to see workers controlling their conditions of work and every man and woman as healthy as possible. He would like to see the end of class subjection and sex subjection. As a socialist he wanted his fellow men and women to enjoy the same advantages as himself. He did not include peace and security among his requirements, because "it is futile to require things which one is most unlikely to obtain." He desired to see universal education and an increasing application of scientific methods in all branches of life.

On Gandhian Principle

The use of complex equipment separates scientists from ordinary people who use ploughs and potters' wheels. Haldane wrote, "I think the physics of ploughing is more important for India than some of the branches of physics which are being investigated in our university laboratories." Following Gandhi, he wrote, "I am not a consistent Gandhian, but I certainly think that Indian scientific research would be the better for adopting a few Gandhian principles, one of which is to regard machines as made to serve men, and never to think of men as made to serve machines."

For several reasons, Haldane urged the use of simple or no equipment in scientific research in India. His own best work was theoretical and mathematical. He firmly believed also, as a socialist, that the use of complicated equipment alienates the scientist from the common man. This was similar to Gandhi's view. It was one of the reasons why Haldane wrote so many popular scientific essays in the *Daily Worker*. He was imparting science education to the average worker who did not have the good fortune of receiving higher education. And he believed that any available funds should be spent on providing better salaries and better living conditions for young scientists and students.[22]

Syadvada System of Predication

Haldane emphasized the uncertainties of so-called scientific principles. He encouraged individuals to think along the lines of the Jaina philosopher Bhadrabahu (?433–357 BC), whose Syadvada system of predication,[23] a system of logic, includes the following stages:

> May be it is.
> May be it is not.
> May be it is and is not.
> May be it is indeterminate.
> May be it is and is indeterminate.
> May be it is not and is indeterminate.
> May be it is, is not, and is indeterminate.

These systems of certainty/uncertainty would be conducive to original thought, as those pursuing scientific inquiry are constantly waiting for surprises. Haldane wrote: "There may, of course, be 'laws of nature.' But if so, I don't know what they are, and I also know that nobody else knows them. There are, of course, good approximations to them." Haldane predicted that in a few centuries all the sciences may reach the stage when their principles will be known and the finer details, applications, and mathematical theories will be left for posterity. If that happens, life will go on until something unexpected happens when scientific research will once again flourish.

He added, "It is indeed presumptuous to attempt to analyze the principal growing point of the human spirit." There are perhaps many who agree with this comment. However, he noted that there are many more who wrote a great deal of "arrant nonsense" about scientific method.[24]

On Being Finite (Or Dealing with Death)

One of Haldane's most charming and poignant essays deals with his thoughts on death and dying.[25] This essay, "On Being Finite" was written by Haldane for the *Rationalist Annual* (1965), just after his operation for cancer of the rectum was performed. He wrote, "I have reached the age of seventy-one, and been operated for a cancer near the hind end of my intestine. These events lead me to attend rather more closely than I did twenty years ago to the fact I shall die within a few years—perhaps one year if the cancer has sent a colony of cells to another part of my body, perhaps twenty-five if it has not, and the rest of my cells behave unusually well." He died on December 1, 1964. Both this essay and the verse he titled "Cancer Is a Funny Thing" are fine examples of the courage he displayed throughout his life.

Haldane went on to make a brief review of the recognition of death throughout human history, pointing out that death as something unavoidable was a recent discovery. Primitive man very rarely saw anyone dying of old age. Most deaths were accidental. As the communities were sparsely populated, infectious diseases were rare. According to Haldane, the notion that "I must die" was a very much later development in human culture. He ruled out the possibility of life after death, stating that very little evidence exists in favor of this belief. He wrote that the trouble about great causes is that they tend, in practice, to be great quarrels. They also give scope for power addicts.

Historically, which human achievements have proven to be permanent? The most permanent have been advances in technology. Strangely, some of the ancient inventions seem to be outlasting the modern. As Haldane noted, the inventors of the lamp, the loom, and the saddle would recognize the descendants of their prototypes. Modern inventions are shorter-lived. Haldane noted that inventors bring forth social revolutions that are rarely achieved by any politicians.

On the subject of credit for scientific discoveries, Haldane noted that some of his more important achievements, such as the estimation of the rate of human genes, are taken for granted, while others, not so important, are carefully acknowledged. He wrote, "No greater compliment can be paid to a scientist than to take his original ideas for granted as part of the accepted framework of science during his lifetime."[26]

The survival of literature and the arts is more chancy, whereas sculpture and mosaics had a better chance of surviving during the last two thousand years. Paintings have a much smaller chance, except some paleolithic murals that have been preserved in caves in some parts of the world, such as in western France and southern India. Today, paintings and sculpture may not survive modern weapons, but music has a much better chance of survival because it appears to be recorded in so many indestructible forms. So Stravinsky is more likely to be directly known two thousand years hence than Picasso.

Death

Writing shortly before his own death in December 1964, Haldane reflected on his death, "I should find the prospect of death annoying if I had not had a very full experience mainly stemming from my work. I missed many possibilities because I got a severe wound in my right arm in 1915. However, adapting to this wound has been experience denied to most people."

Haldane explained that most of the joyful experiences of his life have been byproducts: "For example, I was one of the first two people to pass forty-eight hours in a miniature submarine, and one of the first few to get out of one under water. I doubt whether, given my psychological make-up, I should have found many greater thrills in a hundred lives. So when the angel with the darker drink at last shall find me by the river's brink, and offering his cup, invite my soul forth to my lips to quaff, I shall not shrink."

In an essay written for a radio broadcast in 1929, "My Philosophy of Life,"[27] Haldane reflected on his own death: "But even if I am blown to pieces in the destruction of London during the next war, or starved to death during the next . . . revolution, I hope that I shall find time to think as I die, 'I am glad that I lived when and where I did. It was a good show.'"

Notes

1. J. B. S. Haldane, *Daedalus, or Science and the Future* (London: Kegan Paul, 1923).
   Haldane's first book in which he made several scientific predictions, notably the impact of in vitro fertilization and genetic manipulation on society; ideas used by Aldous Huxley in his *Brave New World*.
2. Freeman Dyson, "*Daedalus* after Seventy Years," in K. R. Dronamraju, ed., *Haldane's Daedalus Revisited* (Oxford: Oxford University Press, 1995), 55–63.
3. Arthur C. Clarke, "Haldane and Space," in K. R. Dronamraju, ed., *Haldane and Modern Biology* (Baltimore, MD: Johns Hopkins University Press, 1968), 243–48.
4. C. Haldane, *Truth Will Out* (London: Vanguard Press, 1950), 75.
5. "On Being the Right Size" is an excellent example of Haldane's popular writing in science. It was first published in a collection of his popular essays, *Possible Worlds and Other Essays*, 20–28 (London: Chatto & Windus, 1927) in 1927. It was reprinted in a later collection of Haldane's popular essays, *On Being the Right Size and Other Essays* (Oxford: Oxford University Press) in 1985.
6. D'Arcy W. Thompson, *On Growth and Form* (Cambridge: Cambridge University Press, 1942).
7. T. A. Davis, "Biology in the Tropics," in *Haldane and Modern Biology*, ed. K. R. Dronamraju (Baltimore: Johns Hopkins University Press, 1968), 327–33.
8. Haldane's popular article "How to Write a Popular Scientific Article," which was first published in the 1930s, was reprinted in K. R. Dronamraju, ed., *What I Require from Life: Writings*

on *Science and Life from J.B.S. Haldane*, with a Foreword by Sir Arthur C. Clarke and Preface by James F. Crow (Oxford: Oxford University Press, 2009), 3–8. Haldane began the article with the following advice: "The first thing to remember is that your task is not easy, and will be impossible if you despise technique. For literature has its technique, like science, and unless you set yourself a fairly high standard you will get nowhere. So don't expect to succeed at your first, or even your second, attempt."

9. J. B. S. Haldane, "The Starling," in *Everything Has a History* (London: Allen & Unwin, 1951), 190–92.
10. J. B. S. Haldane, *An Indian Perspective of Drawin, (1959), Centennial Review* (of Arts and Sciences, Michigan State University), 3, 357–62.
11. Haldane's article on the future of science and its applications was published in the *New York Times Magazine*, April 19, 1964, on the occasion of the New York World Fair, New York.
12. James Watson's, *The Double Helix* (New York: Atheneum Publishers), which was published in 1968, became an instant bestseller. It was dedicated to Haldane's sister, Naomi Mitchison. Watson wrote, "I did not sit through the Christmas holidays in Cambridge. Avrion Mitchison had invited me to Carradale, the home of his parents, on the Mull of Kintyre. This was real luck, since over holidays Av's mother, Naomi, the distinguished writer, and his Labour M.P. father, Dick, were known to fill their large house with odd assortments of lively minds. Moreover, Naomi was a sister of England's most clever and eccentric biologist, J.B.S. Haldane" (71).
13. J. B. S. Haldane, "The New Caste System," *Hindu*.
14. From *The Rationalist Annual* (London: The Rationalist Press Association, 1960); reprinted in J. B. S. Haldane, "On Expecting the Unexpected," in *Science and Life, Essays of a Rationalist*, with an introduction by John Maynard Smith (London: Pemberton, 1968), 135–44.
15. J. B. S. Haldane, "What I Require from Life," in *What I Require from Life: Writings on Science and Life from J.B.S. Haldane*, ed. Dronamraju (Oxford: Oxford University Press, 2009), 48–50.
16. K. R. Dronamraju, ed., *What I Require from Life: Writings on Science and Life from J.B.S. Haldane* (Oxford: Oxford University Press, 2009), 191–94.
17. Parsec: One parsec equals about 3.26 light years (30.9 trillion kilometers or 19.2 trillion miles). All known stars lie more than one parsec away, with Proxima Centauri showing the largest parallax of 0.7687 arcsec, making the distance 1.3009 parsecs (4.243 light years). Most of the visible stars in the nighttime sky lie within 500 parsecs of the Sun. The parsec was introduced to make quick calculations of astronomical distances without the need for more complicated conversions, namely, knowing the true speed of light to calculate light years. Parsec is named from the abbreviation of the parallax of one arcsecond, and was first suggested by the British astronomer Herbert Hall Turner in 1913.
18. Babur (1483–1530) was the founder of the Moghul Empire in India. He was a descendent of Genghis Khan and Timur, but sought to establish a more lasting civilization based on religious tolerance and promotion of the arts. Babur was born in Farghana in Turkestan. At the age of twelve he became ruler, following the death of his father. However, he was soon usurped by his uncles. In his early life, Babur had to fight many battles against his enemies and was frequently struggling to rule over a territory. He moved to Afghanistan and then to India. It was in India that he was able to cement his domination, laying the foundations for the modern Moghul rule of India. He allowed people to continue with their Hindu religion and customs. Babur promoted the arts and was instrumental in bringing Persian culture into India. Babur's son was Humayun. Babur's grandson was Akbar the Great.
19. Haldane, "What I Require from Life," 173–76.
20. J. B. S. Haldane, *Science and Life, Essays of a Rationalist*, with an introduction by John Maynard Smith (London: Pemberton, 1968).
21. Haldane, "What I Require from Life," 19–22; 20.
22. J. B. S. Haldane, personal communication to the author, 1961.
23. The Syadvada system of predication is a system of logic developed in India by the Jaina philosopher Bhadrabahu in the fourth century BC. Haldane thought that it facilitated a more logical and more accurate system of classifying scientific results where it is not always possible to

assign absolutely positive or absolutely negative values. The Syadvada system allows various grades of truth or falsehood, which is often the situation in certain experiments.
24. These views appeared in a series of articles by Haldane that were published in *The Rationalist Annual*, 1929–65, and were reprinted in Haldane, *Science and Life*.
25. J. B. S. Haldane, "On Being Finite," *The Rationalist Annual* (1965); reprinted in Haldane, *Science and Life*, 192–203.
26. Haldane, "On Being Finite," 201–2.
27. J. B. S. Haldane, *The Inequality of Man and Other Essays* (London: Chatto & Windus, 1932), 211–24.

14

Haldane and Huxley

J. B. S. Haldane (1892–1964) and Julian Huxley (1887–1975) had known each other all their lives. Haldane's sister Nou recorded that they knew each other while attending the Eton school.

She wrote in "Beginnings"[1] that, while her brother was being severely harassed by his fellow students, "He was to some extent protected by his fag masters[2] of whom one was Julian Huxley and the other was Geoff Wardley, who was in turn my husband's best friend—this bringing them together." Haldane recalled years later that Julian gave him an apple while they were students at Eton, a mark of exceptional favor from a senior to a junior. Nou herself acknowledged that she and her brother, Jack, as well as the Huxley brothers—Julian and Aldous, all grew up together in Oxford. She had a crush on Aldous and hoped longingly that he would kiss her!

Both Haldane and Huxley attended Eton and Oxford. Huxley was the senior by five years. Both came from distinguished scientific families. Julian was deeply influenced by his famous grandfather Thomas Henry Huxley, the great nineteenth-century zoologist and evolutionist ("Darwin's bulldog"). Haldane's early research interests were guided and fostered by his father, John Scott Haldane, the eminent Oxford physiologist. The two families were closely entwined in friendship. When Julian's first child, Anthony, was born in 1920, Haldane's father, John Scott Haldane, and his wife Kathleen provided much parental guidance and advice, which was of great comfort to the young mother, Juliette (later Lady Huxley). Juliette had no mother-in-law, as Julian's mother died when he was still in school.

In the years following world War I, both Julian Huxley and J. B. S. Haldane were fellows of New College, Oxford—Huxley in zoology, and Haldane in physiology. Haldane was a frequent visitor to the then newly married Huxleys. In his *Memories*, Huxley wrote, "One of our most frequent visitors was Jack Haldane, he too a Fellow of New College, teaching physiology, though he had taken a First in Greats. He was another odd character. He dropped in whenever he liked—which was usually at tea-time—and devoured plates of biscuits, protesting that he couldn't eat a crumb, while reciting Shelley and Milton and any other poet you chose, by the yard. He had a fantastic memory and knowledge of the classics, and enjoyed displaying them."[3]

In *Memories*, Huxley recalled that young Haldane, during his teatime visits, often asked him if he read such and such a paper and Huxley had to admit he had not. What humiliated Huxley even more was that the topics which Haldane was discussing were in Huxley's field of activity! This went on until one day fate caught up with Haldane in the form of Julian's wife, Juliette, who asked him, after one of those questions, "Have you read them?" Haldane was chagrined to admit that he had not read them but he did note the titles and authors! There was friendly rivalry between Haldane and Huxley. It was never antagonistic or hostile in any way.

When I was a student of Haldane at the Indian Statistical Institute in Calcutta (now Kolkata), Sir Julian and Lady Huxley visited us. Their first visit was on a Sunday afternoon. There was no one around except Haldane and myself.

During the day Haldane told me that the Huxleys would be visiting and I should let him know in the library upon their arrival. They arrived about 4:00 pm, and I informed Haldane as instructed. Gyan, the Bengali cook, served us tea and samosas as we sat on the open terrace outside Haldane's flat in the institute. It was a lovely warm day with a gentle breeze. As I sat there, I was enjoying the fragrance that wafted from the jasmine and frangipani flowers that were blooming downstairs. I was impressed by the cordial friendship between these two great scientists. What was more impressive was that they remained friends for almost seventy years through various turbulent years of their lives, including Haldane's difficult years at Eton; his service with the Black Watch in World War I when he almost got killed; the notorious divorce of his first wife, Charlotte, from her previous husband and the resulting loss of Haldane's readership at Cambridge University; conversion to communism and the Lysenko controversy; and finally his move to India. Huxley had his own ups and downs too, including his mother's early death while he was a student; his frequent nervous breakdowns; his marriage to Juliette; his affair with an American schoolgirl and the resulting marital problems; his difficult collaboration with H. G. Wells and his son G. P. Wells; his conflicts as director of the London Zoo; his own stand on Soviet science and Lysenko, which was quite different from Haldane's position; and his controversial appointment as the director-general of UNESCO. But that was all behind them.

What I saw that afternoon was a pleasant and peaceful discussion between two old friends, mostly about problems of biological research that interested Haldane in India. Still very young and shy, I had just started my research with Haldane, but had initiated my research on the preferential visits of butterfly species to varieties of *Lantana* differing in flower color. Julian expressed great interest in my research.

Haldane talked at length about his plans for research in India. Juliette was silent during much of this discussion. But they both spoke a lot a few days later, when they returned one evening for dinner with the Haldanes, who had invited all our research colleagues. Each member of our team made a brief presentation of his research and answered a few questions from Julian. At one point, everyone was presenting so much data and talking at the same time that Helen (Mrs. Haldane) was constrained to explode: "Let the poor man eat!" But Julian was happily snacking on fragrant *Jeera papads* while sipping wine throughout all this commotion while paying close attention to the data.

In his *Memories II*, Julian recorded Haldane's response when asked what message they (Julian and Juliette) should give Haldane's mother, Louisa, upon their return to Oxford. In Julian's words: "'Tell her,' said this stormy petrel of British science, that at long last I am able to do some constructive work, without being hamstrung by bloody red tape.' Alas, he did not escape trouble, usually of his own making, but finally found a refuge at Bhubaneshwar, under the wing of our friend Patnaik."[4]

It was in 1920 that Huxley conducted his famous experiments on the metamorphosis of the axolotl,[5] which is seldom seen in its adult stage. His discovery, which indicated the transformation of the larva by feeding it on thyroid gland, received much attention in the popular press. It was also during those Oxford years that Haldane's sister Naomi "dragged" them into acting.[6]

After the Oxford years, the lives of Haldane and Huxley diverged.

In 1923 Haldane left for Cambridge to become the Dunn Reader in Biochemistry under F. G. Hopkins. In 1925 Huxley resigned from his position as demonstrator in zoology at Oxford and accepted the chair of zoology at King's College, University of London. However, he resigned from that post as well shortly afterward, when invited by H. G. Wells to collaborate with himself and his son G. P. Wells in writing *The Science of Life* (1929–1930).[7] Haldane, in the meantime, began his series of papers on the mathematical theory of natural selection,[8] which formed the foundation (along with the works of R. A. Fisher and Sewall Wright) for theoretical population genetics.

In the 1920s, Haldane and Huxley collaborated in writing an excellent textbook called *Animal Biology*.[9] The book was designed to provide what was considered to be lacking in the science books at that time. The introduction stated, "If it is the scientific point of view, and not merely a collection of facts, that we wish to impress on those we teach, then it becomes increasingly necessary to cut out needless detail, to concentrate on fundamentals, to arouse interest from the outset."

Haldane's mother was much amused by the frontispiece, which she thought bore a remarkable resemblance to the authors: two deep-sea angler-fishes, a fat one and a skinny one, representing Haldane and Huxley, respectively!

Popular Writing

Both Huxley and Haldane first came to the public attention in the 1920s. Huxley's initial fame due to his axolotl work was soon followed by his *Essays in Popular Science* (1926) and *The Stream of Life* (1928) and, above all, *The Science of Life* (1929–1930) with H. G. and his son G. P. Wells.

Haldane's first book, *Daedalus* (1923), was soon followed by *Callinicus, A Defense of Chemical Warfare* (1925) and the collected essays titled *Possible Worlds and Other Essays* (1927), establishing once and for all Haldane's brilliance as a first-rate scientist as well as a superb popularizer.

Haldane's popularization covered a great range of subjects, encompassing geology, chemistry, astronomy, philosophy, and statistics, as well as a number of biological sciences. Collections of essays by Haldane appeared as books with such titles as *The Inequality of Man and Other Essays* (1932), *Heredity and Politics* (1938), *Keeping Cool and Other Essays* (1940), *A Banned Broadcast and Other Essays* (1946), *Science Advances* (1947), and *Everything has a History* (1951). From the late 1930s onward, Haldane's writings had a decidedly Marxist flavor.

Huxley's prolific writings included biological topics, colonial education in Africa, social planning, Soviet genetics, wildlife, humanism, and travel notes. His books include *A Scientist among the Soviets* (1932), *Africa View* (1931), *The Uniqueness of*

Man (1941), *New Bottles for New Wine* (1957), *The Humanist Frame* (1961), *Essays of a Humanist* (1964), *Memories* (1970), and *Memories II* (1973).

One of Huxley's most influential books was *Religion without Revelation*, published in 1927.[10] In arriving at his religious philosophy, Huxley was influenced by the essays of Lord Morley, which he came across while browsing in a public library at Colorado Springs. This book and his aunt Mary Ward's book, *Robert Elsmere*, together made a deep impression on Huxley's mind, converting him to religious humanism. In *Religion without Revelation*, Huxley stated his belief that "any religion which stresses the need for propitiating an external Power will be diverted away from the more essential task of using and organizing the spiritual forces that lie within each individual.... I believe in the religion of life."

Biological Interests

Huxley's biological interests included animal behavior, amphibian metamorphosis, allometry,[11] and embryogenesis. His twin interests in taxonomy and Darwinian evolution were skillfully interwoven in his influential book *Evolution: The Modern Synthesis* (1942).[12] Huxley's early observations on the courtship behavior of the great crested grebe[13] opened up an entirely new aspect of vertebrate ethology. He introduced the term "ritualization" to describe the evolutionary process by which such behavior patterns as feeding or feather-preening might become dissociated from their original functions and used in self-exhausting ceremonies. His work provided important evidence that Darwinian sexual selection did not account for some of the ceremonies of sexual behavior by birds, because these were not performed until long after their mates were chosen. Huxley's most important contribution was in the measurement of relative growth.

Inspired by the early work of his Oxford tutor, Geoffrey Smith,[14] on differential growth rates, Huxley compared the growth in width of the abdomen of female fiddler crabs (*Uca puqnax*) with the growth in width of the carapace. The abdominal width is expressed as a percentage of the carapace width. In very young specimens, the width of the abdomen was found to be similar in both sexes, but it became broader in adult females by allometric growth. In 1924 Huxley published a brief but important paper on the heterogonic growth[15] of the larger "chela," in the male crab. He derived an expression to measure heterogonic growth: $y = bx^k$, where k is a measure of the differential growth of the "chela" in relation to the rest of the body; y is the weight of the chela in relation to that of the rest of the body, x; and b is a constant representing y expressed as a fraction of x, the latter being considered to be unity for the purpose. Huxley emphasized the relevance of this work to the taxonomist, because animals with heterogonic growth cannot be described meaningfully (with respect to their heterogonic organs), except in terms of k. He continued his studies of relative growth in a number of species (e.g., the size of the forceps of the earwig, *Forficula auricularia*), and summed up his work in *Problems of Relative Growth*.[16] He emphasized the mistake made by systematists when defining species and other groups in terms of percentage relationships of parts of the body. It is of interest to note that Huxley's interests in relative growth may be traced back to his first researches in dedifferentiation and morphogenesis when he studied the varying rates at which the organs underwent dedifferentiation in the marine animal-ascidian *Clavellina lepadiformis*.[17] With G. R. de Beer,[18] he showed that when exposed to poisons, the hydroids of *Obelia qeniculata* and *Campanularia* sp. tended to dedifferentiate and to be resorbed into the stolon,[19] or, if detached, became ovoid, undifferentiated bodies. Various parts of the body dedifferentiated at markedly different speeds.

These few examples can only give very brief glimpses of Huxley's diverse biological interests, which include bird watching as well as experimental embryology. Extensive summaries can be found in the biographical memoir written by his pupil J. R. Baker (1978), and in Huxley's own *Memories*.[20]

In contrast to Huxley's formal scientific training (in zoology), Haldane never took a degree in science. His formal academic qualification was a BA degree in classics and humanities from Oxford. His first experiments and first publication in 1912[21] were the result of his father J. S. Haldane's influence, and were concerned with the physiological aspects of CO poisoning. Other studies dealt with the effects of ingesting various chemicals and of breathing various gaseous mixtures on the blood pH and components of blood and urine. However, Haldane was also involved simultaneously in genetic research from 1912 onward, and spent much of his life in that field. From 1923 to 1933, he was formally identified as a "biochemist" and occupied the Sir William Dunn Readership in Biochemistry at Cambridge University in the department headed by F. Gowland Hopkins. From the 1930s until his death in 1964, he made significant contributions to a number of other fields as well, especially statistics, biometry, and animal behavior.

During the years 1937–1957, Haldane was Weldon Professor of Biometry at University College, London. This was in contrast to Huxley, who resigned from the chair of zoology at King's College, London, in 1927, and was a freelance biologist and intellectual-at-large, with the exception of his appointments at the London Zoo, and later at UNESCO.[22]

Haldane's extensive contributions to genetics are summarized in chapter 4, "Population Genetics."[23,24,25,26]

Among other works, Haldane's papers laid the basis for estimating genetic damage resulting from ionizing radiation, and for the study of infectious disease as a selective agent in relation to the maintenance of certain balanced polymorphisms in human populations.

Haldane was especially noted for conducting painful experiments in diving physiology, with himself as the chief guinea pig, during World War II. He firmly believed that medical personnel and physiologists should pioneer in taking physiological risks. He emphasized the application of diving physiology to space medicine, as it would involve working under no gravitation. Much of this pioneering work of Haldane (conducted in the 1930s and 1940s) has since become incorporated into space research. Haldane's interest in space research and Soviet science was noted by Lederberg.[27]

Common Interests

Huxley and Haldane shared the view that the Darwinian theory of evolution, especially by natural selection, was the most significant aspect of scientific research of their time. Both devoted considerable attention to this problem during their lives. Haldane's approach was mathematical, putting it firmly on a quantitative footing. Huxley was interested in a number of aspects—among them the species concept in taxonomy and sexual selection—and played the role of a grand synthesizer. Both believed in the education of the public, using mass media. Haldane and Huxley were the intellectual products of a most exciting time, and represented the best of English liberal education, which enabled them to indulge in a great diversity of intellectual endeavors. They did their utmost to pass on this knowledge and tradition to the succeeding generations.

From about 1935 onward, Haldane embraced Marxism and wrote very passionate essays on its application to science. However, he became disillusioned when the science of genetics was suppressed in the Soviet Union under the influence of T. D. Lysenko, especially after 1949, and slowly dissociated himself from Marxist philosophy and its applications. Soon after, in the 1950s, he took up another (equally passionate) interest in India and Hindu philosophies, and migrated to India in 1957, where he died in 1964.

Huxley, in contrast, continued to advocate "evolutionary humanism" in its widest sense, applying it to include both the sciences and the arts. He wrote to C. P. Snow[28] that his concept of "evolutionary humanism" could bridge the "culture gap" discussed by Snow: "I would have thought that you should extend this idea of the third culture, and make it everything based on the ideas of evolution and its course, both biological and psychosocial. If so, this would be more than a bridge—it would be a new pattern of thinking about nature and human nature, which would not merely reconcile the existing conflict but transcend it in a new pattern."[29]

Both Huxley and Haldane were major figures in twentieth-century science. Their contributions to our science and culture have been quite profound, in extending the boundaries of our knowledge, in public education, and in the application of science to human affairs.

Huxley's influence on world affairs included his work as the first director general of UNESCO, and Haldane's, through such works as *Daedalus*,[30] which profoundly influenced Julian's brother, Aldous Huxley, in writing his *Brave New World*.[31]

Notes

1. Naomi Mitchison, "Beginnings," in *Haldane and Modern Biology*, ed. K. R. Dronamraju (Baltimore: Johns Hopkins University Press, 1968), 299–305. Haldane's sister, Naomi was a prolific and distinguished author of many books on history, literature, travel, politics, and fiction.
2. Fagging is the system in English public schools whereby younger pupils act as servants to the older boys. Originally an emulation of domestic household task distribution and paternal authority, fagging formerly included harsh discipline. Facing public scrutiny, the practice of personal fagging was gradually discontinued during the 1970s and 1980s, but in most schools

it has been replaced by a system where junior boys are required to do tasks for the benefit of the general school community.
3. J. S. Huxley, *Memories* (London: Allen & Unwin, 1970), p. 137.
4. J. S. Huxley, *Memories II* (New York: Harper & Row, 1973), 163. Biju Patnaik was Chief Minister (equivalent to a Governor in the United States) of the State of Orissa in India where Haldane worked for two and half years until his death on December 1, 1964.
5. The axolotl, also known as a Mexican salamander (*Ambystoma mexicanum*) is closely related to the tiger salamander. Although the axolotl is colloquially known as a "walking fish," it is not a fish, but an amphibian. Axolotls are unusual among amphibians in that they reach adulthood without undergoing metamorphosis.
6. Mitchison, "Beginnings."
7. H. G. Wells, J. S. Huxley, and G. P. Wells, *The Science of Life* (London: Amalgamated Press, 1929–1930).
8. J. B. S. Haldane, "The Mathematical Theory of Natural and Artificial Selection: Part I," *Transactions of the Cambridge Philosophical Society*, 23 (1924): 19–41.
9. J. B. S. Haldane and J. S. Huxley, *Animal Biology* (Oxford: Clarendon Press, 1927).
10. J. S. Huxley, *Religion without Revelation* (London: Benn, 1927).
11. Allometry is the study of relative growth of a part in relation to an entire organism or to a standard. The term "allometry" was coined by Julian Huxley and Georges Teissier in 1936.
12. J. S. Huxley, *Evolution: The Modern Synthesis* (London: Allen & Unwin, 1942).
13. The great crested grebe (*Podiceps cristatus*) is a member of the grebe family of water birds noted for its elaborate mating display. The great crested grebe and its behavior was the subject of one of the landmark publications in avian ethology: J. S. Huxley, "The Courtship-Habits of the Great Crested Grebe (*Podiceps cristatus*); with an Addition to the Theory of Sexual Selection," *Proceedings of the Zoological Society of London* 84, no. 3 (1914): 491–562.
14. Geoffrey Smith (1881–1916), Oxford zoologist and tutor of Julian Huxley.
15. Heterogonic growth: relating to, or marked by allometry.
16. J. S. Huxley, *Problems of Relative Growth* (London: Methuen, 1932).
17. The *Clavellina lepadiformis*, sometimes called the light-bulb sea squirt, is a floating, transparent marine organism. They vary in size, shape and color. Essentially they are a bag containing intestines and gills with two openings, siphons, through which water is drawn or expelled. They are a subphylum of the Chordates containing fish and mammals. The link is in the presence of a notochord, but they lack a backbone.
18. Gavin Rylands De Beer (1899–1972) was a British evolutionary embryologist. He was director of the British Museum of Natural History, was president of the Linnean Society, and received the Royal Society's Darwin Medal for his studies on evolution. De Beer's early work at Oxford was influenced by J. B. S. Haldane, Julian Huxley, and E. S. Goodrich. *The Elements of Experimental Embryology*, written with Huxley, was the best summary of the field at that time (1934).
19. Stolon is a branching structure in lower animals, the anchoring rootlike part of colonial organisms, such as hydroids, on which the polyps are borne (from the Latin *stolō*, shoot).
20. Huxley, *Memories*; Huxley, *Memories II*.
21. C. G. Douglas, J. S. Haldane, and J. B. S. Haldane, "The Laws of Combination of Haemoglobin with Carbon Monoxide and Oxygen," *Journal of Physiology* 44, (1912): 275–304.
22. UNESCO (United Nations Educational, Scientific and Cultural Organization), is a specialized agency of the United Nations (UN). Its purpose is to contribute to peace and security by promoting international collaboration through education, science, and culture in order to further universal respect for justice. UNESCO has 196 member states and nine associate members. Julian Huxley was its first director general (1946–1948).
23. Linkage is the likelihood of certain genes to be inherited together.
24. Map distance: Distance between genes on a chromosome, expressed in centimorgans (cM), which was suggested by Haldane in 1919.
25. Mapping function: A mapping function was introduced, first by Haldane in 1919 and by others later, to correct the errors in the construction of gene maps. It is a mathematical relation

between the probability of genetic recombination and map units. Traditional gene maps underestimated the recombinant fraction, which distorted the genetic map. J. B. S. Haldane, "The Combination of Linkage Values, and the Calculation of Distances between the Loci of Linked Factors," *Journal of Genetics*, 8 (1919): 299–309.

26. J. B. S. Haldane, "Some Recent Work on Heredity," *Transactions of the Oxford University Junior Scientific Club*, 1, no. 3 (1920): 3–11.
27. Joshua Lederberg, "Sputnik + 30," *The Scientist*, October 5, 1987.
28. Charles Percy Snow (1905–1980), later raised to the peerage as Baron Snow, was an English chemist and novelist who also served in several important positions in the British Civil Service and the UK government. He is best known for his series of novels known collectively as "Strangers and Brothers," and for "The Two Cultures," a 1959 lecture in which he laments the gulf between scientists and literary intellectuals.
29. Huxley's letter to Sir Charles Snow dated July 29, 1959. Rice University Archives, Houston, Texas.
30. J. B. S. Haldane, *Daedalus, or Science and the Future* (London: Kegan Paul, Trench, Trubner, 1923; New York: E. P. Dutton, 1924).
31. Aldous Huxley, *Brave New World* (New York: Penguin, 1932). Aldous Huxley was a close friend of Haldane's sister, Naomi Mitchison. The Huxleys and the Haldanes grew up together in their youth in Oxford.

PART 1950S

15

Relations with Other Scientists

Ce n'est pas un homme, c'est une force de la nature!
—Boris Ephrussi

Contrary to his popular image in the lay press, which portrayed him as an abrasive and explosive personality, Haldane was extremely careful in cultivating his friendships with other scientists. He was careful not to offend his loyal friends and colleagues, whose friendship he nurtured and treasured with much skill and forethought. Haldane possessed a deep sense of duty to show his appreciation to those who inspired and guided his own life and career, such as his father, J. S. Haldane, the Cambridge University biologist William Bateson,[1] and the distinguished biochemist F. G. Hopkins. He spoke of them in his later years with a deep sense of gratitude and affection. This attitude was also evident in his letters to colleagues and former students as well as in the personal recollections of others who knew him well. One such example is the biographical memoir of Haldane, written by his former disciple and junior colleague N. W. Pirie[2] for the Royal Society of London. A look at Haldane's associates, colleagues, and friends tells us much about his own interests, his attitude toward others, and his character.

Friendship

Haldane was not the kind of man who developed close intimate friendships. The nearest instance might be his friendship with the French biochemist Rene Wurmser and his scientist-wife, Sabine, whom he knew well from 1926 until his death in 1964. That bond was based on both common scientific interests in biochemistry and Haldane's admiration for French science and culture. Wurmser first met Haldanei n 1926 at the International Congress of General Physiology in Stockholm, where Haldane presented some experiments on tetany by overbreathing, with a personal demonstration on his own body. The audience was at first alarmed and certainly impressed as they watched and learned of Haldane's habit of experimenting on himself. Referring to their friendship, Wurmser wrote, "Haldane did indeed show the apparent rudeness and Etonian self-assurance about which so much has been said. It is also true that he did not take the trouble to make a conquest of people who did not interest him, but he could be intensely kind to those he loved. For them, he had the most delicate attentions. His generosity and fidelity toward friends knew no limits."[3]

Some friendships were based on science or politics. His friendship with the Indian anthropologist Nirmal Kumar Bose[4] in Calcutta was based on his interest in Indian anthropology and Gandhian politics. Bose was a follower of Mahatma Gandhi,[5] and was, in fact, secretary to Gandhi for some years during the 1940s. Haldane and Bose shared a keen interest in the Gandhian ideals and approach to life. Bose used to accompany Haldane and his foreign visitors, such as Ernst Mayr of Harvard University, on their visits to ancient Indian temples of Orissa, providing historical and archeological commentary. He possessed an immense knowledge of Indian anthropology, especially of tribal people, and archeology of Hindu temples, which interested Haldane.

There were former colleagues and students in Great Britain with whom Haldane maintained a friendly correspondence for many years, but they were not close friends by any definition. They include Joseph Needham, J. H. Quastel,[6] and N. W. Pirie from the biochemical years during the 1920s and later, at Cambridge University. There were others, such as the botanist C. D. Darlington from the John Innes Horticultural Institution (and later Oxford University) and the famous statistician R. A. Fisher from University College London (and later Cambridge University), with whom Haldane at first had a friendly relationship but fell out in later years. This was not entirely Haldane's fault. Both Darlington and Fisher were notoriously difficult people, and I have written more about the Haldane–Darlington relationship elsewhere (see chapter 7, "Chemical Genetics"). Aldous Huxley's brother, Julian, was Haldane's senior at Eton and they maintained a lifelong friendship. Haldane was five years younger than Huxley. They

Figure 15.1 JBS with visiting colleagues, M. J. Sirks (Netherlands), extreme right C. D. Darlington.

collaborated in writing a textbook of zoology. Haldane's relationship with Huxley was complicated by an intense competition between them in their youth. Both were biologists, both came from famous families, both were educated at Eton and Oxford, and both became prolific popularizers of science in the public media.

Haldane possessed a deep sense of loyalty and responsibility toward his students, or "junior colleagues" as he called them. In contrast to the critical or caustic attitude he showed toward his senior colleagues and equals, Haldane showed much kindness and consideration toward his students, encouraging them to ask questions about research matters. Occasionally, he provided some financial assistance from his own pocket to meet students' travel to a scientific meeting, or to purchase books and pay for publication costs.[7] He gave top priority to research discussions with students, allowing maximum possible time. His notoriety for rudeness or feigned deafness was frequently evident when dealing with journalists or strangers who intruded on his privacy, wasting precious time in frivolous conversations. Consequently, such incidents figured prominently in the popular press. One example was a sensational incident in Calcutta in 1957, when Haldane angrily threw out a reporter from his apartment at the Indian Statistical Institute. It was widely reported in the press.

As the captain of a Black Watch regiment in World War I, Haldane exemplified leadership qualities, courageous in battle but also very protective of his men on the battlefield. A man of exceptional intelligence, of a caliber rarely found on the battlefield, he quickly saw the futility of war (see *Daedalus*, 1–3).

Joseph Needham (1900–1995)

Needham was born in London on December 9, 1900. His father was a Harley Street physician, and his mother a music teacher. After attending Oundle School he studied biochemistry at Cambridge University, where his mentors were F. G. Hopkins and his second-in-command, J. B. S. Haldane. Needham's early research was in chemical embryology. His major works[8] of this period are his *Chemical Embryology* (1931) and *Biology and Morphogenesis* (1942). But by the time this second book appeared, he had already started his research into the science and civilization in China, which occupied the rest of his life, resulting in the publication of many volumes. After his move to India in 1957, Haldane recalled how remarkable it was that his own interest in India paralleled Needham's great interest in China.

Needham's relationship with Haldane and the esteem he showed are evident in the following excerpt:

> Haldane was my immediate predecessor as Sir William Dunn Reader in Biochemistry at Cambridge, in the laboratory founded and presided over by our great teacher and friend Sir Frederick Gowland Hopkins. For nearly a decade Haldane was an outstanding source of inspiration and provocative stimulus to all his colleagues and students—one of those who did most to give the place and the period its unforgettable brilliance. Experimental work on eggs and embryos in those days had benefit from his pipe-puffing criticism; his Johnsonian dicta at tea and tea-club talks sharpened many

ideas in the writing of *Chemical Embryology*. Then, in the same year, 1933, he departed permanently for London, and we, with C.H. Waddington, for that period in Berlin which started for us the fusion of biochemistry and *Entwicklungsmechanik*.... Many occasions later on gave opportunity for meeting, and Haldane's conversion to Indian culture had the deepest sympathy and appreciation of one who had become, as it were, an honorary Chinese. Now I am honored indeed in contributing this paper to the ancestral altar of so old a friend.[9]

R. B. Fisher

Professor R. B. Fisher of the Department of Biochemistry at Edinburgh University recalled his memories of Haldane:

He [Haldane] was one of my examiners for my doctorate. In the *viva*, his colleague [fellow member of the committee] objected very strongly to my use of regression analysis to work out the relation between nitrogen ingestion and uric acid excretion in pigeons because I had not made duplicate analyses. I couldn't get over to him that attempts to sample the extracta introduced more error and that the analysis of the results of single estimations produced an acceptable measure of random error. Jack [Haldane] took over and spent half an hour giving his colleague an excellent tutorial on the subject of regression analysis which saved me.

The other story concerns his [Haldane's] duty as a Reader in Biochemistry. Biochemical Society meetings used to be delightful when he attended because he would get up at least once and congratulate Dr. X on his paper and say he wondered if Dr. X was aware of the findings of Dr. Y—working on some apparently quite different topic—because there were some striking similarities in the two pieces of work, which he would proceed to explain. These interventions were usually sufficiently cogent to persuade me that I really ought to read the *Biochemical Journal* the whole way through.

I suppose you know the story about Jack [Haldane] and Kinematic Relativity. Milne in Oxford published a model in which space was supposed to expand steadily as time passed. Jack wrote a letter to *Nature* in which he wondered whether it had occurred to Professor Milne that if this were true the solar system could have arisen at a sufficiently remote epoch as a result of the collision of a single photon with sun. *Nature* carried a series of congratulatory letters for some weeks.[10]

Cyril Darlington (1903–1981)

In 1927, Darlington was a 24-year-old botanist at the John Innes Horticultural Institution in Merton, near London, when Haldane agreed to join that institution as a "Part-time Officer-in-Charge of Genetical Investigations." The two got along well. Haldane was eleven years older than Darlington and was already an accomplished and famous scientist. Darlington, on the other hand, was a novice, a naïve young man who was in awe of Haldane's brilliance and accomplishments. In contrast to Haldane's distinguished aristocratic ancestry and intellectual dynasty, Darlington came from a poor and undistinguished family. His father was a schoolmaster of no particular distinction. In contrast to the upper-class education that Haldane received at Eton and Oxford University, Darlington's educational background was common and undistinguished. But Haldane was generous to a fault. He took Darlington under his wing, providing him generous advice and guidance. Although Darlington was jealous of Haldane's distinction and ancestry, he was an ambitious opportunist, who quickly realized that he could further his career by exploiting Haldane's generosity. Haldane wrote a flattering Foreword for Darlington's first book, *Recent Advances in Cytology* (1932). However, their friendship ended a few years later, partly because of political differences but mainly because Haldane became convinced that Darlington "stabbed him in the back" to advance his own career. This is discussed in greater detail in chapter 7, "Chemical Genetics."

Lionel S. Penrose (1898–1972)

In contrast to the Haldane-Darlington relationship, Haldane and Penrose[11] enjoyed a lifelong friendship that was both cordial and mutually beneficial. Penrose was a trained physician who got interested in genetics. His early research was on the genetics of mental defect. This early work was published under the title "A Clinical and Genetic Study of 1,280 cases of Mental Defect: Colchester Survey" in 1938. He emphasized that there were many different types and causes of mental defect and that normality and subnormality were on a continuum. Early in his career, Penrose advanced the study of schizophrenia and developed a test for its diagnosis.

Later, Penrose carried out pioneering work on Down's syndrome. He was the first to demonstrate the significance of maternal age. Penrose was born in London and educated at Cambridge. During World War II, Penrose and his family moved to Canada, where he was director of psychiatric research for Ontario (1939–1945). He wrote to Haldane in London at that time, requesting his help to obtain a professorship at University College London (UCL). With Haldane's support, Penrose returned to London as the Galton Professor of Human Genetics at UCL, a position he held from 1945 until his retirement in 1966. The archives at UCL indicate that Penrose's appointment at UCL itself was engineered by Haldane.

In his Foreword to Penrose's book *The Biology of Mental Defect*, Haldane wrote, "he has weighed the arguments in each case very carefully, and I know of no one better qualified to form a considered judgment. I hope therefore that his book will not merely be used as a text-book by specialists, but will be recognized as a contribution both to thought and to humanism. (1). Haldane and Penrose collaborated in estimating the first mutation rates in the human species.

Haldane and Penrose shared a similar outlook with respect to the goals of human genetics. They were especially united in their opposition to the aims and ideals of the eugenics association, which was supported by some prestigious scientists such as Julian Huxley. They were opposed to the sterilization of mentally defective individuals that was advocated by some eugenic groups. Both shared the belief that the science of human genetics was still in its infancy and could not justify such far-reaching decisions. Penrose was a genial and helpful man who gave his time freely to help his students and colleagues.

There were other colleagues In London who were close to Haldane. Hans Gruneberg was a Jewish refugee from Germany who was hired by Haldane. He later became a well-known expert on the genetics of the mouse, which plays a big part in biomedical research. He was reputed to be a very difficult man, an impression first conveyed to me by Helen Spurway but repeated by others as well. He was the author of *Animal Genetics and Medicine* and several other books. Hans Kalmus,[12] who was also a Jewish refugee, was a Czech doctor who worked in Haldane's department at UCL but never practiced in medicine in London. He was a dilettante of sorts, with varied biological interests, especially in sensory physiology and genetics. He was the author of a fine book on genetics, which was quite successful for a while. Haldane once stated that Hitler was indirectly responsible for staffing his department!

Cedric A. B. Smith (1917–2002)

While at UCL, Haldane attracted several bright young scientists who were keenly interested in studying and collaborating with him.[13] They included Hans Kalmus, Cedric Smith, Harry Harris, Jimmy Rendel, Hans Gruneberg, Ursula Philip, and later John Maynard Smith. They were all members of the combined department of Galton Laboratory and the Department of Biometry. For many years, the Galton Professor of Human Genetics was Lionel S. Penrose and the Head of the Biometry Department was Haldane. Because of his seniority, Haldane was head of the combined department, although he left much of the administration to Penrose. However, Penrose referred any new appointments to Haldane for final approval.

Cedric Smith was interested in the mathematical and statistical applications in genetics. He is best known for his collaboration with Haldane in applying the likelihood approach to estimate the linkage between the loci for color blindness and hemophilia on the human X chromosome. Their paper, published in 1947, was an extension of Haldane's earlier collaboration with Julia Bell in 1937. Many years later, when I edited a memorial tribute to Haldane, Smith contributed a fine chapter on testing segregation ratios in family data, an analysis that was pioneered by Haldane and others.

Smith continued to make important contributions to genetic statistics, especially genetic epidemiology. Among Smith's achievements is the most powerful test for mimic loci, which produce apparently same disease but are located in different chromosome regions. His best ideas were incorporated into genetic mapping, whereby many disease genes were localized as a necessary first step to sequencing. This provided a useful tool for clinical genetics and the development of the Human Genome Project. He is best known for his excellent book *Biomathematics*.

Personal Aspects

Cedric Smith was an unforgettable character of kind and gentle nature. When we first met at UCL in 1961, we found out that we were both vegetarians, which bonded us immediately. Smith was a Quaker and a pacifist. During World War II, he was a conscientious objector and spent the war years working as a porter at Hammersmith Hospital in London. Coincidentally, Dennis Mitchison, Haldane's eldest nephew, was a medical resident at the same hospital during wartime.

Smith was fond of recalling how Haldane interviewed him for his job at the Galton Laboratory. He was first interviewed by the Galton Professor, Lionel Penrose, who informed him that he would be hired only if Haldane gave his permission. In his first meeting, Haldane said, "Prof. Penrose seems to think we have a job for you, but I don't think so." Haldane then asked Smith, "Have you read my paper, so and so?" Smith answered, "No." Again Haldane asked, "Have you read another paper of mine, so and so?" Again Smith answered, "No." This went on for several minutes, Haldane asking Smith if he read any of his papers and Smith saying no. After a while, Smith feared that he would not get the job, as he had not read any of the papers mentioned by Haldane. As he was planning to leave quietly, Haldane finally said, "That's all, you are hired!"

Many years later, long after Haldane moved to India, Smith was appointed to the position that was previously occupied by Haldane—Weldon Chair of Biometry at University College London. That appointment was a matter of great pride and satisfaction to Smith. It was a well-deserved recognition!

John Maynard Smith (1920–2004)

The evolutionary biologist John Maynard Smith was trained in aircraft engineering but decided to study biology later. His early research was on the genetic variation in the fruit fly *Drosophila*, but he was best known for his concept of the evolutionary game theory. It has enriched the way we think of evolutionary stable strategies (ESS), in particular the evolution of characteristics whose reproductive advantage or disadvantage to an individual depended on the response of other individuals (summarized in his book *Theory of Games*, 1982). Whether Smith's mathematical theory can be converted to a predictive model in nature remains to be seen. Smith, who was a Marxist in his youth, has written that Haldane's Marxism had little impact on his scientific contributions. Unlike Smith, Haldane's students and associates during his next phase in India were apolitical. Their common bond with Haldane was in science, not politics.

Jacques Monod (1910–1976)

Haldane and the French biochemist Jacques Monod were both noted for bravery in wartime. Haldane was a fierce fighter in World War I and Monod was a brave resistance leader in occupied France in World War II. Both were distinguished scientists and decorated soldiers as well. And both were interested in enzyme chemistry and genetics. Monod admired Haldane greatly.

Jacques Monod shared a Nobel Prize with François Jacob and Andre Lwoff for their discoveries concerning genetic control of enzyme synthesis. Monod is widely regarded as a founder of molecular biology. He was head of the Department of Cellular Biochemistry at the Pasteur Institute in Paris when he was awarded the Nobel Prize. In 1971, he was appointed director of the Institute itself. He improved the financial condition of the institute and initiated many reforms to raise the quality of research.

In 1961, Monod visited Haldane in Calcutta and lectured at the Indian Statistical Institute about his operon model. I had the pleasure of meeting both Jacques Monod and his wife, Odette Bruhl, at that time. I met them again in the following year during a visit to Paris and enjoyed a lovely lunch at their Paris apartment. She was famous for her photographic book on Indian temples.

Jacques Monod was a suave and sophisticated gentleman who liked to dress well and spoke excellent English with a slight American accent. His mother was an American from Milwaukee, and his father was French. Monod was a talented musician who spoke several languages. He was also known for his philosophical writings, especially for his excellent book *Chance and Necessity* (1971). He was particularly noted for his bravery as a resistance leader in occupied France during World War II. Monod was awarded many honors, including both the Croix de Guerre and the American Bronze Star. He died of leukemia in 1976. A recent book narrates and analyzes the friendship between Monod and the writer Albert Camus.[14]

Sir Arthur C. Clarke (1917–2008)

Among Haldane's many admirers was the well-known writer of science fiction, Arthur C. Clarke,[15] whose work *2001: Space Odyssey* was a bestseller. Clarke was well known for his brilliant prediction of communication satellites. An interest shared by Haldane and Clarke, space flight, first brought them together. In 1951, as chairman of the British Interplanetary Society, Clarke invited Haldane to speak on the biological aspects of space flight. Haldane readily agreed, although he was invited at short notice as a substitute for J. D. Bernal. Haldane discussed three problems in his lecture: how men would live in spaceships, how they would live on other planets, and what sort of life they might find there. As Clarke had remarked, these were not subjects with which many reputable scientists cared to be associated at that time. Furthermore, Haldane himself was far too conservative in his early writings on the prospects for space flight. In his futuristic essay titled "The Last Judgement," Haldane set the first landing on Mars in the year 9,723,841 and an expedition to Venus "half a million years later!" Clarke remarked how hard it is for even the most farsighted scientists to anticipate the future. Haldane's address to the Society contained ideas that are still of interest today. He was one of the first to suggest that space voyages should be made during periods of minimum solar activity to avoid the dangers of solar flares.

Curiously, both Haldane and Clarke migrated to the East in 1957; Haldane to India and Clarke to Sri Lanka (or Ceylon). Their paths crossed again in 1960, when the Ceylon Association for the Advancement of Science invited Haldane to Colombo to address its annual meeting. During our visit to Colombo, Sri Lanka, in 1960, Arthur C. Clarke came to see us. But he did so with much trepidation. Clarke noted that experience in the article "Haldane and Space," which he contributed to my book *Haldane and Modern Biology*:

> I debated for a considerable time before calling on him. In the intervening years, I had heard rumors of his ferocity—some reports of his behavior to journalists made him sound similar to Conan Doyle's Professor Challenger—and I had no idea whether he remembered our last encounter, still less whether I was *persona grata*. Nevertheless, quaking slightly, with my partner Mike Wilson to give me moral and (if necessary) physical support, I called at his hotel and sent up my card. His first words when he arrived on the scene, dressed in his white gown and looking as a Hindu patriarch would, were not very reassuring. 'Oh, my god!' was distinctly discouraging, and a real or feigned deafness made any further communication appear hopeless. I was about to leave with as little fuss as possible when I suddenly realized that, far from being exasperated at the intrusion, he genuinely was glad to see me. I was not so surprised to discover that he had read most of my books, for Haldane, of course, had read *everything*.[16]

Clarke invited our party to dinner at his house in Colombo, which he shared at that time with Mike Wilson. Upon arriving, Haldane leaped on Clarke's technical library "like a starving man" (in Clarke's words).

Figure 15.2 JBS enjoyed children; here he is in a relaxed mood with the children of Hans Kalmus in the1940s, in Harpenden near London.

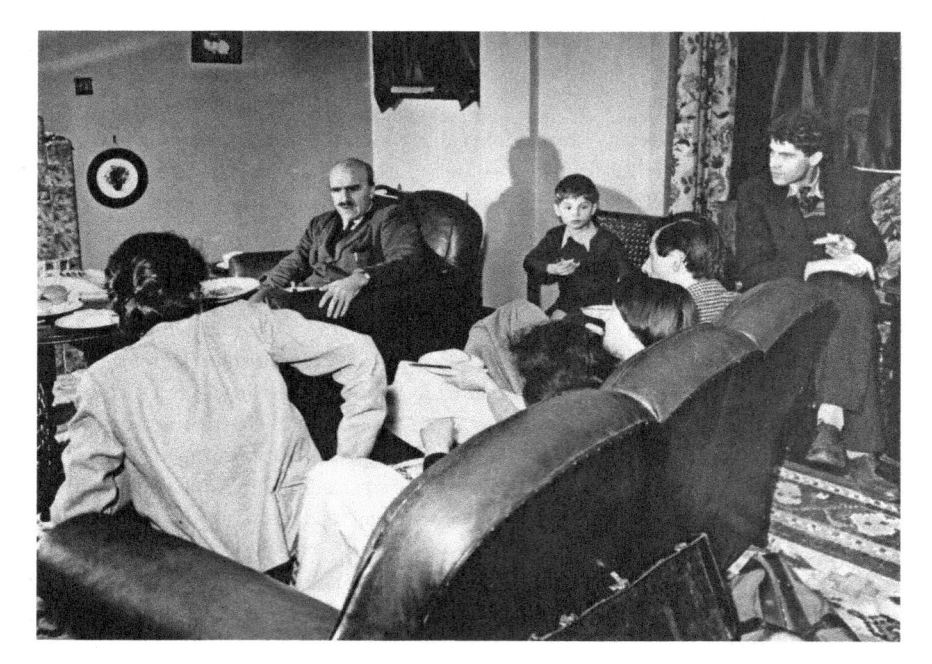

Figure 15.3 JBS enjoyed children; here he is in a relaxed mood with the children of Hans Kalmus in the1940s, in Harpenden near London.

After the dinner, Clarke's partner, Mike Wilson, screened his underwater movie, *Beneath the Seas of Ceylon,* showing the behavior of the teeming population of the Great Basses Reef, and in particular, recording the intelligence of a family of black groupers, *Epinephelus fuscoguttatus.* Clarke later recalled, "The spectacle of these giant fish cooperating as movie extras impressed Haldane so much that he frequently gave vent to a surprisingly schoolboyish 'Golly!'—a term which, for all its *naivete,* expresses the sense of wonder that is the hallmark of the great scientist."

Haldane and Clarke corresponded with each other until Haldane's death in 1964. Haldane's ideas in those letters were all related to space flight, astronautics, and life on other planets. Describing his plight after colostomy, Haldane remarked, "I (and a million other surgical cases) would be quite satisfied with lunar surface gravitation (1/6 g)."[17]

Finally, Clarke wrote, "It was a great shock to hear of his death.... And to realize that communication had at last broken with the finest intellect it has ever been my privilege to know."[18]

Many years later, shortly before his own death, Clarke contributed a fine Foreword to a book of Haldane's popular essays, which I edited. Indeed, his last question delivered to me by his assistant was about the publication of that book!

I was truly saddened upon hearing of his demise. Ever faithful Nalaka Gunawardene kept me informed of the last moments. Arthur was one of the most intelligent people I had ever met. He was an extremely sophisticated conversationalist with a sparkling wit and great imagination—a truly delightful and unforgettable experience!

Ernst Mayr (1904–2005)

Several colleagues and friends of Haldane told me that he was the most intelligent person they had ever met. The Harvard University biologist Ernst Mayr[19] and science fiction writer Arthur C. Clarke were among these. Mayr admired Haldane greatly but also questioned the value of the mathematical theory of evolution (later called population genetics), which was founded by Haldane, Fisher, and Wright. He called this early work "bean bag genetics," which I discuss in detail in my book *Haldane, Mayr, and Beanbag Genetics* (2011). Their early work, especially by Fisher and Haldane, considered, for the sake of mathematical formulation, only the simplest situations where genes were considered in isolation and interaction between genes was ignored. However, in later investigations they considered more complicated genetic situations. In response to Mayr's criticism, Haldane wrote an excellent, combative essay titled "A Defense of Beanbag Genetics."[20]

Mayr's early biological interest was in ornithology. Like Salim Ali, Mayr too studied under Erwin Stresseman in Berlin. Moving to the United States, he worked for twenty years at the American Museum of Natural History before joining the Harvard faculty in 1953. When Mayr became the editor of the journal *Evolution*, he invited Haldane to join its advisory board and contribute a paper on the quantitative aspects of evolution. Several years later, Mayr visited Haldane in India. Haldane showed him around the Calcutta zoo and took him to see the ancient temples of Orissa. Mayr enjoyed that visit very much, and was happy to include an early morning bird watching walk in Bhubaneswar.

T. Dobzhansky (1900–1975)

Theodosius Dobzhansky was an immigrant from Ukraine. He came to the United States on a Rockefeller Foundation Fellowship in 1926 to continue his genetic research. His specialty was fruit fly genetics, to which he made extensive contributions. Among many honors received during his lifetime, he was a recipient of the US National Medal of Science and several honorary degrees from major universities. Dobzhansky's career in the United States is divided among three universities: Columbia University and the Rockefeller University in New York, and the California Institute of Technology. After his retirement, he moved to the University of California at Davis, close to his former pupil Francisco Ayala. While he was working at the California Institute of Technology in 1932, Haldane joined him there as a visiting professor. He also met the president of Caltech, Thomas Hunt Morgan, who was awarded the Nobel Prize in 1933 for his pioneering research in mapping *Drosophila* genes. Dobzhansky's book *Genetics and the Origin of Species* (1942) was one of the foundations of the synthetic theory of evolution.

Dobzhansky and his wife, Natasha, visited the Haldanes in India in 1959. Haldane suggested his name as a possible foreign delegate to the Indian Science Congress in a letter to the prime minister of India Jawaharlal Nehru. Dobzhansky had originally intended to collect fruit flies from the Himalayas for his research and invited me to join him as a graduate student at Columbia University in New York. However, my plans changed when Haldane recommended me to work with Professor Pontecorvo at Glasgow University in Scotland. Dobzhansky recorded his travel experiences in India and elsewhere in his *Travel Letters*.[21] Dobzhansky died of leukemia in 1975.

Peter Medawar (1915–1987)

Peter Brian Medawar was born in Rio de Janeiro. His father was a businessman who was a naturalized British subject, born in Lebanon. Medawar was educated at Marlborough College and Magdalen College, Oxford, and studied zoology under Professor J. Z. Young. During World War II at the Burns Unit of the Glasgow Royal Infirmary in Scotland, he carried out research on tissue transplants, particularly skin grafting. That work led him to recognize that graft rejection is an immunological response. Medawar's discoveries in immune tolerance related to skin grafting earned him a Nobel Prize in Medicine in 1960.

After the war, Medawar continued his transplantation research and learned of the work done by the Australian immunologist Frank Macfarlane Burnet, who first advanced the theory of acquired immunological tolerance. According to that hypothesis, during early embryological development and soon after birth, vertebrates develop the ability to distinguish between substances that belong to its body and those that are foreign. The idea contradicted the view that vertebrates inherit this ability at conception. Medawar lent support to Burnet's theory when he found that fraternal cattle twins accept skin grafts from each other, indicating that certain substances known as antigens "leak" from the yolk sac of each embryo twin into the sac of the other. In a series of experiments on mice, he produced evidence indicating that, although each animal cell contains certain genetically determined antigens important to the immunity process, tolerance can also be acquired because the recipient injected as an embryo with the donor's cells will accept tissue from all parts of the donor's body and from the donor's twin. Medawar's work resulted in a shift of emphasis in the science of immunology from one that assumed a fully developed immune mechanism to one that attempts to alter the immune mechanism itself, as in the attempt to suppress the body's rejection of organ transplants.

Medawar's books include *The Uniqueness of the Individual* (1957), *The Future of Man* (1959), *The Art of the Soluble* (1967), *Pluto's Republic* (1982), and his autobiography, *Memoir of a Thinking Radish* (1986).

Medawar's and Haldane's careers overlapped briefly at University College London. Haldane was fond of recalling that his protege, Peter Gorer (1907–1961),[22] had discovered quite early the genetic basis of histocompatibility, the H-2 locus in the mouse, which became the foundation for all transplantation research.

Salim Ali (1896–1987)

Haldane greatly admired Sálim Ali, an Indian ornithologist and naturalist who became the legendary "birdman of India." He studied ornithology under the great biologist Erwin Stresseman at the University of Berlin. Salim Ali was among the first Indians to conduct systematic bird surveys across India, and his books on birds of India helped develop ornithology. He was closely associated with the Bombay Natural History Society and used his personal influence to seek government support for the organization, creating the Bharatpur bird sanctuary (Keoladeo National Park) and preventing the destruction of what is now the Silent Valley National Park. He was awarded India's second highest civilian honor, the Padma Vibhushan in 1976. Haldane and Salim Ali got along well. Ali's work met several criteria that Haldane considered essential for ideal scientific investigations of field ecology: it involved local species, it did not use expensive and complicated apparatus, it involved the appreciation and conservation of biodiversity and ecological resources, and it yielded data of great scientific interest. Salim Ali accompanied Haldane on several field expeditions in India. He was often mentioned by Haldane as a perfect example of what Indian scientists ought to be doing—using their own natural resources for studying science instead of imitating Western countries that were far better funded.

In his autobiography, *The Fall of a Sparrow*, Salim Ali wrote, "One of the most remarkable men it has been my good fortune to be associated with, even for the few short years he lived and worked in India, was Professor J.B.S. Haldane—a veritable giant both physically and intellectually. It was a fortuitous coincidence also that my earlier work on the ecology of Indian birds had come to the notice of Professor Haldane before he migrated to India, and had caught his fancy and won special approbation." Ali also noted that Haldane had emphasized the need to use local species and resources for research. He wrote, "Haldane lost no opportunity to emphasize that no advantage was being taken by Indian scientists of the matchless opportunities for biological field studies available at their doorstep—opportunities that were the envy of western biologists."[23]

During his last years, Ali collaborated with S. Dilon Ripley, secretary of the Smithsonian Institution, in Washington, DC, in writing several volumes of the *Handbook of the Birds of India and Pakistan: Together with Those of Bangladesh, Nepal, Sikkim, Bhutan and Sri Lanka*, a ten-volume set and a monumental contribution.

When Ali invited Haldane to contribute a paper to the *Journal of the Bombay Natural History Society*, Haldane wrote "The Non-Violent Scientific Study of Birds," suggesting several interesting ideas that could be used for research without killing birds. For instance, he wrote,

> What is the next step? . . . Ultimately we should look forward to a time when there will be an ornithologist for every hundred or so square miles of India capable of enumerating the local species, and a central organization such as the Bombay Natural History Society to make maps showing the distribution of each species in India. As, however, this would require ten thousand or so ornithologists it is not immediately possible. But a start can be made.[24]

Sir Ronald Fisher (1890–1962)

Ronald Fisher was a cofounder of population genetics with Haldane and Sewall Wright. Fisher's book *The Genetical Theory of Natural Selection*,[25] Haldane's *The Causes of Evolution*,[26] and Sewall Wright's[27] extensive paper on the shifting balance theory are widely acknowledged as the foundations of population genetics. Fisher was born in London on February 17, 1890. Because of his poor eyesight, he was tutored in mathematics without the aid of paper and pen, which developed his ability to visualize problems in geometrical terms.

Among his early publications was the groundbreaking paper of 1918, "The Correlation between Relatives on the Supposition of Mendelian Inheritance," which laid the foundation for biometrical genetics. It introduced the methodology of the analysis of variance, which was a considerable advance over the correlation methods used earlier. This paper showed that the inheritance of traits measurable by real values (i.e., continuous or dimensional traits) is consistent with Mendelian genetics. In addition to his work in population genetics, Fisher made extensive and very important contributions to the foundations of statistics.

Haldane and Fisher had a difficult relationship. In their younger years, there was an intense rivalry between them while they were laying the foundation for the mathematical theory of evolution in the 1920s, which later came to be called population genetics. During the war years they cooperated harmoniously. Both were professors at University College London. Many departments were evacuated to the countryside to escape the London blitz. Fisher was able to find space at the Rothamsted Experimental Station in Harpenden, where he had previously been employed for many years. He offered to share his space with Haldane and his staff, which was most welcome. Unfortunately, that cordial relationship between Haldane and Fisher did not last long. After the war, they took opposite sides in the Lysenko controversy, and their political differences made it impossible to maintain friendly relations.

After their retirement, Fisher moved to Australia, and Haldane to India. Ironically, their paths crossed again—in India, where Haldane was a research professor at the Indian Statistical Institute in Calcutta. Fisher occasionally stopped by to see its director, P. C. Mahalanobis, who was a close friend of both Fisher and Haldane. But they never spoke to each other again. I have described that period in my book *Haldane: The Life and Work of J.B.S. Haldane with Special Reference to India*.[28] I used to see Fisher occasionally at that time. I remember him as a short-tempered and difficult man who enjoyed walking around the campus of the institute. We used to chat about my research in plant breeding.

Sewall Wright (1889–1988)

Sewall Wright is best known as a cofounder of population genetics with R. A. Fisher and J. B. S. Haldane. Wright was born in Melrose, Massachusetts, to Philip Green Wright and Elizabeth Quincy Wright. His parents were first cousins, which may have led to Sewall Wright's choice of inbreeding for his early research while he was working for the USDA in Beltsville, Maryland. He discovered the inbreeding coefficient methods of computing it in pedigrees. He extended this work to populations, computing the amount of inbreeding of members of populations as a result of random genetic drift, and he and Fisher pioneered methods for computing the distribution of gene frequencies among populations as a result of the interaction of natural selection, mutation, migration, and genetic drift. The work of Fisher, Wright, and Haldane on theoretical population genetics was a major step in the development of the modern evolutionary synthesis of genetics with evolution. Wright also made major contributions to mammalian genetics and biochemical genetics.

Wright proposed the concept of *genetic drift*, which involves cumulative stochastic changes in gene frequencies that arise from random births, deaths, and Mendelian segregations in reproduction. Wright believed that the interaction of genetic drift and the other evolutionary forces was important in the process of adaptation. In order to evolve to another, higher peak in the adaptive landscape, the species would first have to pass through a valley of maladaptive intermediate stages. This could occur by genetic drift in small isolated populations. If there was some gene flow between the populations, these adaptations could spread to the rest of the species. This was Wright's shifting balance theory of evolution. Wright had a long and bitter dispute with R. A. Fisher, who believed that most populations in nature are too large to facilitate random genetic drift. Wright was a shy, retiring man who did not enjoy small talk. He was frugal in his lifestyle but generous in helping others. In contrast to his difficult relationship with Fisher, Wright got along well with Haldane. He contributed a fine article for my book *Haldane Modern Biology*,[29] a memorial tribute to Haldane. Wright outlived both Haldane and Fisher by many years and was working and publishing almost until his death at the age of ninety-nine years.

Norbert Wiener

Norbert Wiener (1894–1964) was a professor of mathematics at Massachusetts Institute of Technology (MIT). He was a famous child prodigy, who is considered the originator of cybernetics, with implications for engineering, systems control, computer science, biology, and philosophy. For several years during the 1930s and 1940s, Wiener and Haldane enjoyed a close friendship, Wiener attributing his major idea of cybernetics to a paper by Haldane on quantum theory.[30] They also shared a passionate interest in the cultural and intellectual traditions of India. Wiener was a vegetarian, a choice with which Haldane sympathized, and Haldane himself became a vegetarian after he moved to India in 1957.

During a visit to Cambridge University, Wiener spotted Haldane at the Philosophical Library in Cambridge and introduced himself after recognizing him from a photograph in a popular magazine that included a thriller called "The Gold-Makers" written by Haldane. Haldane welcomed Wiener's friendship. The Wieners and the Haldanes became close friends, and the four of them used to play bridge often. Wiener recorded in his autobiography, *I am a Mathematician*, "I have never met a man with better conversation or more varied knowledge than J.B.S. Haldane." Wiener wrote,

> We continued to see a lot of the Haldanes, and I used to go swimming with him in a stretch of the river Cam, which passed by his lawn. Haldane used to take his pipe in swimming. Following his example, I smoked a cigar and,

as has always been my habit, wore my glasses. We must have appeared to boaters on the river like a couple of great water animals, a long and a short walrus, let us say, bobbing up and down in the stream.[31]

Wiener visited Haldane again in the 1940s, after World War II. This time Haldane was professor of biometry at University College London, and was married to his second wife, the geneticist Helen Spurway, who was a lecturer in zoology at the same College.

Biju Patnaik (1916–1997)

Biju Patnaik was the chief minister (governor) of the State of Orissa in India when he invited Haldane to build and direct the Genetics and Biometry Laboratory in the state capital, Bhubaneswar. Haldane moved there from Calcutta in 1962, but death came too soon in 1964 for any long-term research association. Bijoyananda Patnaik was born into a princely family in Orissa on March 5, 1916. His reputation in India rested on his daring exploits as a pilot in the years before and soon after India's independence in 1947. Enlisting in the Royal Air Force, he combined derring-do on behalf of the British forces fighting the Japanese in Burma with secret missions on behalf of the independence movement.

In the first decade after independence of India, Mr. Patnaik concentrated on business, building an industrial empire in Orissa that included airlines, textile mills, iron ore and manganese mines, a steel mill, and plants manufacturing domestic appliances. All the while, he kept up his flying exploits, winning renown for daredevil flights that carried Indian soldiers into battle in Kashmir in 1947, and with a mission in 1948 at the direction of Prime Minister Jawaharlal Nehru, in which Mr. Patnaik rescued two key Indonesian independence leaders from a remote hideout in Indonesia and flew them to India, outraging the Dutch colonialists then ruling Indonesia. "Resurgent India does not recognize Dutch colonial sovereignty over the Indonesian population," he said.

Patnaik's name is permanently remembered for the Kalinga Prize, which he established, given for the best popularization in science each year by the UNESCO in Paris.

Notes

1. William Bateson (1861–1926) was a Cambridge University biologist and a leading founder of genetics. Bateson was regarded by Haldane as one of his three mentors: his father John Scott Haldane in physiology, Bateson in genetics, and F. G. Hopkins in biochemistry.
2. Norman Wingate Pirie (1907–1997), a British biochemist, discovered with Fredrick Bawden that a virus could be crystallized by isolating tobacco mosaic virus in 1936. This was an important milestone in understanding DNA and RNA. His early training in biochemistry at Cambridge University was provided by F. G. Hopkins and J. B. S. Haldane. Pirie remained a close friend of Haldane and visited him in India. He wrote the obituary notice of Haldane for the Royal Society of London. During World War II, Pirie investigated the possibility of extracting edible proteins from leafs.

3. Rene Wurmser, "Haldane as I Knew Him," in K. R. Dronamraju, ed., *Haldane and Modern Biology* (Baltimore: Johns Hopkins University Press, 1968), 313–17.
4. Nirmal Kumar Bose (1901–1972) was a leading anthropologist of India, who played a significant role in building an Indian tradition in anthropology. In India, Bose was a close confidant of Haldane, who never made a major decision without consulting Bose. Bose's student Ajit Kishore Ray collaborated with Haldane in publishing a paper on toe anomalies in India. J. B. S. Haldane and A. K. Ray, "The Genetics of a Common Indian Digital Anomaly," *Proceedings of the National Academy of Sciences, USA* 53 (1965): 1050–53. Bose was also a humanist scholar and a leading sociologist, Gandhian, and educationist. Bose served as Gandhi's secretary in the mid-1940s, and developed a broad appreciation for his ideology, which is reflected in his analytical *Studies in Gandhism* (1940). This period is covered in his *My Days with Gandhi* (1953). In 1929 he started the journal *Cultural Anthropology*, presenting a developing worldview of anthropology and culture. He was also the editor, from 1951 until his death, of the journal *Man in India*. He was the director of the Anthropological Survey of India from 1959 to 1964.
5. Mohandas K. ("Mahatma") Gandhi (1869–1948). Haldane was a great admirer of Gandhi. After he moved to India, Haldane tried to follow some of the principles enunciated by Gandhi. He adopted a vegetarian diet, wore *khadi* (handspun) clothes and sandals, and promoted nonviolent methods in biological research that did not involve the killing of animals. He expressed disappointment that many modern Indians were not adopting at least some of the major precepts preached by Gandhi. Haldane's personal library contained several books by Gandhi as well as many books and articles about him.
6. Juda Hirsch Quastel (1899–1987) was a British-Canadian biochemist who received his early training under F. G. Hopkins and J. B. S. Haldane at Cambridge University. In 1941, when Britain's wartime food supply emerged as a strategic concern, the Agricultural Research Council asked Quastel to lead a new research unit focused on improving crop yield at the Rothamstead Experimental Station. By analyzing soil as a dynamic system, Quastel was able to quantify the influence of various plant hormones, inhibitors, and other chemicals on the activity of microorganisms in the soil and assess their direct impact on plant growth. Best known among his discoveries is the compound commonly labeled as 2,4-D, one of the first systemic or hormone herbicides, a class of chemicals responsible for triggering a worldwide revolution in agricultural output and still the most widely used weed killer in the world. Haldane kept in touch with Quastel during his last years in India.
7. When Haldane was a research professor at the Indian Statistical Institute (ISI), during the years 1957–1961, the institute did not make it clear to him that the costs for publications and reprints would be paid, leaving some doubt about their intentions. Normally, such expenses are covered for many professors and their students in Indian Universities. I recall that Haldane wrote letters to the director seeking clarification, and when none was provided for a long time, he paid from his own pocket all the publication costs of our group, including his own.
8. J. S. Needham, *Chemical Embryology* (Cambridge: Cambridge University Press, 1931); *Perspectives in Biochemistry, thirty-one essays presented to Sir Frederick Gowland Hopkins* (Cambridge: Cambridge University Press, 1937); *Biochemistry and Morphogenesis* (Cambridge: Cambridge University Press, 1942).
9. J. S. Needham, "Organizer Phenomena After Four Decade: A Retrospect and Prospect," in *Haldane and Modern Biology*, ed. K. R. Dronamraju (Baltimore: Johns Hopkins University Press, 1968), 277–98.
10. This letter was kindly made available to me by Haldane's nephew, Professor J. Murdoch Mitchison (1922–2011) of Edinburgh University in Scotland. A cell biologist, Mitchison developed the yeast *Schizosaccharomyces pombe* as a model system to study the mechanisms and kinetics of growth and the cell cycle. The Nobel laureate Paul Nurse was one of his postdoctoral fellows.
11. Lionel S. Penrose (1898–1972) was the Galton Professor of Human Genetics at University College London and a loyal friend of Haldane. Penrose came from a family of distinguished

intellectuals; his son, Sir Roger Penrose, was a distinguished professor of mathematics at Oxford University. Among his numerous contributions to human genetics, Penrose discovered the maternal age effect on the occurrence of Down's syndrome. He was the editor of the *Annals of Human Genetics* for many years.

12. Hans Kalmus was among the Jewish refugee scientists from Berlin whom Haldane helped to find a position at University College London. Kalmus was originally trained in medicine in Prague but switched to research in genetic research in London. As a member of Haldane's Biometry Department, Kalmus was engaged in research in the genetics and sensory physiology of insects including the fruit fly *Drosophila*. He was the author of a bestselling text *Genetic* (London: Penguin Books, 1964), and an autobiography titled *Odyssey of a Scientist* (London: Weidenfeld and Nicholson, 1991).
13. Cedric Smith, "Testing Segregation Ratios," in *Haldane and Modern Biology*, ed. K. R. Dronamraju (Baltimore: Johns Hopkins University Press, 1968), 99–130.
14. Sean B. Carroll, *Brave Genius* (New York: Crown, 2013).
15. Arthur C. Clarke, "Haldane and Space," in *Haldane and Modern Biology*, ed. K. R. Dronamraju (Baltimore: Johns Hopkins University Press, 1968), 243–48.
16. Clarke, "Haldane and Space," 245.
17. Haldane's letter to Arthur C. Clarke, January 8, 1964, reproduced in *Haldane and Modern Biology*, ed. K. R. Dronamraju (1968).
18. Clarke, "Haldane and Space," 248.
19. Ernst Mayr, "Haldane's *Daedalus*," in K. R. Dronamraju, ed., *Haldane's Daedalus Revisited* (Oxford: Oxford University Press, 1995), 79–89.
20. J. B. S. Haldane, "A Defense of Beanbag Genetics," *Perspectives in Biology and Medicine* 7 (1964): 343–59.
21. B. Glass, ed., *The Roving Naturalist: Travel Letters of Theodosius Dobzhansky*, Memoirs, Vol. 139 (Philadelphia: American Philosophical Society, 1980). Dobzhansky recorded his visit to India and the Indian Science Congress that was held in Bombay (now Mumbai) in January 1960. Haldane and the rest of us flew there from Calcutta, as Haldane wanted us to present papers on our respective research projects. An incident at the opening session caught Dobzhansky's attention. The prime minister, Jawaharlal Nehru, as was his annual custom, attended the Congress and made a few welcoming remarks to all the delegates. This tradition, which was initiated by Nehru to indicate an interest in science at the highest level of the Indian Government, continues today; the prime minister always attends the Indian Science Congress on the first day and makes a brief speech. At the Bombay Science Congress (and at other conferences also) the audience is separated from the stage by a wide gap, for security reasons. The sunlight was streaming freely on the podium, especially on the speakers. Dobzhansky wrote in his *Travel Letters*:

> Then came Nehru, who proceeded to give hell to the officials. Could they not see that the sun was making the speakers and those in the semicircle (on the stage) miserable? ... After all this, Nehru delivered the speech which he had intended in the first place, a quite eloquent plea for science and scientists to help combat poverty and human misery of all kinds. It was a pleasure to listen, too, this man is so vibrant with emotion and so sincere! ... But I feel satisfied that good luck allowed me to see the Trimurti aspect of one of the greatest men of our times!" (276–77).

Dobzhansky's participation in the Indian Science Congress was engineered by Haldane himself, who urged his inclusion in his letter to the prime minister himself! *Trimurti* is the name given to a concept in Hinduism "in which the cosmic functions of creation, maintenance, and destruction are personified by the forms of *Brahma* the creator, *Vishnu* the maintainer or preserver, and *Shiva* the destroyer or transformer."

22. Peter Gorer (1907–1961) was a British immunologist, pathologist, and geneticist who studied genetics under J. B. S. Haldane at University College London. He was a codiscoverer of histocompatibility antigens and the elucidation of their genetic regulation. With George Snell,

he discovered the murine histocompatibility 2 locus, or H-2, which is analogous to the human leukocyte antigen. Gorer also identified antigen II and determined its role in transplant tissue rejection. He was elected a fellow of the Royal Society in 1960.
23. Salim Ali (1896–1987) was an Indian ornithologist who received his early training with Erwin Stresseman in Berlin. Salim Ali was among the first Indians to conduct systematic bird surveys across India. His books on Indian birds have educated millions of people. He was closely associated with the Bombay Natural History Society for many years and used his personal influence to gather support for the organization. He was awarded India's second highest civilian honor, the Padma Bhushan in 1976. Haldane had a high opinion of Ali's work in ornithology and recommended his books to his friends and colleagues. Ali was one of a small circle of friends who were close to Haldane in India. Ali wrote a charming autobiography titled *The Fall of a Sparrow* (New Delhi, Oxford University Press, 1985), quotes from pages 184–87.
24. J. B. S. Haldane, "The Non-Violent Scientific Study of Birds," *Journal of the Bombay Natural History Society* 56 (1959): 375–82.
25. R. A. Fisher, *The Genetical Theory of Natural Selection* (Oxford: Clarendon Press, 1930).
26. J. B. S. Haldane, *The Causes of Evolution* (London: Longmans, Green, 1932).
27. S. Wright, "Evolution in Mendelian Populations," *Genetics* 16 (1931): 97–159.
28. K. R. Dronamraju, *Haldane: The Life and Work of J.B.S. Haldane with Special Reference to India* (Aberdeen: Aberdeen University Press, 1985).
29. S. Wright, "Contributions to Genetics," in Dronamraju, *Haldane and Modern Biology*, 1–12.
30. J. B. S. Haldane, "Quantum Mechanics as a Basis for Philosophy," *Philosophy of Science* 1 (1934): 78–98.
31. Norbert Wiener, *I Am a Mathematician: The Later Life of a Prodigy* (Cambridge, MA: MIT Press, 1956), 160–62, 206–7, 314–15.

16

Moving to Paradise (1957)

In his profile of J. B. S. Haldane, titled "Cuddly Cactus," which was published in the *New Statesman and Nation*, Kingsley Martin wrote, "The balm of Sanskrit philosophy has assuaged the burns of Marxism."[1] Surely, if it had not been for his deep disillusionment with Lysenko and the gross mistreatment of geneticists under Stalin, Haldane might well have moved to the Soviet Union. Instead, he chose India, which interested him throughout his life from several points of view. Soviet's loss was India's gain!

From the time of his first visit to India in 1917, when he was sent to the British Military Hospital in Simla[2] to recuperate from wounds received in Mesopotamia (Iraq), Haldane took an instant liking for India, its people, philosophy, religions, and social traditions. His letters to his sister and parents from India give a clear picture of Haldane's first impressions of India. He was then twenty-five years old.

Unlike many of his British contemporaries visiting India during that period, Haldane mingled freely with the local people, ate in roadside stalls, chewed *pan*,[3] and drank unboiled water.

In a letter to his sister Naomi (Nou), dated September 1, 1917, from the field hospital in Simla, Haldane described how he was wounded by exploding bombs while trying to save some trucks and other property near an airplane hangar that was on fire. He was sent to Simla in the foothills of the Himalayas to recuperate. Haldane wrote, "I am glad we are giving commissions to Indians at last, but it will be very awkward indeed if they become generals and command British troops some of whom might not like it."[4] What is most interesting is that during those difficult times, when there was much mistrust between the British and the Indians, Haldane joined a group of mixed British and Indian people who met for a friendly chat on some noncontroversial subjects. Those early experiences had a profound impact on young Haldane's mind. He continued to display a keen interest in India and its people throughout his life. In 1923, in *Daedalus, or Science and the Future*, Haldane commented that Hindus appropriately appreciated the physiological relationship with the cow. Even in those far-off days, Haldane was drawn to India, not Canada, Australia, or South Africa or some other part of the British Empire. In his storybook for children, *My Friend Mr. Leakey* (1937), Haldane invented a character, a Hindu magician called Chandrajyotish, cleverly combining a highly imaginative story with some cleverly unobtrusive instruction in science. In his classic work *The Causes of Evolution* (1932), Haldane referred to the sanctity of peacocks in Hindu mythology, possibly because of the highly decorative tail in males, in a discussion of sexual selection in birds. As a biologist, he saw unusual connections between the wildlife, social and religious customs, and the evolutionary process.

During the years following World War II, Haldane began receiving frequent correspondence from Indian scientists. One typical case was that of V. S. Venkateswar, head of the Zoology Department at an undergraduate college (AVN College) in the university town of Visakhapatnam,[5] which is located on the beautiful coast of Bay of Bengal in the state of Andhra Pradesh. In his letter dated February 28, 1948, Mr. Venkateswar informed Haldane that he had been trying to secure a seat in the University of London to work for a PhD. Not having had any response in over two years, he sought Haldane's help. He wrote further that he was keenly interested in working for a PhD in genetics under Haldane's direction. Haldane responded by saying that postwar reconstruction work was progressing very slowly and space was still inadequate to recruit new students. However, he suggested that Mr. Venkateswar might apply to Professor Waddington at Cambridge University.

Another correspondent was S. P. Ray Chaudhuri of Calcutta University, whose laboratory was visited by Haldane during his post–Science Congress tour of Indian laboratories in 1952. Ray Chaudhuri received his PhD in genetics under the direction of H. J. Muller at the University of Edinburgh in the early 1940s. Muller later received the Nobel Prize for his discovery of the mutagenicity of X-rays in the fruit fly *Drosophila*. In his letter of June 12, 1952, Ray Chaudhuri requested Haldane to write in support of expanded research facilities in animal cytogenetics. The letter was to be sent to Dr. S. S. Bhatnagar, secretary of the Ministry of Natural Resources and Scientific Research, who was deeply involved in building several national laboratories in various sciences. One of the first ones was the National Physical Laboratory in Delhi. Haldane replied at once, stating, "I am particularly glad that you are asking for facilities for genetical work. It is badly needed." In his letter to Bhatnagar, Haldane requested that "in the interests of the development of Indian genetics, he may be given facilities for genetical work which, incidentally, requires little inexpensive apparatus" (see chapter 17, "Life in Paradise") A problem of a different sort was presented by Dr. J. J. Chinoy of the University of Delhi, in the field of agriculture. Chinoy was another scientist who had met Haldane during his visit to India in 1952, this time at the International Statistical Conference in Calcutta. The research was concerned with wheat breeding within the context of Indian climates, and the investigator requested Haldane to write to the Indian Council of Agricultural Research (ICAR) in support of expanded facilities and research programs. Haldane's response was predictable: "research in India may be expected to show characteristics of the wheat plant which could not be developed in cooler climates [i.e., in the United States and Sweden].... Unlike a good deal of other research which is being carried out in India, yours could hardly be carried out elsewhere." Haldane thought such research under varying climates would provide an opportunity to verify Lysenko's claims in the Soviet Union that at least some of the changes brought on by the climactic changes would be fixed in the descendants of the treated plants.

Haldane had resigned from the British Communist Party, stating that Lysenko's ideas were not biologically sound. Now he welcomed the opportunity to prove the falsehood of Lysenko's claims by promoting research in India. Furthermore, he thought that a genetical study of that nature should go hand-in-hand with research on the underlying biochemical variation as well. Haldane was in the habit of quoting the great Cambridge biochemist F. G. Hopkins, saying that as biochemical explanations are more fundamental than others that were in vogue at that time, they are essential to an understanding of biological variation. He wrote, "The biochemistry even of one variety of wheat under standardized conditions is very imperfectly known.... It is not hard to estimate sugars, total nitrogen, and so on, but very much harder to estimate vitamins or enzymes."

These exchanges were among several that are now part of the J. B. S. Haldane archives at University College London. Naturally, he was not able to help most of the correspondents from India, but he always responded politely, suggesting other possibilities. What was more important are the frequent contacts and bonds that were quickly drawing Haldane closer to India.

During one of his visits to India, before his final move in 1957, Haldane was taken to see a traditional Brahmin village in Orissa called Narsinghapur. Haldane translated that name as Man-lion-bury, which is named after one of the incarnations of Vishnu, a man-lion chimera who destroyed an evil demon called Hiranyakasipu. Haldane commented that Manlionbury is not unlike St. Pancras, where Haldane lived in London, or the Boulevard St. Michel in Paris, where some of his French colleagues lived. St. Michael, like Narasimha, is a chimera and a conqueror of demons. Haldane recalled his visit to the village of Narsinghapur. It was designed like the letter H. The central street was occupied by the higher caste, Brahmins and their families, and their servants lived in the other two streets. There was a village school and the library, and the meeting place of the Council on a platform under a tree. They were served a small meal, while the lower castes, sitting at a slight distance, provided some music. Haldane wrote,

> And then suddenly I thought of a College at Oxford, and everything became clear to me. A pious king had founded Man-lion College, as it might be St. John's or Magdalen. He had endowed two chapels and provided for a number of learned Fellows and Scholars. He had also given the College lands and hereditary College servants. I reflected that at Oxford the College servants did not dine at the High Table. Nor do the tenants of the College estates, who provide it with money rather than services. Nor did the servants at Man-lion College. Had they done so at Oxford, they and the Fellows would have been equally embarassed.[6]

Although Haldane had admired the Brahmins for their scholarship, he was aware of the past injustices when the Brahmins denied education for the lower castes. The Indian government later redressed the situation, reserving certain number of seats in universities for the children of lower castes. This is what is called "affirmative" action in the United States, but this too has come under fire and is being challenged in courts. Haldane wrote that these measures did not outweigh the huge injustices of the past.

University College London

Haldane's sister, Naomi Mitchison, recorded that he was not happy at University College (UCL), which raises the question: Was JBS, in fact, a happy man anywhere? At first glance, two aspects stand out about JBS's life—his brilliance and his complaints! His sister wrote,

> But JBS was not happy at University College. When I visited him there, he swept me, scowling, past colleagues of whom he disapproved. Helen, meanwhile, had seen some improbable, but apparently true, happenings among small fishes. New importances arrived in biology.... After all these wars, not quite knowing where we were, he and I played the dangerous game of half-believing in things like religions, as slippery as politics. In India that was very close to everyday life. JBS allowed himself to half-believe in such Indian religious ideas as fitted into his own non-religious ones.

Haldane was a member of the faculty at UCL for twenty years, from 1937 until 1957. As "Prof." at UCL, Haldane became a major scientific and social figure in British society. He promoted science through his lectures, scientific papers, popular essays, radio broadcasts, personal contacts, and travels in Britain and abroad. They were also difficult years because of his conversion to Marxism and the Lysenko controversy in the Soviet Union. At UCL, many colleagues of Haldane were genuinely sad about Haldane's departure. His colleagues arranged a farewell dinner, but Haldane would not sit in the place of honor.[7]

His portrait was painted and hangs today in the faculty common room. Long after his death, a lecture hall at UCL was named after him. An annual lecture in his memory was instituted by the Genetical Society of Great Britain and continues today.

Invitation from India

During the 1930s and 1940s, Haldane's support for the independence movement of India was particularly noted with gratitude by the Indian community in Great Britain and by the scientific community of India. He was a frequent visitor to the India League meetings in London. The first high commissioner of independent India to Great Britain (equivalent to an ambassador in the British Commonwealth), V. K. Krishna Menon,[8] was a close friend of Haldane who made it a point to include the Haldanes for all diplomatic receptions at the Indian High Commission.

Several years later, Krishna Menon became the defense minister of India at the time Haldane moved to the Indian Statistical Institute in Calcutta. On one occasion, Menon sent a plane to pick up Haldane, who was requested by Menon to advise Indian soldiers on problems of acclimatization to higher altitudes and cold temperatures in the Himalayas.

Indian Science Congress

Haldane had been thinking of visiting India for some time, more earnestly since his intellectual crisis during the Lysenko controversy. An excellent opportunity arose in 1951, when he received an invitation to attend the annual meeting of the Indian Science Congress,[9] which was to be held in January 1952. In a letter of invitation dated August 20, 1951, Menon, wrote, "I have very great pleasure in extending to you, on behalf of the Government of India and the Indian Science Congress Association, an invitation to attend the 39th Session of the Indian Science Congress Association to be held at Calcutta from the 2nd to the 8th January, 1952." Another letter came from the general secretary of the Indian Science Congress Association, B. Mukerji, inviting Haldane to stay "for a period of about one month after the session is over and deliver lectures in important academic and research centres in India." Haldane promptly responded, "In principle I am delighted and honoured to accept your invitation for the Indian Science Congress in January 1952."

On September 26, 1951, Haldane sent a letter to the provost of UCL, Mr. B. Ifor Evans, requesting leave to go to India. He wrote, "I should be glad if you would tell me whether the college would be prepared to pay my salary in my absence. You will, I am sure, excuse me for raising this point, as the college would very clearly be justified in discontinuing my salary whilst I am not doing the work which I have agreed to do, but if this view were taken, I do not think I could accept the Indian invitation." In response, the provost stated that this kind of matter should be presented to the college committee but personally he was in favor of Haldane's visit to India and he would recommend to the college committee to approve the visit. He added further that Haldane's visit would help the most important purpose of bringing Indian and British scientists together.

It is customary for the foreign guests of the Indian Science Congress to be taken around the national laboratories and other places of scientific interest. It is expected of each visiting scientist to deliver several lectures during those visits. To say that Haldane greatly enjoyed his visit would be an understatement. He not only had an excellent opportunity to learn much about Indian science but also was generous with advice for future research in India. After his return from India, in a letter to his friend Norbert Wiener, Haldane wrote, "I have been in India, and greatly enjoyed it. You ought to visit Bhabha, Mahalanobis, Saha and Raman. They are very civilized people, and you could get some of the juniors interested. Indian scientists, except at the top, are horribly specialized, and you would do them good."[10]

Letter to Prime Minister Nehru

Immediately upon returning from the Indian Science Congress, Haldane made a number of suggestions for improving research in India in a number of scientific disciplines. He prepared a report, which he forwarded to Prime Minister Jawaharlal Nehru with a cover letter (February 15, 1952).[11]

Haldane was aware that Nehru had earlier studied natural sciences at Cambridge University. One important point that Haldane emphasized was that India presents unique opportunities for nonviolent biological research. He stated further that he felt an obligation to assist a country "whose government is doing as much for world peace as the present Indian Government." In particular, he emphasized research in animal genetics and physiology—two fields to which Haldane made fundamental contributions. He concluded his letter with the following words: "However, I fully realize that the time has ceased when an Englishman can claim any right to advise Indians. If such a view is taken, I can make no complaint. If it is not, perhaps I may be of some service to India."

Haldane's admiration for Nehru can be readily seen from his article "A Rationalist with a Halo," which appeared in the *Rationalist Annual* (1954). Haldane wrote,

> Nehru is a rationalist, in spite of which he is an object of definitely religious veneration. All over India one can buy prints of artistic merit comparable with those of the Sacred Heart in Europe, in which Gandhi is shown ascending into heaven with a large halo and other paraphernalia, while Nehru, with a smaller halo, watches from the ground. But in spite of this halo Nehru has succeeded in making India, by its constitution, a secular state. This is a very great achievement.

Haldane commented further that the prospects of humanism in India are perhaps brighter than in some parts of Europe and America. Of Nehru, Haldane wrote, "The special cheers, which, as several people have told me, greeted him in the coronation process this year, show that he stands, in the minds of many British people, for something which the other Prime Ministers of the Commonwealth do not. He is a Republican, and an atheist, or at least a *nastika*."[12]

Haldane wrote to the provost of UCL with a brief account of his impressions of Indian science. Of particular interest was his suggestion that two other professors from UCL would benefit from a similar visit to India: Kathleen Lonsdale and Thomas Webster (art history). The provost, in turn, arranged a meeting with senior faculty, where Haldane narrated a brief account of his trip to India, which he enjoyed immensely.

Haldane thanked various people for arranging his visit to India, especially P. C. Mahalanobis in Calcutta, who initiated his participation in the Indian Science Congress. He addressed a letter to the Indian high commissioner (ambassador) in London, V. K. Krishna Menon, suggesting that a longer visit to a single research institution would be more beneficial to both sides instead of short visits to several institutions.

The Suez Crisis

It is obvious from a glance at Haldane's life that he was not the sort of man who would go quietly in any situation. He had been thinking of leaving Great Britain for some time. A combination of several political, social, and scientific factors led to that momentous decision.

Figure 16.1 Haldane announcing in 1956 that he would be moving to India soon.

In 1956, an opportunity arose when Israel, Great Britain, and France attacked Egypt in retaliation for the nationalization of the Suez Canal by the president of Egypt, Gamal Abdul Nasser. Haldane's friend Krishna Menon, who was then India's foreign minister and head of India's delegation to the United Nations, condemned the attack on Egypt in no uncertain terms. He saw it as a reversion to the old imperialist and colonial ways of Great Britain.[13]

Haldane promptly issued a press statement to the effect that he did not wish to live in a police state, a criminal state that had attacked Egypt. He added, for a good measure, "I want to live in a free country where there are no foreign troops based all over the place." He was referring to the presence of American troops on British soil. Tensions were high. Much of the British public was opposed to the attack on Egypt.

Police Dog Incident

Although Spurway was warned by the provost of UCL that if she chose to go to jail instead of paying the fine, she would have to resign her lectureship, she refused to pay the fine in Court next morning, as was required. On June 13, she was taken to Holloway Women's prison, where her treatment was far better than she had expected.[14]

In reality, for several years both JBS and Helen had been contemplating going to India after JBS's retirement, which was to happen in two years. Now other, fortuitous, events speeded up their plans. Far from being backed into a corner, Haldane welcomed these events, expressing pride in Helen's actions. Both sent in their resignations to UCL. Haldane wrote, "I wrote to India asking whether it would be possible for my wife and myself to obtain employment there next year . . . wish no longer to be a subject of a state which has been found guilty of aggression by the overwhelming verdict of the human race. I believe that we shall find in India opportunities for research and teaching in a country whose Government, by its active work for peace, gives an example to the world."[15]

In the meantime, Helen learned that a man called "Mr. Waller" paid the fine and arranged her release from Holloway. She was indignant. She thought it might be a reporter, hoping for an interview, in which case he was disappointed.

Haldane's protest against British policy in the Suez crisis and his move to India were widely reported in the press. However, Haldane had been known for his frequent public protests and controversial or even unpopular positions on matters ranging from the trivial to the profound. Initial shock and response soon gave way to tedium in some circles, when one learned that it was yet another crisis involving "Prof. Haldane." In his preface to Clark's biography of Haldane, Peter Medawar wrote that Haldane's extravagances became self-defeating; "he became a 'character', and people began laughing in anticipation of what he would say or be up to next. It is a sort of Anglo-Saxon form of liquidation, more humane but politically not much less effective than the form of liquidation he condoned. In the Russia of Haldane's day . . . Haldane would have been much more offensive and with very much better reasons, but he would not have lasted anything like so long."

While there is some truth in what Medawar wrote, it is also true, however, that Haldane acquired a loyal following during his long career, and many sympathized with his opposition to the British attack on Egypt. After four years in India, when the Haldanes returned for a brief visit to Great Britain in 1961, it was obvious that the outpouring of affection and respect displayed by many was quite genuine. While Haldane's comic opera, as some saw it, had its detractors, in scientific matters as well as in some political and social issues such as nuclear disarmament, it was the same "Prof." whose ideas and comments commanded great respect as always. The fact is that it was not hard to learn that one must make a distinction between the serious side of Haldane and the entertaining episodes with the inevitable press attention that seemed to occur periodically. It was the same Haldane in India, but this time with even greater latitude in behavior, which made life with him such an exciting and interesting experience!

University of Calcutta

Both Haldane and Spurway were looking forward to their life in India. With much foresight, Haldane had earlier agreed to an invitation by Calcutta University to give some lectures during his first visit to that University, suggesting the following topics:

Animal communication—two or three lectures
Animal population—two or three lectures
Natural selection—one lecture

Spurway had offered to give a course of five lectures on vertebrate genetics with special reference to results of captivity.

A Passage to India (1958)

In "A Passage to India," Haldane wrote,

> If I had accepted a post in the US, no one would have been surprised. They would not have called me a traitor because I was leaving a monarchy for a Republic, as I am on migrating to India. They would presumably have guessed that I was in search of a higher salary. This is commonly regarded as an adequate excuse. On February 21, 1957, the Archbishop of Canterbury (whose salary is more than thrice my own) denounced the desire for economic betterment as materialism. I am a materialist, but I am not particularly interested in making money. A bank balance is highly immaterial and idealistic. As a materialist I would sooner see ten people eating good dinners, which are material, than one with a good dinner and a large bank balance, and nine inadequately fed. But not being a materialist in the Archbishop's peculiar sense, I don't want to go to America. For one thing I prefer Indian food to American.[16]

Several different reasons ultimately motivated Haldane's move to India. The foremost among these was his deep-seated fascination for India, an inner desire to live in India some day, and finally an opportunity to realize those dreams. He was interested in almost every aspect of life in India, including the religions, philosophies, languages, literature, history, and social traditions.

Haldane had long believed that the tropical and subtropical fauna and flora of India offered excellent opportunities for research of a unique sort, such as the study of competition and cooperation among neighboring species and the regulation of biometrical variation. In 1958, his presidential address to the Centenary (Darwin-Wallace) and Bicentenary (Linnaeus) Congress, which was held at the University of Malaya in Singapore, Haldane stated, "Many ecological questions are fairly simple in the arctic regions, where a biotic community may consist of only twenty or thirty species; more difficult in the temperate zone, where a hundred or more species must be considered; and extremely complicated in a tropical rain forest, where it might be necessary to consider thousands of species, many as yet undescribed."[17,18] In a letter to Professor K. N. Bahl[19] of the University of Lucknow, who was a student of Haldane in physiology at Oxford University in 1919, Haldane wrote that the descriptions of Indian types would be of great value and would help to train the new generations of Indian biologists.

It seemed that India was his last frontier. His sister, Naomi Mitchison, wrote, "These were the last decades of my brother's life. He was no longer, on the whole, experimenting. He had come to the point where he was bound to try and see science as a whole, to calculate what might happen next, to develop a philosophy which could make sense of it. He had to look into the dark glass of the future."[20]

Sir Julian Huxley stated in his *Memories* that Haldane's choice of India was indicative of the treatment of science and technology in that country. Huxley wrote, "That Jack should have chosen India for his final refuge, rather than the U.S.A. or some European country, is a measure of the liberal treatment accorded here to science and technology. He was able to make some important contributions and initiate many projects, and found happiness and fulfillment in his Indian home until he died."[21]

Besides scientific and cultural reasons, Haldane mentioned from time to time that his move to India was motivated by several other factors. Some are as simple as "sixty years in socks is quite enough," which he often mentioned as one of the reasons. He was happy and relaxed in an Indian attire, which is more suited for a warmer climate than British or European-style clothes. He enjoyed wearing open sandals, which many Indians wear. No need for socks and shoes! Another reason was his desire to visit Hindu temples and shrines as much as possible. His scientific lectures were often punctuated by references to ancient Hindu scriptures and verses, which he compared with Greek mythology. He was fond of quoting such ancient Latin poets as Catullus, Dante and Virgil, proudly displaying his classical education at Oxford.[22]

In his later years, Haldane recalled that several of his classmates chose to become civil servants for the British Empire, which seemed, in those far-off days, to offer the endless promise of a golden future. As he settled comfortably in independent India in 1957, Haldane reminisced that many of those he had known in his youth had now become a bunch of disillusioned old men and women, deeply disappointed in the empire that no longer existed.[23] The end of the empire came sooner and faster than any one had expected. No other statement has been proved so false as the old adage: "The sun will never set on the Empire."

Notes

1. Kingsley Martin was editor of *The New Statesman and Nation* in 1956, when he published a biographic profile of Haldane, which also included a cartoon of Haldane drawn by Vicki (*The New Statesman and Nation*, January 7, 1956). It was reprinted in the collection *New Statesman Profiles* (London: Phoenix House, 1957).
2. Simla (or Shimla) is the capital city of the Indian state of Himachal Pradesh, located in northern India. Its name has been derived from the goddess Shyamala Devi, an incarnation of the Hindu goddess *Kali*. As a large and growing city, Shimla is home to many well-recognized colleges and research institutions in India. The city has a large number of temples and palaces. Owing to its scenic terrain, Shimla is home to the legendary mountain biking race MTB Himalaya. The event was started in 2005 and is now regarded as the biggest event of its kind in South East Asia.
3. *Pan* is betel nut wrapped in betel leaf with some other condiments such as cloves, fennel seed, and elachi, which people chew after a meal, to help the digestive process.

4. J. B. S. Haldane UCL Archives and information conveyed by Naomi Mitchison, 1985.
5. Vizag or Visakhapatnam: Large port city of southeastern India, located on the Bay of Bengal. It is noted for its large ship-building yard and several colleges and universities including a medical center.
6. J. B. S. Haldane, "A Passage to India," *Rationalist Annual* (1958). Reprinted in *Science and Life* (London: Pemberton, 1968), 127.
7. Haldane was usually reluctant to sit in a place of honor at dinners and banquets; instead he chose to sit with students. His colleague at University College London, Hans Kalmus, in his autobiography, *Odyssey of a Scientist*, wrote that Haldane refused to sit in the chair placed to honor him on the occasion of his move to India.
8. V. K. Krishna Menon (1896–1974) was an Indian nationalist, diplomat, and statesman, described as the second most powerful man in India. Noted for his eloquence, brilliance, and forceful, highly abrasive persona, Menon inspired widespread adulation and fervent detraction in both India and the West. The Indian president K. R. Narayanan eulogized him as a truly great man; decades after his death, Menon remains an enigmatic and controversial figure. Menon befriended Haldane while Haldane was still a professor at University College London. Their friendship continued over the years until Haldane's death in 1964. Haldane supported Menon's leadership in fighting for India's independence, which occurred in 1947. Menon continued to consult Haldane on various matters after Haldane's move to India.
9. Annual meetings of the Indian Science Congress are organized by the Indian Science Congress Association (ISCA), which is a premier scientific organization of India with headquarters at Kolkata. The association started in the year 1914 in Kolkata and it meets annually in the first week of January every year. Today, it has a membership of more than 30,000 scientists. The Association was formed with the following objectives: to advance and promote the cause of science in India; to hold an annual congress at a suitable place in India; to publish such proceedings, journals, transactions, and other publications as may be considered desirable; to secure and manage funds and endowments for the promotion of science including the rights of disposing of or selling all or any portion of the properties of the Association; and to perform any or all other acts, matters, and things as are conductive to, or incidental to, or necessary for, the above objects. The annual meetings are held at a different university each time and are attended by several thousand delegates. A small number of foreign delegates are invited each year from several countries.
10. J. B. S. Haldane, letter to Norbert Weiner at MIT in Cambridge, Massachusetts, May 6, 1952.
11. J. B. S. Haldane Archives, University College, London.
12. *Nastika*—Astika ("orthodox") and Nastika ("heterodox") are terms in Hinduism that are used to classify philosophical schools, according to whether they accept the authority of the *Vedas* as supreme revealed scriptures or not. In nontechnical usage, the term *astika* is sometimes loosely translated as "theist" while *nastika* is translated as *atheist*. However this interpretation is distinct from the use of the term in Hindu philosophy. Notably even among the *astika* schools, *samkhya* and the early *mimamsa* school do not accept a God while accepting the authority of the *Vedas*.
13. Michael Brecher, *India and World Politics* (Oxford: Oxford University Press, 1968), 62.
14. Helen Spurway, personal communication to the author, 1958–1959. Spurway was imprisoned in the Holloway Prison for women because she refused to pay the fine, which would imply an admission of guilt. Originally constructed by the City of London and opened in 1852 as a mixed prison, Holloway Prison became all female in 1902. Regime includes both full-time and part-time education, skills-training workshops, British industrial cleaning science (BICS), gardens, and painting. There is a fully integrated resettlement/induction strategy, which identifies individual needs and provides a structured approach for advice and guidance on such issues as housing, benefits, training, and community volunteering programs, as well as anger management, assertion training, and domestic violence. Other programs include desktop publishing, and individual needs-based work with a variety of partnership agencies.
15. Haldane, "A Passage to India," 124.
16. Haldane, "A Passage to India," 124–34.

17. J. B. S. Haldane, "An Autobiography in Brief," *Illustrated Weekly of India, Bombay*, (1961). Reprinted in Krishna R. Dronamraju, ed., *Selected Genetic Papers of J.B.S. Haldane* (New York: Garland, 1990), 19–24.
18. J. B. S. Haldane, "Presidential Address: The Theory of Natural Selection Today," in *Proceedings of the Centenary and Bicentenary Congress of Biology*, Singapore, December 2–9, 1958, ed. R. D. Purchon (Singapore: University of Malaya Press, 1960), 1–7.
19. J. B. S. Haldane, letter to Karam Narain Bahl, September 15, 1943. Bahl was a former student of Haldane at Oxford University, zoologist, professor of zoology, Allahabad University and other universities in India.
20. Naomi Mitchison, "Foreword," in K. R. Dronamraju, ed., *Haldane: The Life and Work of J.B.S. Haldane with Special Reference to India* (Aberdeen: Aberdeen University Press, 1985).
21. Julian Huxley, *Memories* (London: Harper & Row, 1970), and *Memories II* (London: Harper & Row, 1973).
22. "Greats"—*literae humaniores* is the name given to an undergraduate course focused on Classics (Ancient Rome, Ancient Greece, Latin, ancient Greek and philosophy) at the University of Oxford and some other universities. Haldane was proud of his classical education at Oxford University. He enjoyed quoting from the Classics during his scientific lectures.
23. J. B. S. Haldane, personal communication to the author, 1960.

PART 1960S

17

Life in Paradise (1957–1964) (Death)

> I fell in love with India in 1917, but I saw no point in going there till it was independent, as I could not associate with Indians on a basis of equality.
>
> —J. B. S. Haldane, "A Passage to India,"
> *Rationalist Annual* (1958, 132)

Haldane did not wish to move to India as a representative of the British Empire. In closing his famous essay "A Defense of Beanbag Genetics," Haldane wrote, "Meanwhile, I have retired to a one-storied 'ivory tower' provided for me by the Government of Orissa[1] in this earthly paradise of Bhubaneswar and hope to devote my remaining years largely to beanbag genetics."

JBS relished adventure and controversy all his life. He loved to argue and fight. Where many others would prefer a quiet alternative, JBS preferred confrontation. He chose the ultimate adventure in his last years, to live and die in far-away India. And life for Haldane was a continuous adventure with its many ups and downs. Soon it became evident to his former colleagues in England, it was the same Haldane with the usual crises, now living in India. In his short life of six years in India, Haldane went through two resignations, fasted as a protest in the Gandhian tradition, wrote numerous popular essays, traveled widely in several countries, became an Indian citizen, taught courses in genetics, and trained several research students and associates. In addition, he found time to serve on international advisory boards, participate in several scientific conferences, and advise the Indian Government and other bodies on matters related to diving physiology, space flight, and agriculture, among others. Not surprisingly, he was involved in several controversies and debates. Shortly before his death in Bhubaneswar, Haldane wrote that as he lay dying he hoped that he would have time to think of his life and say: "It was a good show!"

And indeed it was. As soon as he arrived in India, Haldane did not waste a moment; he gathered a group of young researchers at once and started his research program at the Indian Statistical Institute in Calcutta. I was one of those youngsters, just barely twenty years old when I first wrote to him and joined his team at the end of 1957. However, before I met him in person I had already heard his voice on All India Radio, when he was invited to deliver the annual lecture in memory of the former deputy prime minister of India, Sardar Vallabhbhai Patel,[2] much respected leader of India's independence movement. Before I met Haldane, I was interviewed in New Delhi (where I was staying then) by the director of the Statistical Institute, Prasanta Chandra Mahalanobis.[3]

The most important reason for Haldane's decision to move to India was scientific. In an article, titled "A Passage to India,"[4] written for the *Rationalist Annual*[5] just before his move to India, Haldane wrote,

> Perhaps my main reason for going to India is that I consider that the opportunities for scientific research of the kind in which I am interested are better in India than in Britain, and that my teaching will be at least as useful there as here. This may seem surprising. It is, however, a fact that the facilities for research now available to me in the University of London are somewhat less than they were twenty years ago, and would in any case have been reduced to zero or near it when I was superannuated in 1960. No doubt if I had devoted half the time which I have spent on research to extracting money from various available sources such as the Nuffield Trustees, the Agricultural Research Council, I could have had the facilities for research instead of the time for it.

About facilities and resources in India, Haldane wrote, "Of course, if my work required electron microscopes, cyclotrons, and the like, I should not get them in India. But the sort of facilities which Darwin and Bateson used for their researches—such as gardens, gardeners, pigeon lofts, and pigeons—are more easily obtained in India than in England." Among other attractions of India, Haldane added, "Economically I shall be better off in India than here ... taxation is lower ... three cheroots for two pence. One does not need a fire, glass windows, or socks. The climate is delightful.... The climate indoors is excellent if one has an electric fan."

Haldane eventually became an Indian citizen in 1961. Although not a religious man himself, he was keenly interested in the Hindu religion. He became a complete vegetarian and started wearing Indian-style clothes. In 1961, Haldane described India as "the closest approximation to the Free World." His American colleague Jerzy Neyman, a professor of statistics at the University of California at Berkeley, objected to this premise. Neyman gave his impression that "India has its fair share of scoundrels and a tremendous amount of poor unthinking and disgustingly subservient individuals who are not attractive." Haldane retorted at once, "Perhaps one is freer to be a scoundrel in India than elsewhere. So one was in the U.S.A in the days of people like Jay Gould, when (in my opinion) there was more internal freedom in the U.S.A than there is today. The 'disgusting subservience' of the others has its limits. The people of Calcutta riot, upset trams, and refuse to obey police regulations, in a manner which would have delighted Jefferson. I don't think their activities are very efficient, but that is not the question at issue." Neyman never responded.[6]

The word "freedom" in this context should be defined in its broadest sense when referring to Haldane's lifestyle, clothing, dietary habits, intellectual pursuits, and scientific interests. In India, it was a slightly different Haldane who made headlines because he fasted to protest against something or other, because he was expressing admiration for some Hindu Gods, and so on. Occasionally, his intentions were misunderstood. On one occasion, Haldane and his wife, Helen Spurway, went to a cattle fair and he started examining closely some cows for genetic variation. The local press reported at once that Professor Haldane was interested in cow worship, which is an important aspect of Hinduism, especially for certain sects in northern India. Haldane quoted freely from memory various passages from Hindu epics and scriptures such as the *Rigveda*[7] and *Upanishads*.[8] He was well versed in the classical literature of India and was often heard drawing parallels between ancient Sanskrit verses and classical Western writings as well as those in ancient Latin and Greek literature. It is thus clear that his intellectual horizons expanded considerably during his Indian period. Haldane's eccentricities blended smoothly into the Indian social scene. He became a nationally recognizable figure with a loyal following of his own. Cartoons and photographs of Haldane appeared in the popular press, such as the *Shankar's Weekly*.[9] A certain latitude in the nature of scientific projects, political views, social or public behavior, style of clothing, and general appearance are not unusual in Indian society. Haldane fully enjoyed this new freedom in India, which was absent in his previous life in Britain. He was obviously very happy to have been accepted into the Indian society.

Haldane enjoyed walking in the countryside, surrounded by his students, occasionally examining the plants and flowers wayside, and observing birds and insects. He was an impressive figure in flowing robes in the Indian countryside, puffing a cigar or pipe, walking with long strides, and keeping us on our toes with his frequent questions about the fauna and flora. We had long conversations about various scientific subjects as well as our individual projects. Mrs. Haldane often accompanied us, collecting caterpillar food, observing birds and butterflies, and to our amusement, occasionally correcting Haldane about something or other in a loud voice. As our small group continued walking, others—total strangers—used to join us. Our numbers swelled, getting bigger and bigger. Some of those people who followed us had no clue who Haldane was. He was impressive, tall and bald headed, wearing saffron or white robes and surrounded by disciples, he was occasionally mistaken for a "holy" man or a sage of the kind one encounters frequently in India. I recall one air hostess approached me on a plane, asking, "Sir, this holy man, which *Ashram* is he from?" I could have replied, "From the *Ashram* of science!"

Nuisance to the State

Five years after Haldane's move to India, he was described in print as a "Citizen of the World" by an American science writer named Groff Conklin.[10] Haldane responded,

> No doubt I am in some sense a citizen of the world. But I believe with Thomas Jefferson that one of the chief duties of a citizen is to be a nuisance to the government of his state. As there is no world state, I cannot do this. On the other hand, I can be, and am, a nuisance to the government of India, which has the merit of permitting a good deal of criticism, though it reacts to it rather slowly. I also happen to be proud of being a citizen of India, which is a lot more diverse than Europe, let alone the U.S.A, the U.S.S.R or China, and thus a better model for a possible world organization.... So, I want to be labeled as a citizen of India.[11]

Haldane's move to India was facilitated by the director of the Indian Statistical Institute (ISI), Prasanta Chandra Mahalanobis, who invited Haldane and his wife, Helen Spurway (who was a scientist in genetics), to join the institute's faculty. Haldane and Mahalanobis corresponded frequently before his move to India in July 1957. Haldane was happy to receive most welcome letters from Mahalanobis while he was still living in London.

On April 2, 1957, Haldane wrote to the director of ISI:

Dear Prasanta:

I have received your two most welcome letters of 27/2/57 and 9/3/57. It seems that we are likely to live in Bengal in a luxury to which we are quite unaccustomed. I am already beginning to imagine how I could spend the income which will accrue to us if the Journal of Genetics pays its way. I have at once thought of mural paintings for our house, particularly of Yudhishtira[12] and the dog being forbidden entry to *svarga*.[13] However, one must have walls before one has murals. And I should like an approximate estimate for a house, including a good deal of space for bookshelves, and at least one room with bath etc., for guests. My wife owns a house here, and should be able to sell it and pay for one in Bengal ... There is a lot of anti-Indian propaganda here. And it is not being properly answered. The Indian official statements about Kashmir are very staid and legalistic, as they probably should be. I shall have the chance of talking to at least one public meeting before I go, and shall make the following points among others. If I were a Shiah, an Ahmadiyya, a member of several other muslim sects, or even a muslim who wished to reduce islam to a rather abstract monotheism, I should be a good deal safer in India than in Pakistan. I shall be willing to support a plebiscite on Kashmir *after* we have had one in Britain on the question of whether we want American armed forces on our soil. I dare say there might be a majority in their favor. But the minority opposed to them, which would include many conservatives, would be too large for the comfort of Mr. Dulles. The first point is quite important. The Indian government can make neither point. I am not qualified. I know I am an inefficient politician. But if nobody else will say such things, I must.

By the way, I am putting "and from the Indian statistical Institute" on some papers which will only be published after I have taken up residence there.... Dr. Carter, at Harwell, is beginning to get quantitative results on mutations in mice provoked by gamma rays, using, among others, a method which I invented.[14] I am sorry to say that the estimate of the rate at which lethal mutations occur which I gave at the ISI last year is being confirmed. More accurately the dose in roentgens needed to produce one is estimated at 400 r where I said 300. But on the available data it could well be 300 to 500. We shall know more in a year. But meanwhile the air will have been still more poisoned

Yours sincerely,
John Haldane

P.S. ... Helen is very worried about her technician. She can hardly start a practical class till one is trained in *Drosophila* work. And as you are spending extra money on microscopes to avoid delay, it might be worth starting training a technician.

One of the topics discussed in their correspondence was a course of lectures in general sciences for students with some "grounding" in mathematics, physics, and chemistry, which Haldane had outlined in a letter[15]:

1. Large scale cosmology. Methods of estimating large distances and times: Estimates exceeding 1 kiloparsec (3×10^{18} cm) based on statistics. Maximum time and distance now measurable are about 1.5×10^{17} sec. and 4×10^{24} cm.
2. Small scale cosmology. Evidence for atomic distances. Smaller distances based on statistical estimation of target areas. Smallest measurable distance about 10^{-12} cm. Smallest time about 10^{-20} sec. The uncertainty principle. Ignorance of processes lasting less than 10^{-5} sec. Mass and Energy measurements.
3. Scientific methods as functions of space and time scales. Range of apparatus 10^4 cm (radio-telescope) to 10^{-16} sec (flash photolysis). The sciences and their inter-relations.
4. Counting and measurement.
5. Interaction of man and other animals with their environment.
6. Dimensional analysis. Models in science and engineering.
7. Human needs, individual and social.
8. Temperature coefficients of chemical reactions.
9. Energetics of living beings.
10. Radiations. "Windows" in upper atmosphere.
11. Touch. Measurement of small forces. Perception and measurement of time.
12. The chemical senses. Some methods in chemical analysis. Biological assay.
13. Chemical structures of living beings.
14. The diversity of living beings.
15. Geology. A brief sketch of its methods.
16. Evolution. The evidence. General conclusions.
17. Communication. Communication within a cell.
18. Communication between cells. Hormones, nervous , and their functions.
19. The internal environment of men and animals. Self-regulating machines. Self-regulation of plants and animals.
20. Communication between animals. Animal societies. Animal behavior. Human languages.
21. Animal technology. (Human technology and science.)
22. Scientific method.

On May 14, 1958, Haldane sent a progress report to Mahalanobis, stating that he wasted a good deal of time in the first year because he had to learn what is easy and what is difficult at the institute. But he promised to be a good deal more efficient next year. He wrote further, complaining about the noise in the neighborhood: "We are tolerating the weather very well. Far worse than the heat are the loud speakers. In the past week we have had one, apparently for a wedding, on the other side of Amrapali [director's residence] but so loud as to interrupt lectures, and one across the Barrackpore Trunk Road which made conversation in my flat difficult. The heat makes me very sleepy, and I can speak for at least 9 hours a day."

Haldane complained that it is not adequately emphasized in teaching the students at the institute that new results in science appear when you notice something you were *not* looking for. He mentioned C. R. Rao, who was in charge of the teaching, in that context to emphasize that the teaching at the institute has only followed traditional lines so far. It has not been designed to prepare the students to expect the "unexpected" results, which is necessary for excellence in research. On another occasion when the famous geneticist T. Dobzhansky was visiting the institute, Haldane received a note from C. R. Rao, "kindly requesting" him to chair a lecture by Dobzhansky. Haldane turned to Dobzhansky, who was sitting in the car with us, and said, "it should say 'inviting' me to chair the session." On other occasions, Rao approached me to convey messages or transfer some papers to Haldane. The institute staff were generally reluctant to contact Haldane directly and were content to use my services for that purpose. Both Haldane and I shared his extensive library, and I was in touch with him on a daily basis. I knew his moods and work habits much better than anyone else. Soon after my arrival, Haldane began entrusting me with various personal and professional matters and was happy to depend on me as a conduit for contacting local people. This was partly due to his lack of familiarity with local traditions and customs but also because he did not know the local languages for communicating with local people and I was able to speak Hindi fluently, which suited most occasions. In research matters, Haldane encouraged me to contact the director Mahalanobis directly, as he did himself, bypassing all others in the institute administration. It saved us precious time, as the director was the only one who made decisions concerning our group.

In another context, Mahalanobis briefly advised Haldane on how communication could be easily misunderstood at points of contact between two cultures. He wrote,

> You would remember that in 1951 I had asked Helen to stay with us; but you told me that you would not like to cause us any inconvenience. This, according to my code of behaviour, implied that you felt it would be inconvenient for you to stay with us. Fortunately, we discovered the mistake in time. If this could happen with me (who knows Anglo-American codes better than most of my countrymen) more serious misunderstandings are, of course, likely to occur with persons who have never gone out of India.

He explained further, "I hope you will now understand why I should like to have all your requests, requisitions, and complaints, if any, ... and suggestions for improvement in writing, directly, from Helen or you. I feel we must eliminate intermediaries completely."[16]

Mahalanobis also drew attention to another interesting facet of cultural differences. It came to his attention that the Haldanes were unhappy with the food at Giridih, where one of the branches of the ISI was located. Helen (Mrs. Haldane) used to go there to buy Tussore silkworm cocoons from the Santal tribal market for her research. Interestingly, neither Haldane nor Helen ever complained about the food, but one of their colleagues mentioned it in a conversation with the local staff. Mahalanobis wrote,

> I am not surprised. This often happens in India. In fact we have many proverbs about persons who wait upon or who work with people of importance.

They are known to be more solicitous about their chiefs than the chiefs themselves. One well-known Bengali proverb is, "Sand is hotter than the Sun." The famous Bengali poet and Nobel laureate, Rabindranath Tagore, has written "What the king, says, his couriers magnify a hundredfold."

I have recorded the details of Haldanes' life in India in my book: *Haldane: The Life and Work of JBS Haldane with Special Reference to India*, with a Foreword written by his sister Naomi Mitchison.

In a clearly defensive article,[17] which was published in the *Rationalist Annual* in 1958, Haldane answered his critics point by point from a rationalist's point of view. He wrote: "But, I am told, I am going to a country riddled with superstition, a country where the Brahmins regard themselves as superior to everyone else, and are only waiting to fix their yoke again on the rest of the population." Haldane wrote that he had a sneaking sympathy with the Brahmins.

Haldane was interested in ideas that are similar in Hindu religion and evolutionary biology. For example, he quoted Professor Birbal Sahni,[18] a distinguished Indian paleontologist, who pointed out that the *avatars* (incarnations) of Vishnu give a rough, though not very accurate, picture of vertebrate evolution—a fish, a tortoise, a boar, a man-lion, a dwarf (perhaps Australopithecine), and then four men. Haldane thought of Narasimha as a rather generalized mammal, perhaps living in the paleocene.

Haldane wrote that he would rather live among adherents of a religion that only tells them what to do and what not to do than another group whose religion also tells them what to believe and what not to believe. He wrote, "I certainly object to a religion like Islam which tells its adherents in great detail both what they must do and what they must believe. A Hindu must, I suppose, believe in some god or gods. But if he chooses, he has scriptural authority for the opinion that the gods only exist in human minds." He quoted relevant passages from the Brihadaranyaka Upanishad (Translated by S. Radhakrishnan, later president of India):[19]

> "This is the highest creation of Brahma, namely, that he created the gods who are superior to him; he, although mortal himself, created the gods who are superior to him, therefore this is the highest creation. Verily he who knows this becomes (a creator) in the highest creation. So whoever worships another divinity (than himself) thinking that he is one and (Brahman) another, he knows not; he is like an animal to the gods; as many animals serve one man, so does each man serve the gods; even if one animal is taken away, it causes displeasure; what should one say of many (animals)? Therefore it is not pleasing to those (gods) that men should know this."

For the author of these texts, Brahma clearly meant the human mind, or what is common to all individual minds. Haldane commented, "Being an idealist, he regarded the creations of this mind as having an existence of their own. The nearest equivalent to this attitude in European literature is perhaps to be found in William Blake's writings, particularly in the last night of *Vala*."

He emphasized further that he was going to India because he was a socialist, and socialism is not only more just but also more efficient than capitalism, that is to say, that it increases the national income and the standard of living more rapidly. He was also going to India because he considered that the recent acts of the British government (in 1957) had been violations of international law. Haldane was referring to the Anglo-French attack on Egypt when Nasser nationalized the Suez Canal. Although that incident acted as a trigger to make his final decision, he had in fact been making plans to move to India for several years.

The Story of Yudhishtira

Haldane's favorite story of Hinduism is from the epic *Mahabharata*. The heroes of the *Mahabharata* are five brothers called Pandavas. They were the victors of the great mahabharata war, in which they had killed all their kinsmen and their friends. They were filled with remorse and great sadness. They renounced the world and set off on a pilgrimage around India. A stray dog joined the party and followed them faithfully. As they climbed the Himalayas to reach *svarga* (heaven), all others except the oldest brother, Yudhishtira, and the dog fell down and died along the way. When they reached the gate of heaven, they were welcomed by the God of Heaven, *Indra*, who told him that his wife and brothers were waiting for him in heaven. However, Yudhishtira was also told that no dogs were allowed in heaven. But Yudhishtira refused to enter heaven without the dog, because a nobleman must not abandon any creature which has put its trust in him. The dog, in fact, turned out to be his father, *dharma*, the god of justice, and they were both allowed into heaven. The moral is that a man must not take any action that he regards as dishonorable, even if ordered by an important god such as *Indra*.[20] Haldane wrote that he knew of no equivalent myth in the ancient or modern religions of Europe. He wrote, "If Abraham had refused to kill Isaac at the divine command, I should have more respect for the Old Testament."[21]

Indian Statistical Institute

In July 1957, I had read in a Delhi newspaper that a great scientist called Haldane and his wife, Helen Spurway, who was also a scientist, were migrating to India to teach and conduct scientific research at the Indian Statistical Institute (ISI) in Calcutta. That article informed me that Professor Haldane would be building a team of young scientists to pursue research in genetics and biology at that institute. Having just obtained my master's degree in genetics, I was keenly interested in finding a research position, and that institute appeared to suit my needs. I was already familiar with the Indian Statistical Institute, as my uncle, V. R. Rao, received his early statistical training there. I wrote a letter to Haldane immediately, expressing my interest to conduct research under his direction. To my great surprise, he replied immediately, asking me to prepare answers to three questions in plant genetics, which were concerned with sex determination in papaya, hybridization method in rice, and propagation in mangoes. As these three topics were well within my area of competence, I was able to prepare a lengthy reply and mailed it within a few days. Again I heard from Haldane within a few days thanking me for my reply and expressing great satisfaction with my answers.

I was, of course, overjoyed to receive such a positive response from the great scientist. We continued our correspondence about further details for a research program that I would want to initiate at the ISI. In the meantime, I received a letter inviting me to meet with the director of the ISI, Professor P. C. Mahalanobis at his office in New Delhi, not far from where I was staying. Haldane, who had never met me at that point, told Mahalanobis that he was interested in hiring me and would like to hear Mahalanobis's personal opinion. I had already heard of Mahalanobis, who was a close friend of the prime minister and a member of the National Planning Commission, and I was tremendously impressed that such an important person would like to see me, a young graduate who was barely twenty years old. That interview was quite brief. Mahalanobis asked me questions about my knowledge of the local resources in genetic research and agriculture, and was happy to send a favorable report to Haldane. After completing some formalities, I finally started my work with Haldane toward the end of 1957.

At the time of my arrival, the ISI was a large sprawling institution occupying several city blocks along Barrackpore Trunk Road, or B.T. Road in the Baranagar suburb of Calcutta. The institute was founded by Mahalanobis, in 1931, as a statistical laboratory at the University of Calcutta, employing one person. It grew larger over the years and moved to the present location. The ISI was recognized as an institution of national importance by the Indian parliament in 1959 and was empowered to grant academic degrees at all levels. The B.T. Road, as it is known locally, is a broad and straight thoroughfare that connected Calcutta port with the army barracks in Barrackpore some 23 kilometers away. During the colonial era the British and the East India Company found it useful for the rapid transport of troops to the seaport of Calcutta. The ISI is divided into several blocks, each serving a specific purpose; 203 B.T. Road was the main part of the institution—the director's office and residence as well as the main administrative and research building were located there. Our research facilities, including Haldane's library, were also located in that building. There were large ponds surrounded by lovely floral gardens and many trees, including tall palm trees, and Haldane was seen swimming in the main pond on most evenings. The trees and flowers served as our research material occasionally. The Haldanes' personal library had just arrived from London when I took up my position at the institute. My desk was placed at the entrance to that library. Haldane sat at the other end of that large room; we were separated by several large shelves containing his books and journals. They included bound volumes of numerous journals such as *Biometrika, Journal of Physiology, Journal of Hygiene, Journal of Genetics, Evolution, Animal Behavior, Hereditas, Proceedings of the Royal Society of London, Annals of Human Genetics, Nature*, and others.

On the grounds of the institute, there was also the publishing house and the press called Eka Press, which published the Journal *Sankhya: The Indian Journal of Statistics*, which was founded by Mahalanobis in 1933 and continues today. At 202 B.T. Road, there were several buildings that served as the residences for statistical trainees who came from all over Southeast Asia, and the offices of the National Sample Survey[22] were located there, as were my experimental plots for growing cowpea (*Vigna sinensis*) and some other plants. There was also another lovely pond surrounded by beautiful floral bushes in that area. At 206 B.T. Road our dormitories and dining hall and my *Lantana* plots for observing Lepidopeteran pollinators were situated. I stayed in the dormitory in my first year at the ISI but moved to share Haldane's flat at the institute afterward. Other administrative and research buildings were located at 204 and 205 B.T. Road. In all, the ISI properties stretched a good distance on B.T. Road, supporting several other businesses along that highway and creating many more jobs in addition to the two thousand employed by the institute. Both Haldane and Helen were familiar figures in that area as they used to walk every day from their home (Basack Villa) to the institute, about half a mile away.

In a recent visit to Calcutta (now renamed Kolkata) to attend the centennial celebration of the Indian Science Congress, I was pleased to see that the major thoroughfare I was traveling on has been named "J.B.S. Haldane Avenue." Such a possibility had never even crossed my mind when I was working with Haldane more than fifty years ago.

Prasanta Chandra Mahalanobis, the founder of ISI, promoted the interaction of statistics with natural and social sciences and its general applicability. The ISI is one of the leading centers in the world for training statisticians and conducting research in statistics—both basic and applied. The institute was most beneficial to Haldane's scientific interests and initial contacts in India. It was a well-established institution with a solid foundation, which was well connected with its chief benefactor—the Indian Government, because of the close friendship Mahalanobis enjoyed with Prime Minister Nehru. The director was also the principal statistical advisor to the central cabinet of ministers and the national planning commission of the government of India. As a research institution, the ISI was highly respected internationally, as the constant flow of visiting dignitaries, Nobel laureates, and distinguished professors from many countries indicated. There was an excellent library and other facilities, which many other institutions in India lacked. Its faculty in the Research and Training School was well qualified and recognized internationally.

Our life and work were also facilitated by the location of the institute in the large port city of Calcutta, which enjoyed the status of being the leading science center of India at that time. The presence of several other research institutes and universities contributed to the wealth of information that was readily available for a scientist. There was the distinguished center of the Indian Association for the Cultivation of Science in Jadavpur, where the famous scientist Sir C. V. Raman[23] conducted his legendary research on spectroscopy ("Raman effect") for which he was awarded the Nobel Prize in Physics in 1930. The University of Calcutta, with its long distinguished record, was located there, as was the "Bose Institute" which was founded by Sir Jagadish Chandra Bose (1858–1937),[24] who was a distinguished polymath in plant sciences, radio physics, and archeology. I used to consult their library with an introductory letter provided by Haldane.

Another advantage was the location of the Dum Dum international airport nearby, which was most convenient for national and international travel by air. There were also several rail links from Howrah station, which were useful for our research when traveling to Giridih in Bihar to collect silkworm cocoons for Spurway's research, and to Visakhapatnam in Andhra Pradesh for my research on consanguineous marriages and Y-linkage of hairy ears.

Now for the drawbacks. The ISI was dominated by one man, the founder and director, Prasanta Mahalanobis. There was indeed a personality cult, which was promoted by

Mahalanobis, over many years, long before our arrival. Any significant decision affecting our work had to wait until the "professor," as Mahalanobis was called by everyone, was available. Unfortunately, he was not, most of the time, as he traveled a lot both in India and abroad. Haldane was often irritated and frustrated because the director was not available for consultation. Of the director's frequent travels, Haldane wrote,

> The journeyings of our Director
> define a novel random vector

The institute lacked a formal structure to address the complaints of a scientist of Haldane's stature. Haldane complained that in certain situations he was treated like the head of a department, and at other times as though he had no status at all. His position was never clearly defined. There was a tendency to keep Haldane at arm's length, perhaps because Mahalanobis was afraid that Haldane would interfere in the inner workings of the institute, which were conveniently managed by Mahalanobis's cronies. Consequently, there were no adequate and timely responses to Haldane's complaints. His many letters to the joint secretary of the institute, Nihar Chakravarthi, Mahalanobis's point man, remained unanswered or received only what Haldane used to call "soothing syrup" replies but no real progress, along the following lines:

Dear Professor Haldane:

Thank you for your kind letter of (date). We will give this due consideration and send you a reply soon. I am sure everything will be settled to your satisfaction. Please write to me again if you have any other concerns or questions.

With my kind regards,
Yours sincerely,
N.C. Chakravarthi,
Joint Secretary

But in many cases, Mahalanobis himself, when he was not traveling, responded to Haldane's letters. They are often lengthy and quite detailed about several points raised by Haldane. Some dealt with essential aspects, whereas others seem very unimportant and trivial.

In one instance, when the institute hesitated to pay my airfare to attend a scientific meeting, Haldane offered to pay the cost from his own pocket. However, Director Mahalanobis objected. In his letter to Haldane dated November 18, 1959, Mahalanobis wrote:
"I am . . . somewhat disconcerted to find that you have offered to pay Dronamraju's travel expenses by air. I am deeply touched by your offer; but I should like you to consider the effect this would have on the institute. Other senior members of the staff, who have a smaller salary than yours, cannot afford to pay for the travel expenses of their juniors by air. If you pay the air fare of your juniors, it may have some adverse psychological effects somewhat similar to those caused by ostentatious disparities of living."

Although I was not aware of this correspondence between Haldane and the Director at that time, Haldane clearly was not convinced by this argument and continued to pay my air fare on several occasions, both in India and abroad.

Figure 17.1 In 1961—Haldane and his team in the Indian Statistical Institute grounds; sitting in the front: Haldane, Pamela Robinson (paleontologist from London), and author.

Figure 17.2 Lecturing at the Indian Statistical Institute in Calcutta in 1960.

Haldane's Birthday Celebration

Haldane's 65th birthday was celebrated with much gusto on November 5, 1957, at the Institute. Mahalanobis himself took great pains to prepare a list of invitees and arrange all the details. Among the ISI archives is a list of fifty-five individuals who were invited, which included several senior scientists, such as the founder of the Bose Institute, D. M. Bose, and the paleontologist from UCL, Pamela Robinson, as well as several scientific workers. Furthermore, Mahalanobis arranged for Haldane to deliver distinguished lectures at national meetings in the capital city of New Delhi. Haldane was resplendent in the Indian garb. At the banquet, Haldane and Mahalanobis sat in the Indian style on the floor, which is decorated with beautiful traditional paintings (see photo).

Research

My primary goal in starting my association with the ISI was to conduct research in genetics under the direction of Haldane. I was aware that the institute's main activity was in statistics and the tasks of most of its two thousand employees were related to statistics in one way or another. I was trained in genetics, but I was satisfied that Haldane's biological interests were sufficiently broad to include the area of genetics in which I was interested. Indeed, this proved to be correct. What is more important, Haldane believed that my presence and that of other nonstatisticians were needed to educate the statisticians on the kind of problems that arise in biological and genetic research and the need for statisticians to develop methods to solve those problems.

During the first three months, I was asked to catalog Haldane's extensive reprint collection, and also read Charles Darwin's books on plants and, of course, his magnum opus the *Origin of Species*. I found Darwin's works most enjoyable, intellectually stimulating and far-sighted. Each of the ideas expressed in his books could be a research project of the future.

Our research projects at the ISI fall into broad categories: those which were suggested to each of us by Haldane, and other projects each of us conceived in the course of our research activities. Our group, including the Haldanes, were members of the Biometry Research Unit of the institute. The term "unit," for some unknown reason, was very popular with Indian administrators at that time. Almost every institution I came across had one or more, often several, units. However, Haldane abhorred the term "unit" and commented, "How I hate that word, it suggests that we shall have, as Blake put it in his able anticipation of totalitarianism [first *Book of Urizen*]:

> One command, one joy, one desire,
> One curse, one weight, one measure,
> One king, one god, one law,
> For Genetics and Biometry [he added].
> Almost anything can be biometry.

That unit had already been established by the director some years ago, with Dr. Bhupen Das and his wife, Dr. Rhea Das (a psychologist from Chicago), and a botanist, Subodh Kumar Roy, as members. As soon as Haldane arrived at the Institute, he suggested some research projects to Roy. I joined Haldane at that point. About a year later, another scientist, T. A. Davis, joined us. The fourth person to join our group was Suresh Jayakar, a young statistician who came to the institute from Lucknow for advanced training. Jayakar joined a small group of trainees who took my course in basic genetics.

Haldane was happily settling into a routine; the mornings were spent either in teaching or answering his vast mail, which was especially large in the first year of his arrival. It appeared that almost everyone in Calcutta wanted to invite him for dinners, give lectures, open exhibitions, or just have a *darshan* (audience) with the famous scientist of whom they had heard so much. His articles in the popular press made him even more popular in India just as he was in England. Unless there were special visitors or meetings, Haldane wrote his mathematical papers in the afternoons and evenings, sometimes working late into the night, taking short breaks for tea and dinner. As I shared a large library with Haldane, I saw him daily, and occasionally he used to leave notes on my desk with ideas for research or books to read.

I describe here briefly some selected research projects that were undertaken by members of our inner circle.

Meristic Variation

Quantitative studies of meristic variation[25] were a big part of our research program at the ISI. Haldane's philosophy toward research in his newly adopted country was defined in his "Bateson lecture," delivered at the John Innes Horticultural institution in July 1957, shortly before his departure for India. Referring to Bateson's "Materials for the Study of Variation,"[26] Haldane wrote,

> When the number of like parts, for example, teeth, vertebrae or petals, can vary, it is usual to find a whole number such parts and unusual to find a miniature or incomplete member of the meristic series.... Gruneberg ... has studied this phenomenon in the third upper molars of a particular pure line of mice. These teeth are sometimes missing. But when they are present they are variable in size and can be decidedly smaller than the normal, though in no way rudimentary or incomplete.... When, at a certain critical stage, the rudiment fell below a threshold, it regressed or did not develop further. Similarly we may suppose that when a rudiment is too large at a critical stage of development it may divide into two or even more parts, giving an extra limb, for example.

Subodh Kumar Roy

Soon after his arrival at the ISI, Haldane suggested to one of our associates, Subodh Roy, several research projects involving quantitative studies of meristic variation.

Roy's research involved the study of variation in like parts in plants, such as petal numbers in flowers in such plants as *Nyctanthes arbor-tristis* and *Jasminum multiflorum* and nipple numbers in goats and cattle. He showed that the petal numbers in *Nyctanthes* vary from four to eight in most flowers. However, when he examined 158,926 flowers, he found that 14 had 9 petals and 1 had 15 petals, with a strong mode at 6 petals. The most surprising finding was that the variance increased toward the end of the flowering season. Haldane commented that if the pots made by a potter became more variable at the end of a day, we should say that he was getting tired, but we do not know what we are to say about a plant. This may well indicate that the developmental canalization[27] (regulation of development) is less rigidly enforced toward the end of the flowering season in plants, but the mechanism that causes it is unknown.

Number of Petals in a flower	4	5	6	7	8	9	15	Total
Number of flowers	451	29147	109345	18956	1012	14	1	158926

Another research program launched by Roy under Haldane's direction was the study of interaction between different varieties (genomes) of rice crops that were planted together in the same plot as compared with those planted in separate plots. Certain varieties, when planted together in the same plot, produced more grain than those planted in separate plots. Roy tested a number of varieties in different combinations and reported that certain varieties, when planted together, in fact depressed the crop yield while others increased the grain production in similar circumstances. Roy was able to make recommendations about the beneficial effects of planting certain crops together.

T. Anthony Davis

Another associate was T. A. Davis, who was working at the Central Coconut Research Station in Kayangulam, in the southernmost state of Kerala, when Haldane visited that institution in 1959. Haldane was impressed by Davis's skill in inventing equipment for climbing coconut trees and to measure root pressure in palm trees, and so forth. He was also interested in the breeding work of Davis on coconut palms. He recruited Davis to join our group at the ISI because the Institute was starting a program in graduate teaching in biology. Furthermore, Davis's research on the leaf spirals of palm trees fitted nicely into our research program on symmetry and biometrical variation in various plant species that interested Haldane. The leaf spirals atop coconut palms are either right-handed or left-handed. No genetic basis for the direction of the spirals was noted. Davis's research involving 45,000 coconut palms all over the world showed that the palms from the northern hemisphere are more often left-handed than those from the southern hemisphere. The difference is significant at the 1.0% level. The radio astronomer Grote Reber (1911–2002) suggested that the foliar asymmetry may be influenced by the magnetic inclination of the earth. However, what interested Haldane was that the palms with the left-handed spirals gave 20.9% more nuts than those with right-handed spirals. Following

Darwin, Haldane called this kind of research, "fool's experiments." In an essay titled "On Expecting the Unexpected," Haldane wrote that expecting the unexpected is an essential part of scientific method.

Under Haldane's direction, Davis extended his research on symmetry to foliar spirals in the plant species *Hibiscus rosa-sinensis* and *Abutilon indicum*, and found an excess of left-handed spirals in their flowers. A similar significant excess of left-handers was found in *Bombax ceiba*, the red silk cotton. Once again, no clear genetic determination was discerned. Unlike Roy's observation that variance increased toward the end of the flowering season, Davis found that variance decreased toward the end of the flowering season in his studies. It is not clear why certain organs may be more rigidly canalized in their development than others, and whether the geomagnetism of the earth might influence the symmetry of various organs in plants and animals.

Davis also measured the pressure in the cell sap of the roots of coconut palms, showing that in some cases the pressure can force the sap to great heights, even higher than the height of the palm tree itself. This interested Haldane greatly.

Suresh Jayakar

Jayakar's training was in mathematics and statistics. He was useful to Haldane, as he checked Haldane's mathematics in his papers, where small errors were found occasionally. Jayakar collaborated with Helen Spurway and myself in studying the nest-building activity of the solitary wasp *Sceliphron madraspatanum*. He collaborated with Spurway in another study—observing the behavior of a pair of yellow-wattled lapwings (*Vanellus malabaricus*). In a letter dated April 6, 1963, Haldane wrote to Ernst Mayr at Harvard University:

> A pair of *Vanellus malabaricus* is now nesting in my garden and can be seen from the veranda with field glasses. S.D. Jayakar and H. Spurway are watching them for 13 hours daily, noting down all journeys. At present they seem to be mainly occupied in keeping 4 eggs cool, both by shading them and by wetting their feathers and hence the eggs. This job will go on for another month or more unless one of the numerous predators such as *Herpestes*, *Varanus*, and *Canis* succeeds in terminating it. However, as another nest not far off has produced some chicks, these birds have a fair chance. I know of no comparable observations on ground-nesters. We are within about 1 km of the area where you saw larks and other birds when you were here. Perhaps you will come again. But you might find us unhospitable during daylight.[28]

Earlier, he conducted a biometrical study of the common edible European mollusc *Cardium edule* in Roscoff, France, where we all spent the summer of 1961 at the Marine Biological Institute directed by Professor Georges Teissier of the University of Paris. Jayakar collaborated with Haldane in publishing several papers in theoretical population genetics. One interesting paper was concerned with an enumeration of all potential relationships in human families that could result in inbreeding.

Krishna Dronamraju

Soon after my arrival, Haldane suggested that I should cross the two common varieties of the prickly weed, Lantana camara L,[29] growing in the suburbs of Calcutta. They look alike, except in flower color; in one, called "pink," the flowers are white and pink, and in the other, called "orange," they are yellow and orange. With the permission of the director, I secured at once a small plot of land in one of the ISI properties near the dormitories. I planted cuttings of both varieties of *Lantana* in alternate rows in my little plot. However, when I tried to cross them, the hybrid seeds (or what looks like them) were inviable. What is more important is the fact that *Lantana* reproduces vegetatively by cuttings, not by sexual reproduction. The crosses were futile because the two varieties differ widely from a genetic point of view, differing not only in chromosome numbers but also in many other aspects. I showed my results to Haldane, who apologized profusely for wasting my time! However, it turned out later that I had not completely wasted my time. During the experiments, I happened to notice that the flowers of each variety were being visited predominantly by certain species of butterflies. In other words, these insects were not visiting the flowers randomly, but were selective in visiting some colors more often. In the case of some butterfly species the discriminating visits were almost exclusive to one variety. As the observations were made over several months, we were sure that several individuals of each butterfly species were involved. It was a consistent behavior that would impact on the plant populations and hence was of considerable importance from an evolutionary point of view.

I was encouraged by Haldane to obtain quantitative data on the preferential visits. I planted the bushes with orange and pink flowers in alternating positions as shown below and recorded the visits of different species of butterflies.

```
Orange   Pink     Orange
Pink     Orange   Pink
Orange   Pink     Orange
```

Visits of Butterfly Species to Flowers of *Lantana*

Scientific name	Duration of observations*	Number of feeds — Orange	Number of feeds — Pink
Precis almana	11 months	218	13
Danais chrysippus	2 months	142	152
Papilio polytes	17 days	15	31
Papilio demoleus	2 months	42	98
Catopsilia pyranthe	12 months	40	603
Baoris mathias	2 months	1	108

*The long durations over which these observations were made indicates that several individual butterflies were involved. This was further confirmed by marking individuals.

My point is that, although in the case of *Lantana* the selective visits of lepidoptera are not important from the viewpoint of sexual reproduction and seed production, such selective pollination represents a very important example of how reproductive isolation in nature can minimize cross-pollination between related varieties of a plant species, causing further genetic divergence between them over a prolonged period. This could, in turn, lead to their evolution into distinct species. Such models are well known when the individuals of a species of plant or animal are subdivided into smaller communities by some geographic barrier such as an earthquake, a flood, or a volcano, leading to speciation (i.e., emergence of new species). That is to say, natural barriers obstruct or discourage pollination between populations that were previously members of the same species. Reproductive isolation from each other will lead to further genetic divergence between two such populations, leading eventually to the process whereby one species is divided into two or more new species.

The evolutionary biologist who emphasized this phenomenon was the German biologist Ernst Mayr,[30] who received his training in zoology under Erwin Stresseman[31] in Berlin. He migrated to the United States and worked at the American Museum of Natural History and later at Harvard University. Mayr expounded his views in many books and papers, such as *Animal Species and Evolution* (1963). However, the model of (potential) isolation that the *Lantana* case represented did not involve geographic or environmental isolating factors. Speciation resulting from an isolation of that kind which arises within a population, for instance, by the appearance of a new flower color mutant in a population, has been called "sympatric" speciation. Mayr believed that sympatric speciation occurs very rarely, if it occurs at all, in nature. In response to a letter from Haldane, Mayr wrote,

> It seems to me that there is no question about the facts. A definite preference is clearly evident. We must make a distinction between the preference of a social bee which may change from day to day and that of most solitary insects which, in many cases appears to be part of the code of information handed down by previous generations. I gather that most of the insects in the *Lantana* case are of the latter sort. . . . I entirely agree with you that this is a most interesting field.[32]

In a personal discussion later, Mayr pointed out that, in the *Lantana* situation that I investigated, the isolation between the two plant varieties was only partial and far removed from a complete or almost complete isolation that would be more convincing to suggest a potentially "sympatric" situation. Nevertheless, I remained convinced that similar situations arising in nature would create "sympatric" isolating mechanisms that would lead to speciation. The preferential visits of lepidoptera would not be evolutionarily meaningful if they were only temporary. I had to show that the preferential visits were inborn and not likely to change from day to day. In terms of Konrad Lorenz's[33] "imprinting," that is to say, behavior determined by the color of the first flower visited had to be ruled out. Spurway and I had a large cage (8 × 8 × 6 feet) built to design experiments to eliminate that possibility. The cage was big enough for two people to sit inside and watch the pupae to see which color would be first visited by the freshly emerged butterflies and their subsequent behavior. Each butterfly species under these experimental conditions showed the same color preferences as I first observed in nature. These results, by confirming my initial observations, showed that the preferential visits are meaningful from the viewpoint of evolution.

Haldane was elated when I showed him my results. He saw my work and that of others in our group as a vindication of his move to India. He had long anticipated that research on Indian species of plants and animals following Darwin would result in unexpected and novel findings. He himself was careful not to take any credit for the research of his students and associates. But he was proud of the fact that he created an atmosphere that enabled us to conduct our research peacefully without any interference from others. He followed this theme in his paper written in honor of Professor Bernhard Rensch of Munster, Germany:

> I wish to put in on record that I had not suggested any such observations to him. Of course I realized their importance, and encouraged him to continue them.... Mutants which produce flowers which do not attract enough of the available pollinators will be eliminated. And if a species spreads into an area where its usual pollinators are rare, mutants attracting other species will be at a selective advantage. Thus, in areas where *Precis almana* is rare, the orange Lantana might set rather little seed. In the remote past (perhaps the Jurassic[34]) when coloured flowers were first evolved the effects were no doubt more rapid."[35]

Bimodal Distribution in Bauhinia

Haldane suggested a simple biometrical project that I could start soon after my arrival at the ISI. Since the time of Darwin, researchers have studied the genetics of floral style lengths, that is to say, the length of the long slender tube that is the female part of the flowers. While in most species of plants, the style lengths do not vary appreciably among the flowers, in a few others the style varies in length quite significantly, so occasionally we have plants with long styles and others with short styles, for example, *Primula sinensis*. Rarely, we can also see a species with three types of style lengths: long, short and mid. On the grounds of the ISI, there were several bushes of *Bauhinia acuminata*,[36] which produces white flowers with long prominent styles. Haldane suggested that I should investigate the variation in style lengths in these plants. In effect, it involved gathering all the flowers from five bushes each morning and measure the lengths of styles from each flower. In total, I measured 3,427 flowers, after obtaining permission from Mrs. Mahalanobis. The resulting distribution of style lengths is what is called bimodal in statistics, with two peaks, one at 5 millimeters and the other at 20. But what was most interesting is that both short-styled flowers and long-styled flowers occurred *on the same* plants, and the short styled flowers are female sterile, that is to say, functionally they are male flowers with the female part being nonfunctional. So we have two kinds of flowers on each tree; most of the flowers are bisexual, containing both male and female parts, while a few on each tree are only male flowers. What I found was a type of heterostyly that was quite different from the classical type found in *Primula sinensis*. While most plants bear bisexual flowers, about

6 percent have unisexual flowers, that is to say, male flowers and female flowers occur on separate plants. The situation I have found in *Bauhinia* appears to be an intermediate step in the evolution of "dioecy," that is to say, the occurrence of male and female flowers on separate plants, as is the case in most palm trees.

Hairy Ears

This project shows the diversity of research topics that Haldane encouraged. While I was following my *Lantana* project, I became involved inadvertently in a human genetic project. A visitor from the Argonne National Laboratory in Chicago dropped by to see Haldane. It was Dr. Herman Slatis,[37] who studied population genetics under Sewall Wright at the University of Chicago. Herman had earlier visited the Haldanes in London and now wanted to see them in their new home in Calcutta. During that visit, he noted the hairy margins of my ears, explaining that I possess one of the few characters in humans that may be determined by a gene on the Y-chromosome.[38] This was a matter of controversy, Herman explained, because several of the characters that were previously claimed to be Y-linked had been shown not to be Y-linked.

One particularly striking candidate for Y-linkage, scaly skin—*Ichthyosis hystrix*—was shown not to be Y-linked in an extensive paper by two collaborators—Haldane's former colleague at the Galton Laboratory in London, Lionel Penrose, and Curt Stern, author of a well-known textbook in human genetics and a professor at the University of California. Herman Slatis told us that we could see the information ourselves from the *American Journal of Human Genetics*. So we all trooped in to the ISI library, where we found the information Herman was talking about. Having known Haldane, the next step was obvious. How soon could I visit members of my family to gather data on hairy ears of my family and possibly others and test for Y-linkage? The funds required for the investigation were fairly modest, for travel in my native state of Andhra Pradesh where most of my relatives were located. Haldane had no intention of going through the ISI bureaucracy for a lengthy process of applying for funds. He told me that I should go ahead with my investigation immediately and he would be happy to cover my expenses from his own pocket. I was accompanied by Spurway during part of the investigation. We visited all the living relatives, and obtained photographic evidence about those who died. Soon we were able to compile a pedigree of six generations of my relatives, and much of the evidence was supportive of the Y-linkage of hairy ears. As soon as I gathered all the required data for completing the three pedigrees including mine, which I included in my investigation, Haldane encouraged me to write my paper on Y-linkage, which was submitted for publication in the *American Journal of Human Genetics*. However, when the reviewers asked me to make several changes in my manuscript, Haldane offered to publish my version in the *Journal of Genetics* with no changes. And that is where it was finally published.[39] Haldane provided the mathematical analysis, showing that the probability in favor of the most likely alternative explanation, that is, an autosomal[40] dominant gene with expression limited to male sex, is 4.1×10^{-7}. This was typical Haldane. He always considered alternative possibilities even when the evidence in favor of a hypothesis is very strong. A few years later, Herman Slatis collected data on hairy ears from Israel, which supported my work.

Inbreeding

Another research project, on inbreeding,[41] involved collection of data on marriages in Andhra Pradesh, where I was born. I was familiar with the high frequency of marriages between close relatives in that region, and mentioned this fact to Haldane, suggesting that I should collect data on these marriages to find out the degree of inbreeding that was being practiced there.

Haldane agreed enthusiastically and, once again, offered to fund the project from his own pocket. I traveled to Visakhapatnam in Andhra Pradesh and recruited a few local people who were willing to assist in the collection of data on marriages, including P. Meera Khan, a medical student at that time, and a few others including Mangipudi R. Sastry and P. Srihari Rao. These two young men later joined us in Calcutta, where I found jobs for them assisting the Haldanes. They became an integral part of the Haldane household and were so helpful that Haldane found them indispensible.

I started with three generations, the marriages of hospital inpatients and of their parents and their children. To counterbalance this group, I selected another subsample, the healthy parents of schoolchildren in the public schools of Visakhapatnam. As expected, I found a high incidence of consanguineous marriages, the most frequent type being marriages with first cousins. It was common practice in Andhra Pradesh to arrange a boy's marriage with his mother's brother's daughter. Other types of consanguineous marriages that were less frequent than those with first cousins were a girl's marriage with her mother's brother and marriages with more distant relatives, such as a first cousin once removed, second cousin and third cousin, and so forth. Analysis of 2,177 marriages is presented below.

Andhra Pradesh, India

Analysis of 2,177 Marriages (in 1960)

	Number of marriages	Percentage
CONSANGUINEOUS	666	30.6
Aunt-nephew	1	0.05
Uncle-niece	157	7.2
First cousin	362	16.6
Other kinds	146	6.7
NONCONSANGUINEOUS	1,511	69.4

Percentage of Consanguineous marriages

	Uncle	First Cousin	Distant relatives	Consanguineous (total)
Inpatients' parents of	5.7	15.0	4.3	25.1
Inpatients	10.7	19.8	8.8	39.3
Inpatients' children	5.3	14.4	14.1	33.8
Parents of schoolchildren	6.6	16.7	4.7	28.0

Wright's Coefficients of Inbreeding[42]

	1	2
Parents of schoolchildren	0.01939	0.01865
Parents of patients	0.01745	0.01671
Patients	0.02777	0.02598
Patients' children	0.01862	0.01572
Total	0.02093	0.01947

Note: The coefficients in column 1 are calculated taking all the known consanguineous marriages recorded in this study into consideration. The coefficients in column 2 are calculated including only those marriages with first cousins or nearer relatives.

Comparison of Global Inbreeding Rates in 1960

I was aware that, except for a few isolated populations such as the Amish communities in the United States and some isolated islands in the Adriatic, the observed rate of inbreeding in Andhra Pradesh was the highest among large general populations of the world. Japan came closest, but it was a distant second. Many countries in Europe and the United States had always had much lower rates. Consequently, our papers on inbreeding in Andhra Pradesh at once attracted the attention of geneticists worldwide who wanted to exploit this population for genetic research.

There were numerous other activities. We traveled a lot, both in India and Europe, for several scientific meetings where we presented lectures about our research. These are described in detail in my book on Haldane in India.[43]

Haldane recorded in his "An Autobiography in Brief" that "his most important work (at the Indian Statistical Institute) was, beyond doubt, starting S.K. Roy, K.R. Dronamraju, T.A. Davis, and S.D. Jayakar on their Scientific careers, which are likely, in my opinion, to be illustrious."[44]

Resignation from the Indian Statistical Institute

Almost from the beginning of his association with the ISI, Haldane found certain administrative and bureaucratic procedures to be irksome. Curiously, he did not blame the present Indian administrators but the old British colonial administrative policies that had been passed on. Foremost was his constant complaint that his position at the institute was never clearly defined. In response to his repeated complaints, he received a clarification from Mr. N. C. Chakravarti, joint secretary of the institute, stating that "Professor Haldane is considered by us as having all powers of a Head of a Department."[45]

Other complaints arose in due course. One of them concerned signing a register to record the arrival and departure of all scientific workers. Haldane objected to that procedure. He wrote, "During 38 years in Europe and North America I have never known scientists ordered to do this. . . . scientific research, especially in biology, cannot be so carried out. For example today my wife has been compiling statistics on animal behavior since 7 a.m. and Dronamraju helping since 8.30. He may very well fail to turn up here [ISI] before 11.30 a.m." One of the problems was that Haldane had to waste precious time on this kind of correspondence instead of doing research or supervising research. Minor irritations of this kind accumulated over time, leading to his resignation from the Institute. The straw that broke the camel's back occurred in connection with the visit of the Russian leader Alexei Kosygin to the institute. Haldane was told that the Institute was expecting a visit from a VIP the next day and all of us should be prepared to show our work and explain its significance. Accordingly, Haldane made plans to present our research projects in a certain order and told us to get ready. However, Director Mahalanobis arrived late that evening from one of his trips and changed our plans without consulting Haldane. Mrs. Haldane conveyed this disagreeable news to her husband, suggesting that he should not put up with such gross interference in research matters. Thereupon, Haldane, who was already fed up with the Institute administration due to various reasons, submitted his resignation immediately, and Helen followed suit. As the drama unfolded, we were about to close for the day and Haldane was locking up his library. Mahalanobis arrived on the scene, looking clearly agitated, and repeatedly requested Haldane to withdraw his resignation. But Haldane quite firmly refused to do so. We were all standing around watching this confrontation with much disbelief and sadness between two great men. Whatever differences arose later, we remembered that their friendship started decades earlier, and the correspondence that initiated Haldane's move from London to India went back to at least ten years before. Our group included, besides myself, S. K. Roy, T. A. Davis, and S. D. Jayakar; the two assistants I hired, Mangipudi R. Sastry and P. Sri Hari Rao; Spurway's technician, Manik Mukherjee; and Haldane's secretary, B. C. Kundu. We were all wondering about the state of our future employment because we had been hired by Haldane. The correspondence between Haldane and Mahalanobis, which is reproduced in the Appendix, indicates their changing relationship during those years.

Shortly after Haldane's resignation, which he and Helen followed in writing, I left for Glasgow University to spend a year with Guido Pontecorvo[46] in somatic cell research. Later I spent another six months in human genetics with Lionel Penrose at the Galton Laboratory, University College London. During my absence from India, I had been in contact with Haldane, who had moved from Calcutta to the Orissa's capital city, Bhubaneswar. While I was in London, Haldane invited me to join him in Bhubaneswar to continue research in human genetics at the Genetics and Biometry Laboratory that he was planning to establish there.

Research in Orissa

Haldane's move to Bhubaneswar was unwise in the opinion of several colleagues in Indian science. He moved from a large metropolis with excellent facilities to the so-called earthly paradise of Bhubaneswar,[47] which was a small provincial capital with very limited resources. We were provided with buildings and salaries but not much else. The local chief minister (governor) of the State of Orissa, Biju Patnaik (see chapter 15, "Relations with Other Scientists"), was keen on encouraging science in his state and thought that Haldane's presence would stimulate teaching and research in science. He was already well known for the Kalinga Prize for science popularization, which he instituted in 1952 under the auspices of UNESCO. However, the arrival of Haldane in Bhubaneswar coincided with a border war with China that was drawing away precious resources from science and other departments. The Genetics and Biometry Laboratory, which was housed in temporary headquarters, was meant to have its own building, as promised by Mr. Patnaik. However, late in 1962, he visited Haldane to tell him that the construction of the new building would be postponed because of the continuing conflict with China. Later in 1963, while he was en route to India from a conference on the origin of life in Tallahassee, Florida, Haldane was operated on for rectal cancer in London but returned to Bhubaneswar in January 1964 and carried on courageously until his death in December 1964. The new building for the laboratory was never built, and the original contract with the State Government of Orissa expired in 1967. Mrs. Haldane (Spurway) then moved to Hyderabad and continued her studies of animal behavior at home until her accidental death in 1977. Her colleague Suresh Jayakar moved to the University of Pavia in Italy and worked with Luca Cavalli-Sforza until his untimely death in 1988.

In spite of the lack of facilities at Bhubaneswar, Haldane seemed personally happier there than he was at the ISI in Calcutta. There was no bureaucracy to deal with. He felt absolutely free, as there was no authority telling him what to do. There were deep psychological scars in Haldane that made him resent authority—any authority—to an extreme degree. Kingsley Martin commented, in his biographical sketch of Haldane, in the *New Statesman and Nation*,[48] that Haldane never recovered from the brutal treatment accorded him by the senior pupils and staff at Eton.

Among our team from the ISI in Calcutta, Roy and Davis decided not to move to Bhubaneswar. Besides the Haldanes, our new team included myself, Jayakar, and the two assistants Mangipudi R. Sastry and P. Srihari Rao. Another scientist, the anthropologist Ajit K. Ray, joined us. Ray was a student of the renowned anthropologist Nirmal Kumar Bose, who had befriended Haldane at Calcutta. Another collaborator, a young medical student who helped us earlier in collecting data on consanguineous marriages in Andhra Pradesh, P. Meera Khan, used to visit us occasionally. At my suggestion, Haldane sent him to receive training with the renowned Italian geneticist Dr. Marcello Siniscalco,[49] who was conducting research on the association between malaria and deficiency of glucose 6-phosphate dehydrogenase in Sardinia. Haldane had originally asked me to work with Dr. Siniscalco, but I was already occupied in collaboration with Professor Guido Pontecorvo on somatic cell culture research at Glasgow University in Scotland.

Colorblindness

Some of our research at Bhubaneswar was concerned with the frequencies of colorblindness in the tribal and nontribal populations of the state of Orissa. Earlier research by Richard Post at the University of Michigan showed that the frequency of color vision defects were much lower in primitive and tribal-nomadic populations of the world than those recorded for large urban populations. I had suggested that since much of the Orissa population was of recent tribal origin, the frequency should be lower in Orissa than in the population of Andhra Pradesh, which we had studied earlier. Accordingly, we found that 4.96 percent of those tested in Orissa, 6.5 percent of Andhra Pradesh nontribals, and 2.5 percent of Andhra Pradesh tribal groups were colorblind. The difference between the tribal and nontribal groups of Andhra Pradesh was found to be statistically significant.

Deaf-Mutism

A project that interested Haldane was the inheritance of deaf-mutism in the small community of Bengali Kayastha,[50] settlers of Orissa. This community, although avoiding close inbreeding, was in fact highly inbred because of the cumulative effect of marriages within a small community over several generations. A similar situation exists in the Amish community of Pennsylvania, who are highly inbred although avoiding marriages with close relatives. In Orissa, one large pedigree indicating an autosomal recessive type of inheritance for deaf-mutism was noted, with 182 individuals in five generations and five families with at least one affected individual. Two of these families (sibships) were doubly related—their parents though not known to be related, were in fact first cousins of the other parents. Unrelated parents produced seven female and four male deaf-mutes recorded in this family, however, because of the small size of the community it is highly probable that the parents in fact shared recessive genes for deaf-mutism. Three deaf-mute females married to normal males produced six normal sons. Haldane believed that the division of the population into numerous castes and subcastes, which are endogamous, that is to say, marry within their group, offers many unique opportunities for research that is unavailable either in Europe or North America.

Toe Anomalies

Because in India a large number of men and women walk either barefoot or wear sandals, the population offers unique opportunities for rapid screening for foot and toe anomalies. Haldane encouraged one of my associates, the anthropologist Ajit K. Ray, to record the frequencies of toe anomalies in the populations of Orissa and Bengal. Ray examined 2,500 men for the defect known as the short fourth toe, which is due to a short metatarsal, and found that the defect is due to an autosomal dominant gene. This research was published in the *Proceedings of the US National Academy of Sciences*[51] jointly by Haldane (posthumously) and Ray in 1965. It was the only publication by Haldane in that journal.

Of the 2,500 examined, Haldane and Ray found that the anomaly was manifested on both feet in 130 cases, on the right foot only in 41 cases, and on the left only in 35 cases; 104 men and 102 women were affected. Other associated anomalies are not rare. Among 117 persons with short toes, 3 had short terminal phalanx of thumb—2 bilateral, 1 unilateral, and 3 had unusually long index fingers on both hands—2 with short third toes on right foot, 1 on left foot, 1 with patella absent on same side as the short toe, and dental abnormalities. The following abnormalities were found in members of the pedigrees with normal fourth toes (each occurred once only): polydactyly on both hands, polydactyly on one hand and foot, short fourth (ring) finger, short great toe on both feet, short third toe on both feet, long index finger and short thumbs on both hands, short giant toes on both feet, long little finger on left hand, syndactyly of right third and fourth fingers, bilateral club foot. Haldane and Ray suggested that most of these abnormalities were manifestations of the main gene. There were similar reports from Japan, but the gene appears to be rarer in Japan than in India.

Animal Behavior

Haldane's large house and gardens provided a "natural laboratory" that offered numerous opportunities for research on various animal species. Haldane, assisted by Suresh Jayakar and Spurway, carried on investigations of the nest-building activity of wasps, which we had earlier started in Calcutta, and behavior studies of the yellow-wattled lapwing *Vanellus malabaricus*. Haldane wrote,

> We watched a pair nesting in our garden at Bhubaneswar through the hours of daylight for several weeks and noticed many features of their behavior.... The eggs are laid on the ground in a shallow nest, in this case exposed to full sunlight, which can heat the ground up to at least 55°C. The birds' efforts during the daytime are largely concerned with keeping the eggs cool.... Both parents may leave the nest for an hour or so soon after sunrise and before sunset. But during the hot hours one was always shading it. However, this was not all. When one parent relieved the other, the freed parent usually went at once to shade or water, then did some foraging, and finally went to the nearest water, in this case the drip from a neighbour's tap, for the lapwings were too shy to use our bird bath. The bird sat down in the water and wetted his or her feathers. On return to the nest, it sat down on the eggs and wetted them. Without this they would probably have been slowly roasted.[52]

The above excerpt from an essay by Haldane, written during the last year of his life, is an excellent example of his last years in India. He was involved in pure and basic research on ecology and animal behavior, which he enjoyed, with no concerns whatsoever about its applications or any necessity to justify it to any one—a life free in every sense, which he had envisioned when he had planned his last years in India.

Death

Haldane was greatly interested in death, especially his own death, throughout his life. He wrote about his personal views of death repeatedly in several articles. In his earliest collection of popular essays, *Possible Worlds*,[53] published in 1928, Haldane wrote, "When I am dead I propose to be dissected; in fact, a distinguished anatomist has already been promised my head should he survive. I hope that I have been of some use to my fellow creatures while alive, and see no reason why I should not continue to be so when dead." Haldane reviewed various belief systems about life after death. He wrote, "When Jesus tells me to love my enemies he is speaking his own mind, and I am prepared to make the attempt; when he tells me that I shall rise from the dead he is only speaking for his age, and his words no more convince me of immortality than of demoniacal possession." Haldane noted that conditions in the present world have been improved largely by recognizing that the laws governing it are not the laws of justice but the laws of physics. He wrote, "As long as people thought cholera epidemics were a punishment for the people's sins, they continued. When it was found that they were due to a microbe they were stopped." He wrote that if he believed in his own eternity it would be a piece of unwarranted self-glorification, and the desire for it a gross concession to selfishness.

When he was dying of cancer in 1964, Haldane wrote a most interesting article on death and his views on life and death, "On Being Finite," published posthumously in the *Rationalist Annual* in 1965. It should be considered as his final statement or perhaps last testament. Haldane wrote that as he was operated on for a cancer at the of age of seventy-one, he was seriously contemplating the fact that he might die within a few years—perhaps one year if the cancer recurred. In typical Haldanian fashion, he discussed death in a detached manner from different points of view. There was no self-pity. It was Haldane at his best.

Death as something inevitable or unavoidable is probably a recent discovery by man. Primitive men saw many humans die, but not from ripe old age. Judging from the bones of paleolithic men, most died before they reached sixty. Almost all the deaths were accidental. Some died in battle. As the land was sparsely populated, deaths due to infectious diseases were very rare, and when it occurred rarely it was attributed to black magic. Haldane compared that situation with the lives of birds today. Their death rate is intense in eggs and nestlings. Once they are able to fly, their death rate is independent of age. About two-thirds of all adult robins will be dead in a year. About one in two thousand lives to be seven years old. Perhaps one robin in a million may die of "old" age, but we do not know at what age that might occur.

Some animals have a precise life span. Mayflies live for a year, almost entirely under water. They die within a day of their emergence from water. Most moths cannot eat or drink, and die within one or two weeks after their emergence from cocoons. To reach the state of contemplating death as something unavoidable requires a certain degree of social consciousness and intellectual sophistication. Death was hard to define or express in words even after language had evolved. As Haldane put it, "It was extremely difficult to accept, because it is unimaginable. One can imagine oneself playing a golden harp in heaven, in a fire in hell, or reborn as a cockroach, but one cannot imagine nothing." Most human beings believe in some kind of life after death. However, Haldane explained that he had no desire to be reborn. If he is to be replaced he would prefer to be replaced by someone without some of his congenital deficiencies, such as tone-deafness. He wrote, "we must, at least provisionally, accept the notion that we are finite in time as we are in space, and act on this acceptance. This means that we must be, to some extent, Epicureans, simply because Epicurus was the first man who did his best to work out the consequences of his finitude and act on them."

On the subject of happiness, Haldane wrote,

> I doubt if happiness is possible unless one works fairly hard and enjoys one's work. Aristotle taught that happiness is good activity, and in India Krishna taught that we should work for the sake of the work, not of its reward. I am in full agreement. What precisions does the acceptance of finitude bring to this notion? The most satisfying work for most people is probably that which brings obvious and immediate benefits to a number of other people whom the worker meets personally, and for which he or she does not have to extract money or gifts from them—for example, the work of a postman, or a doctor under the National Health Service.[54] ... Scientific research and artistic and literary creation, are, at best, the most satisfying of all pursuits. But scientific research is being more and more debased by team work, in which a large number of workers do what they are told to do, not what they want to do, and the results are remote.

Haldane believed that inventors are doing more for world revolution than many politicians. He predicted that some of his scientific discoveries would outlive him, not only for a human lifetime, but for very much longer, because the more important of his discoveries such as the measurement of the human mutation rate are taken for granted. Haldane wrote, "No greater compliment can be paid to a scientist than to take his original ideas for granted as part of the accepted framework of science during his lifetime." What kind of creativity will be preserved for centuries? In the past, sculpture and mosaics had some chance of surviving two thousand years. But paintings deteriorate, and may be burned or damaged in some fashion. Sculpture will not survive modern weapons. But music can be recorded by various means. Haldane predicted that Stravinsky and Hindemith are perhaps more likely to be directly known two thousand years hence than Picasso and Epstein.

The prospect of death is greatly alleviated by the belief that one's work will be of use after death and that others will continue that tradition. Haldane wrote, "I should find the prospect of death annoying if I had not had a very full experience mainly stemming from my work.... One thing which I am really sorry to have missed is walking to France on the sea bottom, which incidentally would have involved some interesting physiological research beforehand. I only got the money needed for this purpose at the age of seventy. Most of my joyful experiences have been by-products."

Among his life's experiences, Haldane listed his experiments with various hallucinogenic drugs, commenting that the alterations of his consciousness due to these drugs were trivial compared with those produced in the course of his research work. Among other experiences of his life, Haldane mentioned that he was one of the first two people to pass forty-eight hours in a miniature submarine, and one of the first few to get out of one under water. As he was dying of cancer, Haldane summed up his life and concluded with satisfaction, "So when the angel with the darker drink at last shall find me by the river's brink, and offering his cup, invite my soul forth to my lips to quaff, I shall not shrink."

Auto-Obituary for BBC

After his surgery at University College Hospital in London, Haldane was interviewed by the BBC and the interview was broadcasted on the evening of the day he died in Bhubaneswar, India, December 1, 1964.[55] It is remarkable that he went out of his way to praise Lysenko as a very fine biologist.

Poem on Cancer

While he was recuperating from cancer surgery at University College Hospital in London, Haldane wrote a poem on cancer that offended and amused readers in equal parts. It was published in the *New Statesman* in January 1963 and was later reprinted in several magazines and newspapers.

Cancer's a Funny Thing (with apologies to W.H. Auden)
I wish I had the voice of Homer
To sing of rectal carcinoma,
Which kills a lot more chaps, in fact,
Than were bumped off when Troy was sacked.

Yet, thanks to modern surgeon's skills,
It can be killed before it kills
Upon a scientific basis
In nineteen out of twenty cases.

I noticed I was passing blood
(Only a few drops, not a flood).
So pausing on my homeward way
From Tallahassee to Bombay
I asked a doctor, now my friend,
To peer into my hinder end,
To prove or to disprove the rumour
That I had a malignant tumour.
They pumped in $BaSO_4$.
Till I could really stand no more,
And, when sufficient had been pressed in,
They photographed my large intestine,
In order to decide the issue
They next scraped out some bits of tissue.
(Before they did so, some good pal
Had knocked me out with pentothal,
Whose action is extremely quick,
And does not leave me feeling sick.)
The microscope returned the answer
That I had certainly got cancer,
So I was wheeled into the theatre

Where holes were made to make me better.
One set is in my perineum
Where I can feel, but can't yet see 'em.
Another made me like a kipper
Or female prey of Jack the Ripper,
Through this incision, I don't doubt,
The neoplasm was taken out,
Along with colon, and lymph nodes
Where cancer cells might find abodes.
A third much smaller hole is meant
To function as a ventral vent:
So now I am like two-faced Janus
The only* god who sees his anus.

*In India there are several more
With extra faces, up to four,
But both in Brahma and in Shiva
I own myself an unbeliever.

I'll swear, without the risk of perjury,
It was a snappy bit of surgery.
My rectum is a serious loss to me,
But I've a very neat colostomy,
And hope, as soon as I am able,
To make it keep a fixed time-table.
So do not wait for aches and pains
To have a surgeon mend your drains;
If he says "cancer" you're a dunce
Unless you have it out at once,
For if you wait it's sure to swell,
And may have progeny as well.
My final word, before I'm done,
Is "Cancer can be rather fun".
Thanks to the nurses and Nye Bevan
The NHS is quite like heaven
Provided one confronts the tumour
With a sufficient sense of humour.
I know that cancer often kills,
But so do cars and sleeping pills;
And it can hurt one till one sweats,
So can bad teeth and unpaid debts.
A spot of laughter, I am sure,
Often accelerates one's cure;
So let us patients do our bit
To help the surgeons make us fit.

J. B. S. Haldane (1964)

After Haldane's death on December 1, 1964, a meeting of the scientific workers of the ISI was held and a resolution was passed expressing profound sorrow. It was noted that Professor Haldane "made significant contributions to a wide variety of fields in science and humanities and has been a source of inspiration to all scientific workers for more than half a century. He laid the foundations of the mathematical theory of population genetics and his fundamental work on natural and artificial selection has greatly improved our understanding of the theory of evolution."

The institute received a telegram from the Soviet Academy of Sciences:

> We share your deep sorrow over death of Professor John B.S. Haldane, Foreign Member of USSR Academy of Sciences, one of the most outstanding biologists who contributed greatly to development of Indian science. On behalf of Presidium of USSR Academy of Sciences and all Soviet biologists express our Indian colleagues and family of the deceased sincere condolences over this great loss.
>
> Academician Sissakyan, Academy's Chief Secretary.

Impact of Haldane on Indian Science

Haldane left a lasting impact on science and scientists of India. His arrival in India was greeted with much anticipation and optimism by all scientists. Soon his articles in the daily press and popular magazines made him known to millions of readers. His approach to scientific research, such as deriving maximum benefit while using simple or no equipment and making original observations in nature in the Darwinian tradition became well known and appreciated by many. He was particularly inspiring to young students because of his brilliant ideas but also because of his simple, nonpompous demeanor and accessibility to young students, which was rarely found in scientists of his stature in India. My own situation was a case in point. When I wrote to him as an unknown young student, he replied immediately. This was unheard of in India. Famous and important scientists do not respond to letters from young, unknown students.

Haldane introduced other ideas and methods. His constant refrain that quantitative expression of scientific results are much more meaningful and significant than qualitative ones, deeply influenced young scientists. He set an ideal example with his own simple lifestyle and methods that inspired and guided younger scientists. Many people admired his Gandhian lifestyle, vegetarianism, and nonviolent approach to research. In a country that inherited a top-heavy bureaucratic colonial administration and pompous lifestyle, Haldane's brilliant ideas and simple living appealed like a fresh breeze, which was intellectually stimulating.

Most importantly, Haldane emphasized the importance of science and scientists to Indian society. Through his personal contacts, lectures, and numerous writings, Haldane attempted to raise the social standing of scientists, particularly young scientists, in India and restored some dignity to their life and work, which had been lacking before his arrival in India. Two scientists who were supported and promoted by Haldane were the ornithologist Salim Ali and the anthropologist

Nirmal Kumar Bose. Both became famous in India, partly because Haldane praised their work in his popular essays that appeared in the daily press. They became successful scientists, receiving many honors both in India and abroad. Because of his eminent stature and fame, Haldane came into contact with many political leaders in India including Prime Minister Jawaharlal Nehru. At every opportunity Haldane expressed support for the younger scientist, advocating more freedom and independence from bureaucracy. He told Nehru that it is better for scientific research and innovation to support small groups of younger scientists working with a proven leader rather than build large research institutes. But he was not very successful in his efforts because politicians generally liked to build large impressive institutions, which they could showcase as symbols of their success. Nehru was no exception. He supported building a chain of national laboratories across India, for instance, the National Physical Laboratory in New Delhi, the Central Chemical Laboratory in Puna, the Central Lather Research Institute in Madras, and so forth. However, excellence in science itself remained elusive, as attested by the dearth of Indian Nobel laureates since C. V. Raman was awarded the Nobel Prize for Physics in 1930!

Politeness and Unprofessionalism

Haldane explored the reasons for the low level of performance in research and teaching in science in his writings. Here are the opening lines for one of his articles, titled "What Ails Indian Science?":

> I have already come to the conclusion as to why science in India is developing with disappointing slowness. It is not because Indians are stupid or lazy. It is because they are too polite. They spend hours daily in conversation with others, not on professional matters, but on personal topics. In London, I talked with colleagues for an hour or more, daily, but it was mostly about the details of our work. In the Indian Statistical Institute (Calcutta) the same is true. But it is not true in most academic institutions where I have been in India. Again at scientific meetings and usually in ordinary discussion my Indian colleagues are polite about one another's work. In Europe, we are usually polite about the work of juniors, and highly critical of that of men and women of established reputation. At a recent international meeting on genetics, an American got up after a paper read by my wife and said that he could not let her highly misleading views pass without criticism. She felt that she had at least reached the status where one is criticized without mercy. She and I at once formed a friendship with the critic. We had something to talk about. In my opinion, only a few branches of Indian science have reached the stage of maturity where this is possible. I may criticize some of my colleagues as I would criticize British colleagues, and hurt their feelings severely. Once again I am up against the choice between politeness and efficiency. I do not know how I shall resolve this dilemma. I hope that as Indian science grows up, it will become less acute.[56]

New Caste System

Haldane observed that a new caste system, based on academic degrees, was evolving in India. One could not teach a number of subjects such as Bengali, chemistry, history, or any other, without a degree in that subject. And a higher degree given for research was necessary if one applied for a professional chair. Referring to his own situation, Haldane wrote that it was only a matter of time before he would be barred from teaching science or statistics, because he possessed no degree of any kind in a scientific subject. But in terms of the new caste system, he was qualified to teach the classics, since he secured in 1915 a marginal first class in *Literae Humaniores* at Oxford University which mostly consisted of a combination of Greco-Roman classics, philosophy, and ancient Greek and Latin. Haldane cited the example of a man in India who was refused a university post to teach his native language because he had no degree in it, although he had published quite a lot in it and possessed a degree in another subject. He mentioned another case where a student was refused admission to a master's degree course because his undergraduate degree was in a different subject. He ridiculed such a practice as calculated to ensure that Indian scientists are too specialized. Successful Indian scientists like Jagadish Bose, Meghnad Saha, and Prasanta Mahalanobis achieved eminence precisely by bridging the gaps between different sciences.

Teaching and Learning

Haldane was shocked to realize that the methods he followed previously in teaching science did not work in India. One Sunday afternoon, we were walking near the Institute grounds when we heard some chanting. In Haldane's words, "I was listening to some mantras, and asked my companion [present author] if he could identify them. The practice of repeating religious formulae is of course about as common in Europe as in India, and I have little doubt that it has an effect in guiding the thoughts of the chanter in a certain direction, even when the chanting has become quite automatic." But I was amused to tell Haldane that the language of the chant was English and the subject organic chemistry. What he thought was a religious chant was, in fact, a young student of organic chemistry preparing himself for a test. Haldane returned to the scene in disbelief and found that I was right. Haldane wrote in his essay,

> The subject of the chant was the preparation of aliphatic amines, with special reference to various precautions. I have learnt a great deal in this way, and have very considerable stock of poetry, in at least ten languages, and eleven if you consider, as I do, that some parts of the Koran are great poetry. Clearly one must learn poetry exactly. But I have never learned any scientific fact in this way. On the contrary, I try to learn them in as many different ways as I can and to teach them from many points of view.... The knowledge of science is, or rather should be, something quite different from the knowledge of poetry. The kind of knowledge, which is most useful in science, is a very long way from that which gets one a first class in a written examination.

Haldane explained that in scientific research it is important to recognize what is not known. Referring to my research on the selective visits of butterflies to different colored flowers of *Lantana*, he wrote, "When one of my young colleagues made what turned out to be a completely original observation I said that I thought nobody had ever noticed such a thing before, and told him to write to two men in Europe and USA to confirm this. Much of my success in research has been due to my knowledge concerning human ignorance. So far as I know this peculiar kind of knowledge is never taught."

Double Loyalty of a Scientist

Haldane pointed out that every scientist has a double loyalty—to science as well as to his country. He observed that in India junior scientists are not engaged in research but waste their time in filling endless forms about work in contemplation and showing visitors around. Some of those workers could have done some research, even if only mediocre, but were ordered to remain in the laboratory, standing beside an incomplete or inefficient apparatus, and explain the potential project to visitors. These were not isolated cases. Haldane saw them all over India in numerous situations. He wrote that there was widespread jealousy toward younger workers by their superiors. The younger workers, who were at their most productive age, were either discouraged in doing original work or their results were stolen by their supervisor. He mentioned the head of a certain laboratory: "This remarkable man had published over fifty scientific papers in one year. No single human being before him has ever made discoveries at this rate! No doubt junior colleagues had done most of the work, or all of it. But their names were not mentioned. It is not surprising that young men do not care to work under such conditions." They went abroad to continue their research, especially in Europe and North America, in a mass movement that came to be called "brain drain." Another reason why Indian graduate students went abroad was mentioned by Haldane, "They are systematically humiliated by the administrative staff of many institutions, and sometimes by professors, in a manner, which is not tolerated in western Europe or United States. Again I hope to live to see this remedied."

Haldane's outspoken comments and his critical evaluation often offended senior administrators and others in powerful positions in India but endeared him to younger scientists. He championed their cause in a fearless manner when they were unable to speak for themselves. He wrote, "The root cause of all this incompetence and worse is not far to seek. A large number of Indian scientists have no pride in their profession, though they are proud of their salaries and positions.... In India today the unworthy successors of Durvasa[57] and Vishvamitra[58] actually invite governors, vice-chancellors, and the like, to address them. This may be a relic of British Rule. If so, it is a regrettable one." Haldane was often invited to visit various universities and research institutes in India. However, it was an unfortunate practice that he was first required to meet with the vice-chancellor or the director of the institution, having tea and wasting his time in polite and trivial conversation. Haldane used to abandon such meetings quite abruptly and go and mingle with the students or younger scientists to discuss their studies or research programs. This was, of course, considered rude, but Haldane was more interested in helping younger workers and often made his views well known with great emphasis. Soon, especially after his move to Bhubaneswar, there was a sharp drop in invitations to universities and social functions. This was partly due to the fact that he was now living in a small town, but also because it became common knowledge that he did not want to be disturbed unless it was a matter that was directly related to the research work of the laboratory.

Notes

1. Government of Orissa (now named Odisha) is an Indian state by the Bay of Bengal. It is surrounded by the Indian States of West Bengal to the northeast and in the east, Jharkhand to the north, Chhattisgarh to the west and northwest, and Andhra Pradesh to the south. It is the modern name of the ancient kingdom of Kalinga, which was invaded by the Mauryan emperor Ashoka in 261 BCE. The region is also known as *Utkala* when mentioned in India's national anthem, "Jana Gana Mana." Cuttack remained the capital of the state for over eight centuries until April 13, 1948, when Bhubaneswar was officially declared as the new state capital, a position it still holds. Odisha is the ninth largest state by area in India. Oriya (officially spelled *Odia*) is the official and most widely spoken language, spoken by three-quarters of the population. Odisha has a relatively unindented coastline (about 480 km long) and lacked good ports, except for the deepwater facility at Paradip, until the recent launch of the Dhamra Port. The narrow, level coastal strip, including the Mahanadi river delta, supports the bulk of the population.
2. Sardar Vallabhbhai Jhaverbhai Patel (1875–1950) was an Indian leader and statesman, one of the leaders of the Indian National Congress and one of the founding fathers of the Republic of India. He was a national leader who played a leading role in the country's struggle for independence and guided its integration into a united, independent nation. As the first home minister and deputy prime minister of India, Patel took charge of the task to forge a united India from the British Colonial Provinces allocated to India and more than five hundred self-governing princely states by the Indian Independence Act of 1947. Using frank diplomacy, backed with the option and use of military force, Patel's leadership persuaded almost every princely state. Often known as the "Iron Man of India" or "Bismarck of India," he is also remembered as the "Patron Saint" of India's civil servants for establishing modern all-India services, especially the administrative and police services. It was a great honor that was accorded to Haldane when he was invited by the Indian Government to deliver the 3rd Patel Memorial Lectures, upon his arrival in India in 1957. They were broadcasted all over India by All India Radio, and later published under the title *Unity and Diversity of Life*.
3. Prasanta Chandra Mahalanobis (1893–1972) was an Indian scientist and applied statistician. He is best remembered for the Mahalanobis distance, a statistical measure. He pioneered studies in anthropometry in India. He founded the Indian Statistical Institute at Kolkata and the Indian statistical journal *Sankhya*. He contributed to the design of large-scale sample surveys, and advised the Indian Government on national planning at the highest level.
4. "A Passage to India": Haldane's essay, which he wrote in 1957, was published in the *Rationalist Annual*. Haldane wrote, "In July 1957 my wife and I hope to leave Britain to settle down in India for the rest of our lives. By the time this article is printed we shall probably be living there."
5. Founded in 1885 in Fleet Street, London, the *Rationalist Annual* has been "promoting reason, science, and humanism and standing up to irrationalism and religious intolerance ever since." The Rationalist Association was founded, as the Rationalist Press Association (RPA), by the radical publisher Charles Watts at his print works in Johnson's Court, just off Fleet Street, London, in 1885. As a publisher of books and pamphlets dedicated to free thinking, science,

and a critique of organized religion, Charles was following in the footsteps of his father, also called Charles, who was a prominent figure in the Victorian free-thought movement and founding secretary of the National Secular Society. Toward the end of the nineteenth century, Watts & Co started to expand from producing *Watt's Literary Guide* and a range of propagandistic pamphlets, to publishing books, including the celebrated series Cheap Reprints, which made the works of skeptical Victorians like Charles Darwin, Thomas Huxley, and John Stuart Mill available to working people at only sixpence a volume. In 1929 Watts began publishing another famed series, the Thinker's Library, which printed 140 volumes over 22 years, including works by H. G. Wells, Bertrand Russell, Mark Twain, and J. B. S. Haldane. The book-publishing arm became less prolific in the postwar years, but publication of *Watt's Literary Guide* continued, as did publication of the RPA annual. This had started as the *Agnostic Annual* in 1884, becoming the *Rationalist Annual* in 1927, and *Question* in 1968, before ceasing publication in 1980.

Starting in 1929 with his important article "Origin of Life," Haldane was a regular contributor to the *Rationalist Annual*. He wrote "On Being Finite" shortly before his death on December 1, 1964. It was posthumously published in the *Rationalist Annual* for 1965. It was Haldane's final statement, bidding "good-bye" to the world, and sharing his final thoughts on death and dying and his view of his own scientific contributions.

6. Jerzy Neyman, J. B. S. Haldane archives, ISI, Calcutta, 1961, University College, London.
7. The *Rigveda* (Sanskrit) is an ancient Indian sacred collection of Vedic Sanskrit hymns. It is counted among the four canonical sacred texts (śruti) of Hinduism known as the *Vedas*. Some of its verses are still recited as Hindu prayers, at religious functions and other occasions, putting these as the world's oldest religious texts in continued use. The *Rigveda* contains several mythological and poetical accounts of the origin of the world, hymns praising the gods, and ancient prayers for life, prosperity, and so forth. It is the oldest extant text in any Indo-European language. Linguistic evidence indicates that the *Rigveda* was composed in the northwestern region of the Indian subcontinent, roughly between 1700 and 1100 BC.
8. The *Upanishads* form the core of Indian philosophy. They are an amazing collection of writings from original oral transmissions, which have been aptly described by Shri Aurobindo as "the supreme work of the Indian mind." It is here that we find all the fundamental teachings that are central to Hinduism—the concepts of "karma" (action), "samsara" (reincarnation), "moksha" (nirvana), the "atman" (soul), and the "Brahman" (Absolute Almighty). They also set forth the prime Vedic doctrines of self-realization, yoga, and meditation. The *Upanishads* are summits of thought on mankind and the universe, designed to push human ideas to their very limit and beyond. They give us both spiritual vision and philosophical argument, and it is by a strictly personal effort that one can reach the truth. The term "Upanishad" literally means, "sitting down near" or "sitting close to," and implies listening closely to the mystic doctrines of a guru or a spiritual teacher, who has cognized the fundamental truths of the universe. It points to a period in time when groups of pupils sat near the teacher and learned from him the secret teachings in the quietude of forest "ashrams" or hermitages. Historians and Indologists have put the date of composition of the *Upanishads* at around 800–400 BC, though many of the verse versions might have been written much later. In fact, they were written over a very long period of time and do not represent a coherent body of information or one particular system of belief. However, there is a commonality of thought and approach.
9. *Shankar's Weekly* was an Indian political cartoon magazine that was founded and edited by Kesava Shankara Pillai (1902–1989), better known as Shankar, an Indian cartoonist. He is considered the father of political cartooning in India. He founded *Shankar's Weekly*, India's *Punch*, in 1948. Today he is most remembered for setting up the Children's Book Trust, established in 1957, and Shankar's International Doll Museum in 1965. A notable cartoon published on May 17, 1964, just ten days before Prime Minister Nehru's death, showed an emaciated and exhausted Prime Minister Jawaharlal Nehru, with a torch in hand, running the final leg of a race, with party leaders Gulzari Lal Nanda, Lal Bahadur Shastri, Morarji Desai, Krishna Menon, and Indira Gandhi in tow, to which Nehru remarked, "Don't spare me, Shankar."

Shankar loved kids and organized Shankar's International Children's Competition in 1949 and Shankar's On-the-Spot Painting Competition for Children in 1952. He instituted an annual Competition for Writers of Children's Books in 1978. Begun in English, this competition is now held in Hindi too. It later drew children from all over the world. He also founded the Children's Book Trust in Nehru House in New Delhi in 1957. Later in 1965, the International Doll Museum too came to be located here. It has now a children's library and reading room, known as Dr. B. C. Roy Memorial Children's Library and Reading Room and a doll development and production center.

Haldane's cartoon appeared in *Shankar's Weekly* in 1959. At first he was shown wearing European-style suit and tie. However, when Haldane protested that he was no longer wearing European-style clothes, it was redrawn, showing Haldane in Indian-style clothes. Haldane was pleased to have his cartoon included in that weekly, which he considered to be a sign of acceptance in the Indian society.

10. Edward Groff Conklin (1904–1968) was a leading science fiction anthologist. He edited forty anthologies of science fiction, and one of mystery stories, and wrote books on home improvement. His book review column, "Galaxy's Five-Star Shelf," was a key feature in *Galaxy Science Fiction* from its premiere issue (October 1950) until October 1955. During that period, he also edited Grosset & Dunlap's Science Fiction Classics series, which he conceived as an inexpensive alternative to hard-to-find small-press editions of such titles as Robert A. Heinlein's *Beyond This Horizon* and Isaac Asimov's *I, Robot*. In the last three years of his life, Conklin was the staff science editor for the *American Heritage Dictionary of the English Language*. He lived in New York at 150 West 96th Street. At the age of sixty-three, he died of emphysema in his summer home at Pawling, New York.
11. Letter to Groff Conklin, ISI Archives, Calcutta, 1961. J. B. S. Haldane Archives, University College, London.
12. Yudhishtira is a princely character from the Hindu mythological saga *Mahabharata*.
13. *Svarga* is heaven in Hindu mythology.
14. J. B. S. Haldane, "The Detection of Autosomal Lethals in Mice Induced by Mutagenic Agents," *Journal of Genetics* 54 (1956): 327–42.
15. J. B. S. Haldane Archives, University College, London; Haldane's letter to the Director ISI, 1956–57.
16. J. B. S. Haldane Archives, University College, London; Letter from the ISI Director to Haldane, 1956–57.
17. See chapter 16, "Moving to Paradise (1957)."
18. Birbal Sahni (1891–1949) was an Indian paleobotanist, geologist, and archeologist. He founded the Birbal Sahni Institute of Palaeobotany in Lucknow, India. His greatest contributions lie in the study of fossil plants as well as living plants of India. He published numerous papers on these topics and also served as the president of the National Academy of Sciences of India. He was elected an honorary president of the International Botanical Congress in Stockholm in 1950, but died before the Congress was held. He was elected a fellow of the Royal Society of London in 1936.
19. S. Radhakrishnan, *The Hindu View of Life* (London: G. Allen & Unwin, 1961), 5.
20. J. B. S. Haldane (1968), A passage to India, p. 133, In *Science and Life*, London: Pemberton Publishing Co.
21. Haldane, "A Passage to India," 133.
22. The National Sample Survey Office (NSSO) was set up by the Indian Government in 1950, with the idea of having a permanent survey organization to collect data on various facets of the economy. In order to assist in socioeconomic planning and policy-making, the NSSO conducts nationwide sample surveys collectively known as the National Sample Survey (NSS). The NSS is a continuing survey in the sense that it is carried out in the form of successive "rounds," each round usually of a year's duration covering several topics of current interest. The surveys are conducted through household interviews, using a random sample of households covering practically the entire geographical area of the country.
23. C. V. Raman (Sir Chandrasekhara Venkata Raman; 1888–1970) was an Indian physicist whose pioneering work in the field of light scattering earned him the 1930 Nobel Prize for Physics. He was the first nonwhite scientist to receive the Nobel Prize. He discovered that when light traverses a transparent material, some of the deflected light changes in wavelength. This phenomenon is now called Raman scattering and is the result of the Raman effect. Raman and

his student Suri Bhagavantam discovered the quantum photon spin in 1932, which further confirmed the quantum nature of light. Raman was honored, in 1954, with the highest civilian award of India, the *Bharat Ratna*. Raman's early research was conducted at the Indian Association for the Cultivation of Science in Calcutta. Later, he joined the Indian Institute of Science in Bangalore. In 1944 he established the Raman Research Institute in Bangalore and remained its director until his death in 1970.

Raman's sincere advice to aspiring scientists was that "scientific research needed independent thinking and hard work, not equipment." Haldane would have agreed thoroughly with that comment, as he too advised students that high-quality scientific research did not necessarily require expensive or complicated equipment.

24. Sir Jagadish Chandra Bose (1858–1937) was a polymath; he was a physicist, biologist, botanist, and archeologist as well as a writer of science fiction. He pioneered the investigation of radio and microwave optics, made significant contributions to botany, and pioneered experimental science in India. He was successful in his research on remote wireless signaling and was the first to use semiconductor junctions to detect radio signals. However, instead of patenting, Bose made his inventions public in order to allow others to further develop his research. Subsequently he made a number of pioneering discoveries in plant physiology. He constructed recorders capable of recording extremely slight responses to various stimuli; these instruments produced some striking results, such as the quivering of injured plants. His books include *Response in the Living and Non-Living* (1902) and *The Nervous Mechanism of Plants* (1926).

25. Meristic variation: Haldane followed Bateson's explanation of meristic variation. This is explained by examples of variation in like parts, such as teeth, vertebrae, or petals. For instance, teeth can be variable in size or missing. When at a certain critical stage, the rudiment fell below a threshold, it regressed or did not develop further. Similarly, when a rudiment is too large at a critical stage of development it may divide into two or even more parts, giving an extra limb, for example. Haldane compared the process of the formation of vertebrae in a tail to that of drops in a liquid filament, even though the forces involved are of quite a different nature. He thought that Bateson followed a similar reasoning. Haldane encouraged research projects in India that followed similar examples, for example S. K. Roy's research on variation in petal numbers in plants, and variation in nipple numbers in cattle.

26. J. B. S. Haldane, "The Theory of Evolution, Before and After Bateson," *Journal of Genetics* 56 (1958): 11–27; 24.

27. Canalization: The term *canalization* was coined by the developmental biologist Conrad H. Waddington of the University of Edinburgh. It is a measure of the ability of a population to produce the same phenotype under varying conditions of its environment or genotype. Biological robustness or *canalization* comes about when developmental pathways are shaped by evolution. Waddington introduced the *epigenetic landscape*, in which a canalized trait is illustrated as a valley enclosed by high ridges, safely guiding the phenotype to its "fate." Waddington claimed that canals form in the epigenetic landscape during evolution.

28. K. R. Dronamraju, *Haldane, Mayr and Beanbag Genetics* (Oxford: Oxford University Press, 2011), 253.

29. *Lantana camara* is a common prickly invasive species of the verbena family, which grows in the suburbs of Calcutta and many other parts of the world. The two varieties used in my research differ in flower color—"pink," which has pink and white flowers, and "orange," which has orange and yellow flowers. It is native to the American tropics and has been introduced to other parts of the world as an ornamental plant.

30. Ernst Mayr (1904–2005), evolutionary biologist and author, was a German immigrant who moved to the United States after receiving his graduate education in Berlin under the direction of the distinguished zoologist Erwin Stresseman. Mayr spent much of his life in the United States in two institutions—the American Museum of Natural History in New York and Harvard University in Cambridge, Massachusetts. His early specialization was in ornithology, but later his interests grew to include many other subjects related to evolutionary biology. Mayr was one of the leading architects of the synthetic theory of evolution. He was a

strong believer in the role of geographic factors in the formation of new species. Mayr's books include *Systematics and the Origin of Species, from the Viewpoint of a Zoologist, Animal Species and Evolution, Evolution and the Diversity of Life, The Growth of Biological Thought*, and *Toward a New Philosophy of Biology*. Mayr was a close intellectual friend of Haldane. Toward the end of his life, Haldane was involved in a friendly controversy with Mayr regarding the importance of mathematical population genetics, which he defended in an amusing and feisty essay, "A Defense of Beanbag Genetics." Mayr visited Haldane in India shortly before Haldane's death.

31. Erwin Stresemann (1889–1972) was a German naturalist and ornithologist. He obtained a PhD in zoology in 1920 at Munich. He was at first in charge of the bird department of the Berlin Zoological Museum and later became a professor of zoology at the University of Berlin. Stresemann was one of the outstanding ornithologists of the twentieth century. He trained a number of younger scientists, including Ernst Mayr, Bernhard Rensch, and Salim Ali. Stresemann was the long-standing editor of the *Journal fur Ornithologie* from 1922 until his death.

32. Ernst Mayr, letter from to J. B. S. Haldane, April 13, 1959.

33. Konrad Lorenz (1903–1989) was an Austrian zoologist, ethologist, and ornithologist. He shared the 1973 Nobel Prize for Physiology or Medicine with Nikolaas Tinbergen and Karl von Frisch. He is one of the founders of modern ethology. Lorenz's research involved studies of instinctive behavior in animals, especially in greylag geese. Working with geese, he rediscovered the principle of imprinting (originally described by Douglas Spalding in the nineteenth century). In later life, his interest shifted to the study of human behavior. Lorenz wrote numerous books, some of which, such as *King Solomon's Ring, On Aggression*, and *Man Meets Dog*, became popular bestsellers. His last work "Here I Am—Where Are You?" is a summary of his life's work and focuses on his famous studies of greylag geese.

34. J. B. S. Haldane, "Mind in Evolution," *Zoologische Jahrbucher Abteilung fuer Systematik* 88 (1960): 117–25; 119.

35. Haldane, "Mind in Evolution," 117–37.

36. *Bauhinia acuminata* is a flowering shrub native to tropical southeastern Asia. Common names include dwarf white bauhinia, white orchid-tree, and snowy orchid-tree. It grows 2 to 3 meters tall. The leaves are bilobed. The flowers are white and fragrant, with five white petals. The species occurs in deciduous forests. It is widely cultivated throughout the tropics as an ornamental plant.

37. Herman Slatis (1932–1976) was a brilliant geneticist who studied with Sewall Wright at the University of Chicago. I knew him well as a personal friend who suggested my PhD project, hairy ears as a possibly Y-linked trait in man, during a visit to Haldane's department at the Indian Statistical Institute in Calcutta. He was a professor of zoology at Michigan State University from 1963 until his untimely death in 1976, when he was struck by a car while crossing a street. In human genetics, Slatis studied the genetics of hirsutism, inbreeding, radiation effects, and several other topics. He was especially skilled at statistical analysis of genetic data in human families. Slatis served as a director of the American Society of Human Genetics from 1964 to 1966.

38. The Y chromosome is one of two sex chromosomes in mammals. It is the sex-determining chromosome in many species, since the presence or absence of Y determines male or female sex. In mammals, the Y chromosome contains the gene *SRY*, which triggers testes development. The DNA in the human Y chromosome is composed of about 59 million base pairs. The Y chromosome is passed only from father to son, so analysis of Y chromosome DNA may thus be used in genealogical research. So far, over 200 Y-linked genes have been identified. All Y-linked genes are expressed with some exceptions. It represents approximately 2 percent of the total DNA in a male cell. The human Y chromosome is normally unable to recombine with the X chromosome, except for small pieces of pseudoautosomal regions at the telomeres (which constitute about 5 percent of the chromosome's length). These regions are relics of ancient homology between the X and Y chromosomes. Most of the Y chromosome does not recombine. It is called the "NRY" or nonrecombining region of the Y chromosome. The SNPs (single-nucleotide polymorphisms) in this region are used for tracing direct paternal ancestral lines.

39. K. R. Dronamraju, "Hypertrichosis of the Pinna of the Human Ear, Y-linked Pedigrees," *Journal of Genetics* 57 (1960): 230–44.
40. Autosomal: Related to any of the chromosomes other than the sex chromosomes; inheritance that is not sex-linked.
41. Inbreeding: Usually refers to matings in any species (or marriages in humans) within a small population, often involving close relatives, such as first or second cousins.
42. The coefficient of inbreeding is a measure of the degree of inbreeding between two individuals. It was invented by Sewall Wright in 1922. Wright's *coefficient of relationship* is a measure of the level of consanguinity between two given individuals. The coefficient of relationship approaches a value of one for individuals from a completely inbred population, and nearly zero for individuals with arbitrarily remote common ancestors. The coefficient expresses the expected percentage of homozygosity arising from a given system of breeding.
43. K. R. Dronamraju, *Haldane: The Life and Work of JBS Haldane with Special Reference to India* (Aberdeen: Aberdeen University Press, 1986), 112–56.
44. J. B. S. Haldane, "An Autobiography in Brief," *Illustrated Weekly of India, Bombay*, 1961. Reprinted in K. R. Dronamraju, ed., *Selected Genetic Papers of J.B.S. Haldane* (New York: Garland, 1990).
45. J. B. S. Haldane, personal communication to the author.
46. Guido Pontecorvo (1907–1999) was professor of genetics at the University of Glasgow in Scotland. "Ponte" was well known for his brilliant research in microbial genetics. He was born in a Jewish family in Pisa, Italy, and received his education in agriculture. In 1938 he fled Italy to escape the fascist regime of Mussolini. Arriving in Scotland, he was interned as an enemy alien during World War II. Later, he studied under the geneticist Herman J. Muller (later Nobel laureate) at the University of Edinburgh, obtaining a PhD in fruit fly genetics. Much of his academic life was spent at the University of Glasgow. "Ponte" came from a large family of eight children. His brother Gillo was the director of the movie classic *Battle of Algiers*, and another brother, Bruno, a nuclear physicist, became famous when he abruptly abandoned his research position at the UK Nuclear Energy Establishment at Harwell and fled to the Soviet Union secretly.
47. Bhubaneswar is the capital of the Indian State of Odisha, formerly known as *Orissa*. The city has a history of over 3,000 years as a center of economic and religious importance in eastern India. With many Hindu temples, which span the entire spectrum of Kalinga architecture, Bhubaneswar is often referred to as a Temple City of India. Modern Bhubaneswar has a population of nearly one million, and about half a million when Haldane lived there during the years 1962–1964. The modern city was designed by the German architect Otto Konigsberger in 1946.
48. The *New Statesman* was founded in 1913 with the aim of permeating the educated and influential classes with socialist ideas. Its founders were Sidney and Beatrice Webb (later Lord and Lady Passfield), along with Bernard Shaw and a small but influential group of Fabians. After undergoing several changes and mergers, it was relaunched in 2006, and newstatesman.com has forged its own identity. The website has facilitated a new generation of readers, as well as an international readership. The *New Statesman* online and weekly magazine furthers the original aims of Beatrice and Sidney Webb. Haldane's caricature by Vicky and a biographical sketch were published in the issue dated January 7, 1956, when Kingsley Martin was its editor. Later, it was included in a collected volume of biographical profiles titled *New Statesman Profiles*, with a note on profiles by Kingsley Martin (London: Phoenix House, 1957), 185–89.
49. Marcello Siniscalco was an Italian geneticist from Naples. He was a student of the renowned scientist Giuseppe Montalenti. Marcello's early research was concerned with glucose 6 phosphate dehydrogenase (G-6 PD) deficiency in relation to malaria in Sardinia. In fact, malaria research occupied his entire life, partly in relation to thalassemia and sickle cell anemia. Sadly, Marcello was never given the opportunity to establish his research center in Italy. He worked in far-off centers, in the Galton laboratory at University College London, in the University of Leiden, and at the Rockefeller University in New York, but he always returned to Sardinia to collect blood samples. He was invited by J. B. S. Haldane to extend his research on malaria and G-6 PD deficiency in the tribals of Andhra Pradesh, south India, where he worked with me and P. Meera Khan. This project was personally financed by J. B. S. Haldane from the Feltrinelli

Prize Awarded to him by the Italian Academy of Sciences. One of the last international conferences attended by Marcello was at the UK Genome Center in Hinxton (near Cambridge) in 2002, "Infectious Disease and Host-Pathogen Evolution," funded by the Wellcome Trust.

50. *Kayastha* is a caste of Hindus originating in India. Kayasthas are members of the literate scribe caste who have traditionally played the role of record-keepers, writers, and administrators of the state, and have been mentioned traditionally as royal officials engaged in writing state documents, maintaining public accounts, and holding highly responsible official roles. Kayasthas have historically occupied the highest government offices, serving as ministers and advisors during early medieval Indian kingdoms and the Mughal Empire, and holding important administrative positions during the British Raj. In modern times, Kayasthas have attained success in politics, the arts, and various professional fields. In eastern India, Bengali Kayasthas are believed to have evolved from a class of officials into a caste between the 5th/6th century AD and 11th/12th century AD, its component elements being putative Kshatriyas and mostly Brahmins.

51. A. K. Ray, and J. B. S. Haldane, "The Genetics of a Common Indian Digital Abnormality," *Proceedings of the National Academy of Sciences USA* 53 (1965): 1050–53.

52. J. B. S. Haldane, letter to Ernst Mayr, April 6, 1963. Reprinted in K. R. Dronamraju, ed., *Haldane, Mayr and Beanbag Genetics* (Oxford: Oxford University Press, 2011), 253.

53. J. B. S. Haldane, *Possible Worlds and Other Essays* (London: Chatto & Windus, 1930).

54. Since its launch in 1948, the National Health Service (NHS) in the UK has grown to become the world's largest publicly funded health service. The NHS was born out of a long-held ideal that good healthcare should be available to all, regardless of wealth, a principle that remains at its core. With the exception of some charges, such as prescriptions and optical and dental services, the NHS remains free at the point of use for anyone who is resident in the UK. The NHS employs more than 1.7 million people. Of those, just under half are clinically qualified, including 39,780 general practitioners (GPs), 370,327 nurses, 18,687 ambulance staff, and 105,711 hospital and community health service (HCHS) medical and dental staff. The NHS deals with over 1 million patients every 36 hours. Funding for the NHS comes directly from taxation and is granted to the Department of Health by Parliament. For 2012/13 its budget was around £108.9 billion.

55. *The Listener* (BBC), December 10, 1964. Haldane's auto-obituary.

56. *Science and Indian Culture* (Kolkata: New Age Publishers, 1965).

57. *Durvasa*—In Hindu mythology, *Durvasa* was an ancient sage, the son of Atri and Anasuya. He is supposed to be an incarnation of *Shiva*. He is known for his short temper. Hence, wherever he went, he was received with great reverence from humans and gods alike. According to local tradition in modern Azamgarh, Durvasa's Ashram or hermitage, where many disciples used to go to study under him, was situated in the area, 6 kilometers north of the *Phulpur* Tehsil headquarters.

58. *Viswamitra* is one of the most venerated sages of ancient India. He is also credited as the author of most of Mandala 3 of the *Rigveda*, including the *Gayatri Mantra*. The Puranas mention that only 24 rishis since antiquity have understood the whole meaning of—and thus wielded the whole power of—the *Gayatri Mantra*. Vishvamitra is supposed to be the first. Vishvamitra was a king in ancient India, also called Kaushika ("descendant of Kusha").

18

An Indian Perspective of Darwin

It was typical of Haldane to examine Darwinism from a fresh perspective after his move to India, for he never missed an opportunity to reexamine long-held beliefs and theories, including his own publications. Haldane devoted his entire life to confirming the Darwinian theory of natural selection in terms of Mendelian genetics and establishing it on a firmer foundation. His admiration for Darwin bordered on adoration. Over seven decades ago, in his collection of essays titled *Keeping Cool and Other Essays*,[1] Haldane summed up Darwin's accomplishments: "Darwin speculated in the most daring manner. He was extremely cautious in publishing his speculations and extremely honest in weighing the arguments against them. But his mind was dominated by an immense respect for facts, and it is this respect more than any other characteristic which has given him his lasting influence on human thought."

Haldane implored his students in India to imitate Darwin's methods. With rich biodiversity and limited resources in India, Haldane thought it was an ideal place to conduct excellent biological research of the kind that interested Darwin—ecological and field studies that did not require expensive or complicated equipment.

Haldane published his views on an Indian perspective of Darwin in a most unlikely and remote journal, *Centennial Review*,[2] of the Michigan State University in East Lansing, Michigan. The fall 1959 issue was dedicated to the Darwin-Wallace Centennial, 1859–1959. As was his custom in several of his publications, Haldane began with a cautious note, "It is probably too early to assess Darwin's significance for human culture. It is, however, much easier to do so if one has the stereoscopic view afforded by a measure of intimacy with more than one of the main cultures of our planet. I could not have written this article before I became an Indian."

Haldane stated his main point in the first paragraph of the essay itself. It was concerned with the impact of Darwinism in India and China as opposed to Europe and America:

> To Europeans and Americans, it inevitably seems that Darwin's greatest achievement has been to convince educated men and women that biological evolution is a fact, that living plant and animal species are all descended from ancestral species very unlike themselves, and, in particular, that men are descended from animals. This was an important event in the intellectual

life of Europe, because Christian theologians had drawn a sharp distinction between men and other living beings. In view of Jesus' remarks about sheep, sparrows, and lilies, this sharp distinction may well be a perversion of the essence of Christianity. St. Francis seems to have thought so.

In contrast to this situation in Europe, Haldane pointed out that "in India and China this distinction has not been made; and, according to Hindu, Buddhist, and Jaina ethics, animals have rights and duties." Haldane explained further that Hinduism is not a religion, as this term is understood by the adherents of proselytizing religions: "It is an attitude toward the universe compatible with a variety of religious and philosophical beliefs." It is best reflected in Hindu mythology and folklore. Haldane referred to the great Hindu epic, the *Ramayana*,[3] in which the divine hero Ram is aided by an army of monkeys and bears to regain his wife, Sita, who has been abducted by the demon Ravana. In a scene from the other great Hindu epic, the *Mahabharata*,[4] which depicts the life of Krishna, the whole background consists of cow's heads in a stone carving at Mahabalipuram near Chennai in South India.

As Haldane put it,

> For every Hindu, the setting of human experience is alive. Of course, he does not live up to his attitudes and beliefs. Nor do Christians. If even fifty percent of Christians forgave their debtors, from the boss who owes a week's wages to the farmer who has mortgaged his means of livelihood, the economic fabric of Christian civilization would collapse in eight hours. Similarly, many, . . . Indians are cruel to animals; but kindness to animals, including vegetarianism, is commoner in India than forgiveness of debtors is in Christendom.

Botanical Research

Haldane wrote that, in his opinion, Darwin's most original contribution to biology is not the theory of evolution, which had been discussed by others before Darwin, but his great series of botanical works. If Darwin had died young, Wallace would have continued his work on evolution. However, as Haldane pointed out, it would have taken longer for its acceptance because Wallace's arguments covered a smaller field than Darwin's. However, the impact of Wallace's work on evolution would have been less on Western thought than Darwin's because Wallace left loopholes open for supernatural intervention, and Darwin did not.

In his later years, Darwin wrote several books on experimental botany that were concerned with those aspects of plant life that resembled animal life. Two were concerned with climbing plants and insectivorous plants respectively. Three books were devoted to sexuality in plants, especially the humanlike aspects such as the evil effects of incest, and curious modifications of nature that prevent self-fertilization. Benefits of that research and its extensions include the discovery of plant hormones and the invention of weed killers. The economic and social benefits of the systematic outcrossing of maize (corn) are well known. It is one of the great success stories of the North American continent.

Darwin's Attitude

The next question Haldane addressed is what the attitude was that led Darwin to his discoveries.

Clues to the answer can be found in Darwin's autobiography and in the memoir by his son Francis. Darwin was devoted to his experimental objects. His son wrote,

> I used to like to hear him admire the beauty of a flower; it was a kind of gratitude to the flower itself, and a personal love for its delicate form and colour. I seem to remember him gently touching a flower he delighted in; it was the same simple admiration that a child might have. He could not help personifying natural things. This feeling came out in abuse as well as praise—e.g., of some seedlings—"The little beggars are doing just what I don't want them to." His emotional attitude to animals was one of profound aesthetic admiration. One of his favorite words was "wonderful." Here is a typical passage concerning the second stage larvae of barnacles. "They have six pairs of beautifully constructed natatory legs, a pair of magnificent compound eyes, and extremely complex antennae; but they have a closed and imperfect mouth, and cannot feed."

Francis narrated an account of what his father called "a fool's experiment." Francis was ordered to play the bassoon to some seedlings. This did not influence their growth but vibration of the table did. Another classic fool's experiment was to cut several scalene triangles of paper, leave them on the lawn, and later find that the earthworms chose the most acute angle to drag them to plug their holes. Haldane commented, "Darwin did not draw a sharp line between earthworms and the old gentleman who had failed to interest him in mathematics at Cambridge."

Such dedication is seen in Indian sages, but it does not lead to increased knowledge. Usually, it results in a flood of sympathy with animal and human suffering with a new series of ritual prohibitions affecting a few thousand people. Darwin's devotion led him to make scientific observations with great accuracy. So Haldane concluded that from the Hindu point of view Darwin possessed some of the attributes of a saint.

Darwinism and Logic

Aristotle developed a logical system based on similarities. On the other hand, Darwin, in the last chapter of *The Origin of the Species*,[5] anticipated a logic based on differences. Darwin wrote, "Systematists will have only to decide (not that this will be easy) whether any form be sufficiently constant and distinct from other forms, to be capable of definition, and if definable, whether the difference be sufficiently important to deserve a specific name."

Darwin further wrote, "Hence, without rejecting the consideration of the present existence of intermediate gradations between any two forms, we shall be led to weigh more carefully and to value higher the actual amount of difference between them."

Haldane suggested that Darwin's second sentence has formed the foundation for a whole new branch of statistics. In this context, two important questions can be answered: First, "Does population A of animals, plants or humans differ significantly from population B, or could their difference be due to random sampling from the same larger population?" Second, "Does population C differ more or less from population A than from population B?"

The statistical methods devised by Gossett,[6] Pearson,[7] Mahalanobis,[8] and others to answer such questions with respect to biological data have become important for data in physics, geology, and other sciences. Haldane predicted that as all sciences became more and more statistical, as some scientists had predicted, Darwin would be recognized as a pioneer in this process.

Ethics

Haldane considered that Darwinism has so far been responsible for more evil than good in the field of ethics. It has been greatly misrepresented and misinterpreted for various reasons. Haldane thought that Darwin himself was partly responsible for its misrepresentation.

Darwin arrived at his theory of evolution by natural selection after reading Malthus[9] on population (1798). The following account was written by Darwin[10] in his autobiography:

> In October 1838, that is, fifteen months after I had begun my systematic inquiry, I happened to read for amusement Malthus on *Population*, and being well prepared to appreciate the struggle for existence which everywhere goes on from long-continued observation of the habits of animals and plants, it at once struck me that under these circumstances favourable variations would tend to be preserved, and unfavourable ones to be destroyed. The results of this would be the formation of a new species. Here, then I had at last got a theory by which to work.

Darwin was especially impressed by Malthus's observation that in nature plants and animals produce far more offspring than can survive, and human populations are also capable of a similar overproduction. Until recently, this was the situation in India, where many infants and newborn were killed off by infectious diseases of many kinds. Malthus thought that unless family size was regulated, poverty and famine would follow. But he thought that was God's way of preventing man from being lazy. Both Darwin and Wallace extended Malthus's principle in natural terms without invoking divine intervention. They came to the conclusion that producing more children than can survive would create a competitive environment among siblings, and that the variation within a sibship would result in some individuals with a greater chance of survival. However, centripetal selection still operates, that is to say, extremes have fewer offspring than animals or plants near the average. Haldane wrote, "Giant species appear to be much less likely than those of moderate size to leave descendants."

Haldane pointed out that natural selection still operates in a population that is so fortunate to have room for everyone. The example he cited is the population of North America, which has been growing steadily since 1700 AD or earlier. There has been space for everyone (at least until now) but there was also natural selection for fertility, disease resistance, and other traits. The population of some countries such as France has been steady for a long time but it is still subject to natural selection.

In India, Haldane mentioned the example of the development of resistance to DDT[11] by mosquitoes and other insects as an ongoing evolutionary process, and urged local scientists to pursue quantitative studies of that process. No other agency than natural selection can produce rapid evolutionary changes in large populations. But evolution is usually so slow that it is hardly noticeable in a single generation.

Social Darwinism

Darwinism has been misinterpreted by some to mean that the struggle for life during the course of evolution was similar to warfare and economic competition in the human populations. Consequently, Darwinism was used to justify warfare, racism, slavery, and numerous other cruelties. Haldane wrote, "Finally various forms of human violence and injustice have been defended as somehow conforming to Darwinian principles. But in fact the struggle for life is seldom conducted with violence. The fitness of an animal or plant from a Darwinian point of view is measured by the number of progeny which it leaves." Haldane stated that "during the last few thousand years the main agents of natural selection in man have probably been infectious diseases, and the struggle for life has been a struggle with bacteria and other pathogens, and not with fellow human beings."

Haldane pointed out that in much of the world the poor breed quicker than the rich, even when their higher death rates are considered, and "those who are victorious in the struggle for wealth are defeated in the struggle for life. The view that Darwinism is an immoral doctrine is based on ignorance of such elementary facts as this."

It is also a fact of human history that ruling classes die out due to several reasons. They may be massacred in revolutions, or their inbreeding makes them susceptible to various infections, or more likely they may die of infertility. The sex researcher Alfred Kinsey[12] reported that sexual behavior among affluent Americans has resulted in lowered fertility in the richer and better educated Americans. Whatever the reasons were, Haldane concluded that economic success has been usually correlated with biological failure. He noted that a significant fraction of West Africans was content to live meekly as slaves, while other, prouder, races were not. Consequently, their descendants are now a majority in several American cities and neighboring islands.

In *The Descent of Man*,[13] Darwin described how medical advances meant that the weaker were able to survive and have families, and as he commented on the effects of this, he cautioned that hard reason should not override sympathy and considered how other factors might reduce the effect: "Thus the weak members of civilized societies propagate their kind. No one who has attended to the breeding of domestic animals will doubt that this must be highly injurious to the race of man. It is surprising

Figure 18.1 Haldane invited the Italian geneticist Marcello Siniscalco for genetic research among the tribal people of Andhra Pradesh in India.

how soon a want of care, or care wrongly directed, leads to the degeneration of a domestic race; but excepting in the case of man himself, hardly any one is so ignorant as to allow his worst animals to breed. . . . We must therefore bear the undoubtedly bad effects of the weak surviving and propagating their kind."

Notes

1. J. B. S. Haldane, *Keeping Cool and Other Essays* (London: Chatto & Windus, 1940).
2. *Centennial Review*, published by Michigan State University at East Lansing, invited Haldane to contribute an article for the issue celebrating the centennial of the publication of Charles Darwin's *Origin of Species*. The topic chosen by Haldane for his article is how Darwin and Darwinism look from an Indian point of view when viewed against the background of India's cultural and religious traditions. One important point emerges: In contrast to the conflicts that arose between Darwinism and Christianity in Western countries, there is remarkable harmony and compatibility between Darwinism and the Hindu religion. I am pleased to acknowledge the help of Dr. Herman Slatis, who was then on the faculty of Michigan State, in obtaining a copy of the article. "An Indian Perspective of Darwin," *Centennial Review of Arts & Sciences, Michigan State University* 3 (1959): 357–64.
3. The *Ramayana* is a great epic of Hindu mythology. It tells about life in India around 1000 BCE and offers models in life's duties and ethics as well as the ideals and wisdom of everyday life. The hero, Rama, lived his whole life by the rules of dharma. He was an ideal husband to his faithful wife, Sita, and a responsible ruler of Ayodhya. They were an ideal couple, whom Hindus are taught to emulate. The original *Ramayana* was a 24,000-couplet-long epic poem

attributed to the Sanskrit poet Valmiki. It has since been told, retold, and translated throughout Southeast Asia. The *Ramayana* continues to be performed in dance, drama, puppet shows, songs, and movies all across Asia.

The *Ramayana* and *Mahabharata* are the two epics that help to bind together the many peoples of India, transcending caste, distance, and language. Two all-Indian holidays celebrate events in the Ramayana. Dussehra, a 14-day festival in October, commemorates the siege of Lanka and Rama's victory over Ravana, the demon king of Lanka. Divali, the October–November festival of lights, celebrates Rama and Sita's return home to their kingdom of Ayodhya.

4. The *Mahabharata* is the epic narrative of the Kurukshetra war and the fates of the Kaurava and the Pandava princes. It contains philosophical and devotional material, such as a discussion of the four "goals of life." Among the principal works and stories in the *Mahabharata* are the *Bhagavadgita*, the story of *Damayanti*, an abbreviated version of the *Ramayana*, and the *Rishyasringa*, often considered as works in their own right. Traditionally, the authorship of the *Mahabharata* is attributed to Vyasa. The oldest preserved parts of the text are thought to be not much older than around 400 BCE. The title may be translated as "the great tale of the Bhārata dynasty," and it is extended from a shorter version of 24,000 verses called simply *Bhārata*.

The *Mahabharata* is the longest known epic poem, containing 100,000 *shlokas* or over 200,000 individual verse lines (each *shloka* is a couplet), and long prose passages. It is about 1.8 million words in total. Its importance to world civilization has been compared to that of the Bible and the Qur'an.

5. Charles Darwin, *On the Origin of Species by Means of Natural Selection, or, The Preservation of Favoured races in the Struggle for Life* (London: J. Murray, 1859).

6. William Sealy Gosset (1876–1937) was a statistician. He published under the pen name Student, and developed the "student's t-distribution." He attended Winchester College before reading chemistry and mathematics at New College, Oxford, with which both Haldanes—J. S. and J. B. S.—had a long association later. Gosset and Karl Pearson had a good relationship. Gosset was employed by Guinness Breweries. That Gosset wrote under the name "Student" explains why his name may be less well known than his important results in statistics. He invented the *t*-test to handle small samples for quality control in brewing. Gosset discovered the form of the t distribution by a combination of mathematical and empirical work with random numbers, an early application of the Monte-Carlo method.

7. Karl Pearson (1857–1936) was a major contributor to the early development of statistics as a scientific discipline in its own right. He founded the Department of Applied Statistics (now the Department of Statistical Science) at University College London in 1911; it was the first university statistics department in the world. The present departments of Statistical Science and Computer Science, as well as the Genetics and Biometry Department and Physical Anthropology were all part of his legacy to UCL. When Francis Galton died in 1911 and left part of his estate to the University of London for a chair in eugenics, Pearson was appointed to that chair, as desired by Galton. His commitment to socialism and its ideals led him to refuse honors, such as an OBE (Order of the British Empire) when it was offered in 1920, and a knighthood in 1935.

8. Prasanta Chandra Mahalanobis (1893–1972) was an Indian scientist and applied statistician. He is best remembered for the "Mahalanobis Distance," a statistical measure. He made pioneering studies in anthropometry in India. He founded the Indian Statistical Institute at Calcutta, and *Sankhya, the Indian Journal of Statistics*. He contributed to the design of large-scale sample surveys. Because of his close friendship with the newly independent India's Prime Minister Jawaharlal Nehru, Mahalanobis became a member of the National Planning Commission and advisor to the Government. In the second 5-year plan, he emphasized rapid industrialization of India and played a key role in the development of a statistical infrastructure including national sample surveys. Mahalanobis died on June 28, 1972, a day before his seventy-ninth birthday. Even at this age, he was still active doing research work and discharging his duties as the secretary and director of the Indian Statistical Institute and as the honorary statistical advisor to the Cabinet of the government of India.

Mahalanobis was a man of great intelligence and organizational abilities. I knew him personally, as he was the director of the Indian Statistical Institute when I was a research fellow with Haldane during the years 1958–1961. Later, I was happy to receive my PhD diploma, which he signed and the research for which was conducted under Haldane's direction.

9. T. R. Malthus, *An Essay on the Principle of Population; or, A View of its Past and Present Effects on Human Happiness; with an Inquiry into Our Prospects Respecting the Future Removal or Mitigation of the Evils which it Occasions* (London: J. Murray, 1826; reprinted numerous times).

10. Charles Darwin, *Autobiography of Charles Darwin, with two appendices, comprising a chapter of reminiscences and a statement of Charles Darwin's religious views* (London: Watts & Co., 1949).

11. DDT (dichlorodiphenyltrichloroethane) is a colorless, crystalline, tasteless, and almost odorless organochloride known for its insecticidal properties. First synthesized in 1874, DDT's insecticidal action was discovered by the Swiss chemist Paul Hermann Muller in 1939. After the war, DDT was made available for use as an agricultural insecticide, and its production and use duly increased. Müller was awarded the Nobel Prize "for his discovery of the high efficiency of DDT as a contact poison against several arthropods" in 1948.

In 1962, the book *Silent Spring* by Rachel Carson was published. It cataloged the environmental impacts of indiscriminate DDT spraying in the United States and questioned the logic of releasing large amounts of chemicals into the environment without a sufficient understanding of their effects on ecology or human health. Its publication resulted in a large public outcry that eventually led, in 1972, to a ban on the agricultural use of DDT in the United States. A worldwide ban on its agricultural use was later formalized under the Stockholm Convention, but its limited use in disease vector control continues to this day and remains controversial.

12. A. C. Kinsey, *Sexual Behavior in the Human Male* (Philadelphia: W.B. Saunders, 1948), and *Sexual Behavior in the Human Female* (Philadelphia: W.B. Saunders, 1953).

13. C. Darwin, *The Descent of Man, and Selection in Relation to Sex* (London: J. Murray, 1913).

19

Life with Haldane

It has been said that while success has many parents, failure is an orphan. It is a tribute to Haldane's fame and success as a great scientist that I have heard several individuals claiming to be his "pupils, associates, acquaintances or even intimate friends" in several countries. These false claims increased exponentially long after his death. Haldane used to explode occasionally upon reading in the press that someone or other was claiming to be his one of his friends or associates, which turned out to be a complete fraud.

I knew Haldane intimately during his life in India (1957–1964), acquiring a thorough knowledge of his life and work, his former pupils, past associates, and close friends as well his enemies. I met him when I was twenty years old and he was sixty-six. He had just immigrated from London to Calcutta to join the faculty of the Indian Statistical Institute (ISI). I had just obtained my master's degree in genetics and plant breeding from Agra University, which is, of course, located near the world-famous Taj Mahal. I joined Haldane's department in January 1958 and was promptly assigned a large desk and a bookshelf at the entrance to his large personal library. Haldane did not sit at a desk because of the pain he suffered from an injury he sustained in a diving experiment during World War II. He preferred to sit on cushions in a low armchair, surrounded by his books and papers on the floor, arranged in an arc. Haldane sat at the other end of the library, and we were separated by several rows of tall bookshelves. It was part of my de facto duty at the entrance to screen visitors who wished to see Haldane. Facing my desk were several shelves with Charles Darwin's books, which I was recommended to read during my first three months. A year later I was given a separate study of my own next door.

Quite rapidly, I became a close confidant of both Haldane and Spurway. By the end of my first year, 1958, I was invited to share their residence, first an apartment in the Institute and later a house nearby. We traveled together all over India and Europe, attending and presenting research papers at scientific conferences. Memorable visits included those to Sri Lanka, Kashmir, Scotland, France, and Israel, which I have described in detail elsewhere.[1] Both marriages of Haldane were childless. He treated me like a son and depended on me for carrying on various domestic issues related to our daily routine as well as communicating with other colleagues and the local bureaucracy. I hired the staff, including the cook and other servants in the household, and helped Spurway in resolving various domestic crises. Soon Haldane also needed two personal assistants to help him deal with the Institute bureaucracy, look after his personal library, and run errands daily as needed. So I hired two young men from my native state of Andhra Pradesh, Mangipudi R. Sastry and P. Srihari Rao, who proved to be most faithful and hard-working staff members of the Haldane circle, first in Calcutta and later in Bhubaneswar.[2] In addition to my research, Haldane also asked me to help him with the editing of *Journal of Genetics*, which he transferred to India after his move.

Naomi Mitchison

Haldane's sister, the celebrated writer Naomi (Nou) Mitchison (later Lady Mitchison), was the first relative of Haldane who came to see us, shortly after my arrival. Haldane invited her to join us on a short train journey to the Indian Institute of Technology*—a sprawling institute of technology located at Kharagpur. The purpose of our visit was to see the Head of the Genetics and Plant Breeding Department, Dr. Nirad Sen, who was developing different strains of cowpea, *Vigna sinensis*, which would be ideal for simple Mendelian hybrid experiments. Haldane thought it would be a good project for a beginner like me and asked Dr. Sen to give us seeds to start our own breeding program. We were received cordially and obtained the seeds we wanted. What interested Haldane and Nou was the extent to which the Institute was rapidly becoming self-sufficient. Various machine tools were being made on a large scale. New generations of tractors were on display, as the country was being transformed rapidly, replacing old methods by industrialization.

Nou was a prolific author of over ninety books in historical and classical fiction, travel, biography, and political commentary. She was married to Dick Mitchison, Labour Member of the Parliament, who was made a life peer later. Legend has it that Nou and Dick played a historic role in India's independence from Great Britain, by arranging the first meeting between the postwar Labour prime minister, Sir Clement Atlee, and the future prime minister of India, Jawaharlal Nehru, at her home.

Nou died at the age of 101, in 1999, in her long-time residence at Carradale in Scotland. She was a feminist before there were any feminists, although she never used that term herself. One of her best-known books is *Memoirs of a Spacewoman* (1962). She was a generous hostess to generations of distinguished writers, artists, and scientists.

I recall the cheerful Christmas holidays at Carradale in 1961, when I was invited by Nou. That was my first visit to Carradale. I was spending a year at Glasgow University. Nou asked another houseguest, Anne McLaren,[3] to pick me up in her car. The house was full of family and friends. All her children were there: Denis, Murdoch, Avrion, Lois, and Val. There were friends from Africa, Australia, and Canada. The house was beautifully decorated. There were lights and a tall Christmas tree. At the dinner, Nou placed me on her right, clearly a place of honor. There were more than twenty people at that dining table. The children sat at a separate table. Presiding over all this were Naomi and her husband Dick.

Several years before my visit, one of her guests was the young Jim Watson, who later won the Nobel Prize for discovering (with Francis Crick) the molecular structure of DNA. Watson's bestseller *The Double Helix* was dedicated to Naomi Mitchison. She enjoyed the friendship and admiration of many writers, including Aldous Huxley, D. H. Lawrence, J. R. R. Tolkien, and Rebecca West, among others.

Visitors

From the very beginning, Haldane had a steady stream of visitors in India. Joshua and Esther Lederberg were the first, stopping in Calcutta on their return journey from a Fulbright Lectureship in Australia. Lederberg's meeting with Haldane, coming on the heels of the Sputnik launching by the Soviets in 1957, was momentous in stimulating his lifelong interest in exobiology. Discussing the recent Soviet triumph, the two men agreed that if Sputnik were to land on the moon, it would carry with it microorganisms from Earth, which would irreversibly contaminate the lunar surface. On his return to the United States, Lederberg wrote a series of memos that warned of possible contamination and the importance of sterilizing all possible spacecraft thoroughly. Lederberg was awarded the Nobel Prize shortly afterward for his discovery of sexuality and genetic recombination in bacteria.

Lederberg was soon followed by two visiting Australians—the physiologist Derek Denton and the microbiologist Frank J. Fenner. Others included Sir Julian Huxley, P. M. S. Blackett (later Baron), Sir Ronald A. Fisher, Sir Harold Jeffries, Baron Ritchie-Calder, Sir Howard Florey, N. W. Pirie, George Gamow, Ernst Mayr, T. Dobzhansky, Harlow Shapley, Curt Stern, William Cochran, M. S. Bartlett, A. E. Mourant, E. J. H. Corner, J. Konorski, Nathan Keyfitz, and Niels Bohr from Copenhagen, as well as several Frenchmen including Antoine Lacassgne, Rene Wurmser, and Jacques Monod from Paris. Another visitor was the academician Alexander Nesmeyanov, president of the Soviet Academy of Sciences ("the man behind the Sputnik," *TIME* cover, June 12, 1958). Others included the Vietnamese leader Ho Chi Minh, who was invited by Mahalanobis to visit the Institute, the Oxford University philosopher A. J. Ayer, and the Welsh architect Clough Williams-Ellis.

There were, of course, many Indian visitors, such as Mihir Sen (one of the first to swim across the English Channel); the anthropologist Nirmal Kumar Bose, who was Mahatma Gandhi's secretary; the noted ornithologist and writer Salim Ali; and the nuclear physicist Homi J. Bhabha. We met the famous author of science-fiction Sir Arthur C. Clarke (*2001: Space Odyssey*), at a meeting in Sri Lanka, but he was already known to Haldane in London.[4]

Figure 19.1 Indian Statistical Institute director Mahalanobis celebrating Haldane's 65th birthday in 1957 in Indian style at his residence "Amrapali."

Our Daily Life

As expected, the Haldane household was unique, perhaps more so than any other I have seen. A visitor was apt to encounter various animals roaming around the house. Visitors were told not to disturb the wasp's nest in the bathroom. Helen Spurway (Mrs. Haldane) and I were making quantitative observations of all the activities of the solitary wasp, *Sceliphron* sp. We published an extensive paper in the *Journal of the Bombay Natural History Society*, which was then edited by Haldane's friend, the distinguished ornithologist Salim Ali. Biometrical studies of that wasp behavior was very much in keeping with Haldane's earlier statistical analyses of the bee dances and the breathing behavior of the mudskipper *Anabas*. Helen enjoyed keeping and studying various animal species all her life. For several years in Calcutta she kept guppies, which were originally part of her stock at University College London. Her famous paper on pathenogenesis ("virgin births") was based on her earlier research.

Helen and I used to take long walks in the suburbs collecting caterpillars and twigs from their host plants. She kept stocks of tussore silk moths (*Anthraea mylitta*).

Cocoons of different colors were bought periodically from the *Santal* tribals near Giridih in Bihar, where the ISI maintained a field station for crop trials. Helen and I used to go there once in a while, accompanied by two assistants, by night train from Howrah station. We enjoyed visiting the Santal tribal villages in the forest. I recall that at one point Helen wondered whether the Santals realized that she was a woman, because they had some taboos against women trespassing certain sacred areas of their land; or perhaps that taboo did not extend to a white woman from a foreign country. Her hair was short like a man's, and she wore and Indian-style man's white shirt and white pyjamas in public. She wore no makeup and no lipstick. Unless one was particularly observant, it was easy to mistake her for a man (especially in contrast to the Indian women!). This happened quite often in India, where the expected image of femininity followed certain traditional well-defined lines of dress and behavior. Her high-pitched voice was not helpful either; it was neither masculine nor feminine. Spurway always carried several books on the sociology and anthropology of various Indian communities and the tribal people. One of her favorite authors was Nirmal Kumar Bose, author of several books on Indian anthropology and editor of the journal *Man in India*.

There was no clear distinction between home and the lab. We kept experimental animals in both the house and the lab. We talked about our research most of the time. There was an almost religious fervor about any topic related to the maintenance and safety of animals and plants. Helen had always been keenly interested in the nature and causes of domestication of various animal species. And that is what led to her death eventually. She died of tetanus resulting from a jackal bite. A visitor would not have found the usual household decorations, ornaments, pictures, photos, stereo equipment, radio, musical instruments, or flowers. Indeed, Helen considered jewelry to be too ostentatious and thought flowers looked better on the plants. Haldane used to tell visitors that while other women bought jewelry, Helen preferred to buy books. There was no music of any kind. Haldane was tone-deaf, although his sister Naomi recorded that she used to waltz with her brother during their teenage years.

The Haldanes showed no interest in opera, symphony, dance, sports, or a great number of other popular forms of entertainment. We saw Hindi movies and Bengali theater once in a while, because Haldane was greatly interested in the nature of entertainment that was available for the masses and the techniques of mass communication, which could be converted to large-scale science education of the public.

At first, when they arrived at the Institute in July 1957, the Haldanes stayed with the director in his mansion, which has several guest rooms. It is surrounded by beautiful gardens and ponds and is conveniently located within a few hundred feet of the Institute. However, by the time of my arrival in December they were busy moving into an apartment on the fourth floor of the Institute, which I also shared later. I used to call it my shortest commuting distance; we lived on the fourth floor and worked on the second floor! But a year later we moved to a large house, "Basack Villa," which was at a short walking distance from the Institute. It was shared by the Haldanes and myself upstairs and by Haldane's secretary, Asit Bhattacharya, and his mother downstairs. Both the apartment and the house were visited by numerous distinguished scientists from several countries who came to see Haldane. I hired a cook from Kerala, as Haldane was very fond of south Indian food. I still recall the excellent *dosas* that Haldane enjoyed at breakfast. He also had a boiled egg, toast, and coffee.

Our daily routine began about 8:30 a.m., when we gathered in the dining room. Haldane used to read the local newspaper, *The Statesman*, while having breakfast. We talked about scientific matters in an unhurried manner. To a casual observer, we looked like a disorganized bunch, but there was serious purpose and coordination behind our life and work. While almost never mentioning it aloud, Haldane appeared to have had certain broad lines of inquiry in mind. Each of us filled a niche in his intellectual landscape.

Our minds were never far from our research, and much of our conversation was concerned with science. It was hard work being with Haldane. He used to keep us on our toes, asking questions about our research and the Latin names of local plant and animal species. We dined out occasionally on Sundays at one of our favorite restaurants, the Kwality Punjabi restaurant on Park Street, which Helen liked as it was located next to the Oxford Book Shop. Both the bookshop and the restaurant still exist today and are doing well, as I found out during a visit in 2012, after a lapse of about fifty years. Another restaurant we used to frequent was the Chinese *Nanking* restaurant.

Lacassagne's Comment

Our lifestyle was aptly captured by Professor Antoine Lacassagne, a visiting scientist in radiation biology from the Curie Foundation-Radium Institute in Paris, who wrote,

> I could not fail to be struck by the admiration and devotion of these young Indian students for their Master, whose diligence was an example to them all, whose enthusiasm was contagious, and who waged a constant battle for the advance of scientific research in India.... This impression was enhanced in the course of the patriarchal lunch to which we were all convened in the Haldanes' home a few hundred yards from the institute. Not only did Haldane dress as an Indian, not only had he adopted their customs (he was by then a vegetarian), he had taken even Indian nationality. In the living room of his small villa, a variety of dishes, seven or eight of them, were brought in and placed on a side table (with, in addition, some meat and fish solely for me). Each of us helped himself at will, then carried his plate to a chair or sofa, or even squatted on the floor. The whole atmosphere of the house was unpretentious, homely, somewhat disorganized even; and one couldn't help feeling moved by the warmth and friendliness of the hospitality of our host and hostess.[5]

Haldane usually spent the mornings answering his mail, barring lectures or meetings with VIPs, or taking care of some problem that was confronting Spurway that particular morning. The Haldanes were most generous in providing daily lunch for all of us. Their Bengali cook, Gyan, a name that Haldane had particularly enjoyed pronouncing, as it meant "knowledge," was most accomplished in providing us with several vegetarian dishes as well as some with meat and fish. To the great astonishment and shock of his former colleagues in England, Haldane adopted a totally vegetarian diet in India. Occasionally, we had visitors who came to see Haldane and were lucky enough to be invited for lunch. There were many scientists as well as occasional civil servants from Delhi who came to consult Haldane on matters related to science and technology. Haldane carefully arranged local sightseeing for close friends such as T. Dobzhansky and Ernst Mayr, the two "musts" being bird watching at the Calcutta zoo and the Museum of Anthropology. The zoo trips were arranged at sunrise or sunset, when the largest numbers of birds could be seen departing or arriving at the large freshwater lake inside the zoo. Another attraction was the famous Calcutta Botanic Gardens, where the largest and the oldest banyan tree (*Ficus bengalensis*) is located. This was of great interest to one of our visitors, the distinguished botanist E. J. H. Corner from Cambridge University, who spent many years in Singapore and Malaysia, studying tropical flora. He was an assistant director of Singapore Botanic Gardens for many years and was later professor of tropical botany at Cambridge University (1965-1973).

Mantras of Science

Haldanes and I used to walk a good deal in the suburbs near the ISI. We were occasionally joined by, Pamela Robinson, a visiting paleontologist from University College London. During one of our walks, Haldane heard a male voice raised in a monotonous chant. Of that incident, Haldane later wrote:

> I supposed that I was listening to some *mantras*, and asked my companion [present author] if he could identify them. The practice of repeating religious formulae is of course about as common in Europe as in India, and I have little doubt that it has an effect in guiding the thoughts of the chanter in certain directions, even when the chanting has become quite automatic.... But my companion stated that the language of the chant was English, and the subject organic chemistry. We returned, and I found that he was right. The subject of the chant was the preparation of aliphatic amines, ... I have never learnt any scientific fact in this way.... As I see it, science is an attitude, not a set of facts.

Haldane had many interactions with the Indian science establishment.[6] Unfortunately, most of them were negative. From the very beginning, he fought consciously not to become assimilated into that hierarchy. He was invited to the Indian Science Congress, but considered it a waste of time.

In fact, he successfully identified with the scientific workers and students. In numerous visits to universities and research institutes, Haldane was invited to spend precious time with the vice-chancellors and directors, which mostly consisted of having tea and polite and trivial conversations. He detested that practice. I saw him reject such invitations on many occasions and spend that time instead with students and young scientists, discussing their research projects. One example was the time when Haldane walked out of the office of the Delhi University's vice-chancellor (a well-known economist, Dr. V. K. R. V. Rao) and went to meet the graduate students and offer suggestions for their research. His brusque manners and refusal to cultivate friendships with people in power naturally made him unpopular in certain circles. He used to attend various lectures and social functions at the ISI, but often preferred to sit among the students, while ignoring the specially reserved seats meant for the VIPs like himself. He was often seen writing popular articles or scientific papers on long pads while paying close attention to the lecture, which soon became evident when he joined the discussion at the end. He was well known for his ability to divide his attention simultaneously among different mental tasks, including complex algebra, with ease. In his profile of Haldane in *The New Statesman*, Kingsley Martin wrote, "He shares with the late Lord Simon (one of his bêtes-noires) a capacity to divide his mind and his attention. At a conference or a public meeting he conditions himself to work out complicated mathematics and at the same time be master of the speaker's arguments—as he will presently show in discussion."[7]

What was most refreshing about working with Haldane and being with him was his total intellectual honesty and willingness to share his considerable knowledge with no pretense of any kind. Unlike some other professors whom I encountered in several countries, Haldane was not pompous, was always ready to help students, and assumed no airs of self-importance. He was not in the least offended when we pointed out occasional mistakes in his papers, especially in his mathematics. On the contrary, he thanked us profusely for improving his papers. Haldane wrote his papers very fast. He used no calculators or any other aids in performing his calculations. It was all done in longhand. Indeed, he enjoyed working out what some have called "brute force" mathematics. What is surprising is that he made so few mistakes!

Figure 19.2 Author (center) and the Haldanes arriving at the Second International Congress of Human Genetics in Rome in 1961.

Just being with Haldane was a great education, not only about various sciences, but also about ancient classics, philosophy, literature, history, world religions, and civilization. A brief conversation with Haldane saved us hours of work in the library. He was fully conscious of the great intellectual difference between himself and others but tried hard to bridge the gulf when dealing with various individuals. On some occasions he enjoyed displaying his intellectual prowess, especially when he wished to get rid of unwanted visitors or unwelcome intruders. I was amused to see that he achieved the same goal by feigning deafness on some occasions!

In 1960, Haldane was invited to be the chief guest of the Ceylon Association for the Advancement of Science in Colombo, Sri Lanka. Immediately upon arriving, he rejected the accommodation reserved for him thoughtfully by our hosts at the most expensive British-style hotel and chose instead a modest Indian hotel in the poorest part of Colombo. Our hosts, a group of Western-educated snobs, were, of course, shocked and insulted. There was an editorial in the *Ceylon Daily Mail* next morning deploring his ungrateful and rude behavior! But Haldane was not bothered by being called "rude" in such matters. He considered it a badge of honor when he was trying to stick to his principles. After that incident, that visit was a great success. Next morning, he was welcomed most warmly by the governor general of Ceylon, Sir Oliver Goonetilleke, who presided over the inaugural session. The audience was greatly impressed by Haldane's scholarly address, his erudition, and the classical knowledge he displayed in his opening remarks.

In his spare time, which was usually late at night or occasionally at teatime in the afternoons, Haldane used to read fiction of a sort, usually either science-fiction or stories of P. G. Wodehouse, which I and others liked to read in those days. Haldane had first read them as a schoolboy when they were first published around 1905 onward. Occasionally, he used to read the English translations of the great Hindu epics, or of various Hindu classics and philosophical works. One of his favorite authors was Dr. Sarvepalli Radhakrishnan, who held the chair of Eastern Classics at Oxford University, but later served as vice-president and president of India in his last years.

Haldane's Attitude Toward His Associates

It was obvious that Haldane genuinely enjoyed being surrounded by his students and associates. He was intensely loyal and supportive toward his close associates. His attitude in turn elicited our loyalty toward him. As I found out later, not all professors were like Haldane. There is a story about how the distinguished astrophysicist and Nobel laureate Chandrasekhar was ridiculed as a young scientist in a most embarrassing manner by his own professor, Arthur Eddington, at a Royal Society symposium. They disagreed strongly about Chandrasekhar's conclusions regarding stellar evolution, an important discovery that came to be known later as the "Chandrasekhar limit" and for which Chandra was awarded the Nobel Prize in his later years.

I mention this incident to indicate that this could never happen to a pupil of Haldane. Insulting or humiliating his own students in public was never his style.

Figure 19.3 Haldane photographed by Dr. Klaus Patau while visiting the University of Wisconsin in Madison in 1963.

Haldane was visibly proud of our little group and was not above bragging about our achievements. He was most comfortable when surrounded by his close pupils and associates. In marked contrast, he was most cautious when approached by strangers. Often he preferred to state his ideas and comments in writings. I used to find his notes on my desk when I arrived at my desk each morning. He worked long hours after dinner, enjoying the peace and quiet late into night. He disliked answering the phone and used to feign deafness while asking the caller to write a letter instead. I never saw him pick up the phone to make a call. Haldane was particularly concerned, to a degree that I have seldom seen, about the possibility that some one could misquote or misuse his statements. He was easily irritated by any misprints or errors that appeared in his publications. When he wrote popular articles for newspapers, he made it a condition that he should have the opportunity to correct the proofs before their final publication, a practice that is normal for scientific journals, but not newspapers.

During our travels, Haldane always carried a briefcase with some note pads and pens. Whenever we had a break or a waiting period he would start writing a scientific paper or a popular article for the press. With his prodigious memory he did not need to look up references or other resources. Haldane could write quite unperturbed in the middle of chaos, surrounded by other passengers at airports, or sitting in a café in Paris, or in a classroom while listening to a lecture. His powers of concentration were legendary.

In his personal financial matters Haldane was thrifty to a fault. His lifestyle was almost Spartan, involving no luxuries of any kind. Indeed, he regarded excessive wealth as a nuisance that complicates one's life.

In addition to his institute salary and income derived from family inheritance, Haldane earned a substantial sum from his popular writings and books.

Haldane provided generous funds from his own pocket to support my travels to attend scientific meetings and to buy books and journals needed for my research.

Afternoon Teas

One of the most memorable aspects of life with Haldane was the time we spent together during afternoon teas at the ISI in Calcutta. There were two kinds of tea breaks: morning and early afternoon breaks in the tearoom of the Research and Training School of the Institute. These were attended by the faculty and any other scientists who might be visiting the institute at that time. I recall meeting there the physicist and Nobel laureate P. M. S. Blackett, the statistical pioneer Sir Ronald A. Fisher, and two other statisticians Bill Cochran and M. S. Bartlett, as well as the famous demographer Nathan Keyfitz.

I remember one amusing incident when the Harvard University statistician Bill Cochran was introducing himself to Haldane, who asked him where he was from. Haldane's loss of hearing seemed to have appeared suddenly. Thereupon Cochran began shouting, "Harvard! ... Harvard! Harvard! ... in a louder and louder voice. Finally, Haldane responded quietly, "Yes, Yes, I have heard of the place!"

Our longer and more productive as well as much more enjoyable tea breaks took place later in the afternoon, usually after 5:00 p.m. We used to meet on the roof outside Haldane's apartment on the fourth floor of the ISI. These were gatherings of the inner circle, consisting of Haldane, Helen, myself, Subodh Roy, T. A. Davis, and a few younger assistants. Occasional visitors included the anthropologist Nirmal Bose, the ornithologist Salim Ali, Sir Julian and Lady Huxley, Prime Minister Nehru's secretary Pitambar Pant, and so forth. Those were our best times with the "Prof," when he was most relaxed, sipping tea and eating *samosas*, and answering our questions on a great number of topics, mostly related to science. At dusk, Haldane used to identify the stars, referring to both their *Sanskrit* and Greek names and telling us a great deal of information about their dimensions and properties. It was quite an impressive display of knowledge by Haldane, in a subject that was far removed from our research topics. He commented frequently that the sky over Calcutta was much clearer than in London. One of his bedside books was C. W. Allen's *Astrophysical Quantities*.

Notes

* Based on author's train journey with Haldane and his sister, from Calcutta to the Indian Institute of Technology in Kharagpur, 1958.
1. K. R. Dronamraju, *Haldane: The Life and Work of JBS Haldane with Special Reference to India* (Aberdeen: Aberdeen University Press, 1985), 113–39.
2. Suresh Jayakar had not yet joined us at that time.

3. Dr. Anne McLaren (1927–2007), who was a pupil of Haldane, was an embryologist at University College, London. In her later years, she served as the Foreign Secretary of The Royal Society of London. Both Anne and her husband Dr. Donald Michie, died in a car crash in 2007.
4. See K. R. Dronamraju, ed., *Haldane and Modern Biology* (Baltimore: Johns Hopkins University Press, 1968), 243–48.
5. Antoine Lacassagne, "Recollections of Haldane," in *Haldane and Modern Biology*, ed. K. R. Dronamraju (Baltimore: Johns Hopkins University Press, 1968), 307–11; 310.
6. Dronamraju, *Haldane: The Life and Work of JBS Haldane*, 14–15.
7. Kingsley Martin, *New Statesman Profiles* (London: Phoenix House Ltd, 1957), 187.

20

Haldane and Religion

> It is the nature of religion to evolve towards intolerance by natural selection.
> —J. B. S. Haldane

Haldane displayed an ambivalent attitude toward religion. He often stated that he was an agnostic, but that does not even begin to explain the entire story. In their childhood, both JBS and his sister, Naomi (Nou), attended New College chapel, although they were brought up without religious beliefs. His sister wrote, "My mother used to read to us out of Montefiore's *Bible for Home Reading*. . . . We had no religious conflicts with our parents."[1]

Throughout his life, JBS was fascinated by religion, in fact he was curious about all religions, and possessed an impressive amount of knowledge regarding Christianity, Judaism, Islam, Hinduism, and a score of other Eastern religions including Jainism and Buddhism. He returned to the topic of religion repeatedly in his writings.

Science and Theology as Art Forms

Haldane summarized his view of the relationship between science and religion as follows. Religion is a way of life and an attitude to the universe. Statements of fact regarding religion are untrue in detail but may contain some truth at their core. Science is also a way of life and an attitude to the universe. Statements of fact made in its name are generally right in detail. But they do not reveal the true nature of existence. Haldane wrote that the wise man regulates his conduct by the theories of both religion and science, however, "he regards these theories not as statements of ultimate fact but as art-forms."[2]

Haldane began explaining that religion and science are human activities with both practical and theoretical sides. He pointed out that while they differ in some important ways, they also share certain resemblances.

The scientific man starts from experiences that are emotionally flat, however, he may end by producing a theory as exciting as Darwinism, or a practical invention as important as antibiotics or high explosives. The mind of the religious man, on the other hand, works on a descending scale of emotions. The dogma or good works that he may produce are inevitably less thrilling than his religious experience.

The raw material of scientific thought consists of carefully collected facts that can be verified with sufficient patience and skill. The theories they have produced are far from certain; they change from generation to generation, or even from year to year. It is a fact that there is no great stability to be found in scientific theories, which encourages the opponents of science to talk of the intellectual bankruptcy of our age. We must also examine the claims of the theological beliefs that are offered as substitutes for science. Religious and moral experiences are facts, but their interpretation is in a prescientific age. Haldane explained that there is something true in theology, because it leads to right action in some cases. But there is also something untrue, because it often leads to wrong actions and leaves out a great deal such as not being able to explain the origin of evil. But Haldane's main objection is that once religious myths are made, it is hard to destroy them. Science is quite different. Haldane wrote, "Chemistry is not haunted by the phlogiston theory as Christianity is haunted by the theory of a God with a craving for bloody sacrifices."

Another topic of interest to Christians, prayer, was investigated statistically by Francis Galton in England. He tested the efficacy of prayer among those most prayed for, namely, the sovereigns and the children of the clergy. If prayer is effective they should live appreciably longer than other persons exposed to similar risks of death. Galton's investigation led him to the conclusion that the much-prayed-for persons had slightly shorter lives on an average than those with whom he compared them.

Haldane conceded that religious experience is a reality. It cannot be communicated directly, but those who experience it can induce it in others by myth and ritual. This fact is fully recognized in Hinduism. Haldane predicted that religion would become unimportant in human life. However, it might be more likely that religion as we know it will some day be superseded altogether. Haldane suggested that it is possible to become convinced of the supreme reality and importance of the spiritual without postulating another world or even a personal God. And such convictions may be supported by mystical experience. However, he doubted whether most people are capable of the abstraction necessary to adopt such a point of view.

"What Is Religious Liberty?"

In the *Rationalist Annual* of 1939, Haldane wrote "What Is Religious Liberty?" He was lucid, as usual. He quoted Thomas Paine: "Liberty consists in the right to do whatever is not contrary to the rights of others." Under the rights of others, Haldane included the following: Do children have a right to choose their own religion? Does one have a right to wake up one's neighbor in the morning by holding a religious ceremony? Do they include a right to use the streets for processions? Do they include a right to use endowments to propagate a doctrine when the property was originally intended for a different purpose or to support a different doctrine?

If Jesus Lived Today

"If Jesus Lived Today" was included by Haldane in a book of his collected essays titled *The Inequality of Man and Other Essays*.[3] In his preface, Haldane attempted to clear up the confusion between the identities of himself and his father.

> To remove a misconception which has frequently found its way into print, I take this opportunity of declaring my non-identity with my father, Professor J.S. Haldane. It is true that our opinions differ mainly on questions of emphasis and terminology rather than of fact. But they are sufficiently different to have allowed an ingenious American writer to convict us, in our joint capacity, of flagrant inconsistency. I hope that in future my intellectual sins will not be visited on him.

How did an agnostic see Jesus in the 1920s, when Haldane wrote his article? Haldane answered,

> I see Jesus as a man whose perception of spiritual facts was extraordinarily intense. He was far more intelligent, as appears from his sayings, than his disciples. They misinterpreted his words, and as we only see him through their eyes we cannot know how he would appear to our own. If Jesus were born in our time to a poor Jewish mother in capitalist Europe or North America he would receive a far wider education than 1900 years ago, when his reading was probably confined to the law and the prophets. Perhaps it was for this reason that his general ideas were always stated, either in parables drawn from everyday life, or in the terminology of religion.

Haldane explained further that today Jesus could talk in terms of science, psychology, and economics. In his own time he tried to simplify religion, and was accused of blasphemy, whereas today most religious people would probably regard him as an infidel.

"Today, most of us would learn about him through the press. He cures disease in an unprofessional manner. He wants to abolish wealth. After two or three years, he becomes an intolerable nuisance to the authorities. He has seen visions. Two doctors, already jealous of his unprofessional cures, certify him as insane. He is arrested but dies in prison before there was any trial. Some say he committed suicide. But that is not the end of him. Some of his disciples say that he is still with them. Other disciples stress the mystical side of their master."

Haldane concluded, "The future is unknown. Has the Man started the real world revolution, or only another religion?"

The Pre-Christian Religions of Europe

Soon after his arrival in India, in June 1958, Haldane was invited to deliver a lecture at the much respected Ramakrishna Mission Institute of Culture in Calcutta. He took it upon himself at once to draw a table of comparison between Greek, Roman, and Hindu gods.[4]

Greek	Roman	Hindu	Function
Zeus	Jupiter	Indra	Supreme, thunderer
Poseidon	Neptunus	Varuna	Sea, earthquakes
Hades	Dis	Yama	Ruler of dead
—	Pluto	Kubera	Mineral wealth
Hera	Juno	Durga	Ambivalent, wife of Zeus
Athene	Minerva	Sarasvati	Wisdom, born from Zeus' head
Ares	Mars	Skanda	War
Apollon	Apollo	Surya	Sun, prophecy, healing
Artemis	Diana	—	Moon, huntress
Hephaistos	Vulcanus	Visvakarman	Fire, maker of thunderbolts
Hermes	Mercurius	—	Messenger
Demeter	Ceres	Annapurna	Agriculture
Aphrodite	Venus	Rati. Laksmi	Love, beauty, fertility
Castor, Pollux	Caster, Pollux	Asvins	Warriors, help sailors
Aiolos	Aeolus	Vayu	Wind
Herakles	Hercules	Krishna	Hero, deified
Dionysos	Bacchus	Indra	Wine
Asklepios	Aesculapius	Dahanvantari	Medicine, son of Apollo
Hekate	—	Kali	Destruction

The Dark Religions

In his essay "Dark Religions," published in the *Rationalist Annual* for 1961, Haldane was concerned that religion had been used to justify irrational conduct. Haldane wrote that Christianity has the worst record for intolerance of any of the great religions. Islam always permitted Christians and Jews to follow their cults in Islamic states, although they were heavily taxed and occasionally massacred. However, Haldane believed that as a regulator of private lives Christianity is, at its best, better than Islam, at its worst, worse. He suggested that with infantile hygiene and birth control a wife can find time for public activities, and added,

> It is surprising how many of the women fellows of The Royal Society are mothers. And an educated wife can be an asset to an intellectual husband. Whereas a man or woman who has no intimate acquaintance with the other sex is liable to be a menace to society, particularly if he or she is a member of a celibate order. Among other things, celibates generally oppose reforms to marriage, such as birth control, easier divorce, and above all sex equality. This is natural enough if they regard marriage as intrinsically evil.

Haldane wrote that Christianity was mostly confined to Europe until technology and science allowed it to spread, and that "Scientific technology developed in the teeth of Christian opposition. . . . At the present time Christianity is not so inimical to science as are the other great religions."

Buddhism

Haldane regarded Buddhism as a religion that had degenerated abjectly from the ideals and doctrine of its founder. Buddha's doctrines are part of some sect or another of Hinduism. He denounced many of the practices of Hinduism. He was far more philosophical than Jesus or Mohammad. Buddha cast out some of the devils of Hinduism, only to be replaced by far worse devils in their place. He spoke out against idolatry, but his followers continued to worship a new set of deities, especially Buddha himself. And they are not opposed to violence. Haldane mentioned that the former prime minister of Ceylon, Mr. Bandaranaike, was murdered by a Hinayana[5] Buddhist monk.

Haldane opposed the central dogma of Buddhism, which is also the central part of Hinduism. That is, my good or bad fortune in this life is the reward, or punishment, of my actions in my previous life. This doctrine is diametrically opposed to science, especially medical science.

Hinduism

The last of the four great religions, Hinduism, differs from all others. There is no founder or prophet, no dogmas, and no single moral code. There are a whole range of creeds and codes. Haldane compared this situation with the pre-Christian religions of Europe. The various groups tolerate each other. Haldane recognized four levels. At the bottom are some unpleasant female hobgoblins, such as Sitala, who presides over smallpox. On the third level there is polytheism similar to the saint worship of southern Europe. Various deities such as Ganesha for financial transactions and Annapurna for the crops belong to this level. The second level is a monotheistic religion, and its adherents believe that certain beings were avatars (incarnations) conducive to noble lives. At the highest level, Hinduism believes in the existence of a supreme being. But it has no qualities. The great philosopher Sankara[6] denied that the existence of God could be proved by reason.

Haldane admired Hindu atheists, who preserved a great deal of Brahmin ethics and did not eat meat but took scholarship more seriously than money. Their imagination was molded on Hindu mythology. At its highest level Hinduism was more compatible with science than was any other religion. He did not mention other religions that have fewer than ten million adherents, for example Sikhs, Jews, and Jains. He predicted that the African cults might become important as Africa became liberated. He foresaw conflicts between Islam and Christianity as Islam spread southward in West Africa. He wrote that everywhere religion seems to divide people. It is the nature of religion to evolve toward intolerance by natural selection. The success for scientific research appears to be negatively correlated with the success of religion. The least religious states usually have the greatest economic success.

One of our memorable visits was to see the holy temple at Benares or Varanasi, the holiest shrine for Hindus, where foreigners are rarely allowed. A man at the entrance to the temple stopped us and asked me, "Who are these people? Are they also going in?" I answered, "They are devotees, just like you and me." We were allowed inside. It was an occasion of great joy for Haldane to step inside that temple at Benares or Varanasi! It was my impression that he came closest to being a Hindu or what he felt like being a Hindu at that point. A former associate of Haldane at the Cambridge University's Biochemistry Institute, N. W. Pirie, reviewed in *Nature*[7] a previous book of mine in which I discussed Haldane's life in India.[8] Pirie wrote, "I feel that Dronamraju overstates the extent to which Haldane accepted Hinduism. He was, of course, sympathetic to the principle of non-violence." Pirie was seriously mistaken. If he were present at that point when Haldane entered the holy temple at Varanasi, I am sure Pirie would have reached a different conclusion. Furthermore, there is a lot more to Hinduism than nonviolence.

Haldane's Daedalus

At the end of his small and stimulating book, *Daedalus, or Science and the Future*,[9] Haldane considered the type of religion that would satisfy the scientific mind. Haldane wrote that we must not take traditional morals too seriously. He wrote, "And it is just because even the least dogmatic of religions tends to associate itself with some kind of unalterable moral tradition, that there can be no truce between science and religion."

Haldane wrote that there was no reason why a religion should not arise with an ethic as fluid as Hindu mythology, but it had not yet done so. Christianity has the most flexible morals of any religion because Jesus left no code of law, but every Christian church has tried to impose a code of morals of some kind for which it has claimed divine sanction. The only kind of religion that would satisfy the scientific mind would openly admit that its mythology and morals are provisional. However, Haldane wondered whether it could properly be called a religion at all!

The Causes of Evolution

In his classic work, *The Causes of Evolution*,[10] Haldane considered alternative explanations for Darwinian evolution by natural selection. He wrote that the most obvious alternative to this view is to hold that evolution has throughout been guided by *divine power* (italics mine). However, two objections to this hypothesis were stated by Haldane: One is that the fate of most lines of descent is extinction, and usually the end is achieved by a number of different lines evolving simultaneously. Haldane wrote, "This does not suggest the work of an intelligent designer, still less of an almighty one."

The second objection, a moral one, is more serious, according to Haldane. Several earlier life forms have evolved into parasites, causing much pain and death to other species. From an ethical point of view, their loss of faculties combined with an infliction of suffering in the same class as moral breakdown in a human being, which can often be traced to genetical causes. Haldane questioned whether God made the tapeworm. An affirmative response fits neither our knowledge of evolution nor what we believe to be the moral perfection of God.

Haldane provided three possible answers[11]:

1. "We can consider the dark as well as the bright side of evolution as evidence of divine ingenuity, (*Isaiah*): "I make peace, and create evil: I the Lord do all these things."
2. The second answer was provided by Plato. In the "Republic," Socrates says "God therefore, since He is good, cannot be responsible for all things, as the many say, but only for good things." However, we are reminded that the creation of a tapeworm presents as much challenge as does the rose. Haldane added, "We should have to give the Devil credit for a large share in evolution."
3. We can conclude that there is no need at present to postulate divine or diabolical intervention in the course of the evolutionary process."

With reference to the direction of evolution, Haldane wrote that the change from monkey to man might well seem a change for the worse for a monkey. But it might also seem so to an angel.[12]

McOuat and Winsor raised the question whether Haldane took up the study of evolution because he was interested in religion. There is no convincing evidence to support such a hypothesis. He was interested in evolution because of his interest in natural history and his interest in Harry Norton's analysis of natural selection. Several points need to be made in this context:

1. Haldane's attention was drawn (by R. C. Punnett) to Harry Norton's table on the effect of natural selection, which was reproduced in Punnett's book *Mimicry in Butterflies*,[13] published in 1915, and which kindled Haldane's interest to extend Norton's analysis to various situations of Mendelian genetics. Haldane's "Mathematical Theory of Natural Selection" was published in 1924. Previously, however, Haldane had already showed his great interest and knowledge of the evolutionary process and natural history by proposing his Rule (later called "Haldane's Rule") in 1922. William Bateson, whose knowledge of natural history was beyond any doubt, was one of the first to support "Haldane's Rule"[14]
2. As early as 1911, Haldane was already reading a paper on the first case of linkage in vertebrates and its evolutionary implications (based on Darbishire's data) in a seminar organized by Professor E. S. Goodrich at Oxford University, and collaborated with his sister in breeding guinea pigs and mice to obtain original data on linkage.
3. During the years of my association with him in India, Haldane was greatly interested in natural history, took part in frequent field trips, bird watching with Ernst Mayr, and was rarely seen without his field glasses.

The Argument from Design

Haldane was referring to the well-known argument of Paley, who attempted with great skill to prove the existence of a creator from the design of living organisms. A good many of his arguments were met by Darwin's theory of evolution by natural selection. Haldane pointed out that Paley's argument did not lead to Christianity. It forces us to believe in a malignant creator or creators. A biologist who has spent his life studying parasitic animals must inevitably take human misery and injustice for granted. The moral effect of the belief that the world was made by a benevolent and almighty creator is far worse. C. S. Lewis's book *The Screwtape Letters*,[15] provides a good example of its effects on an intelligent man. The book is supposed to be written by a devil. The devil is strongly in favor of modern medical practice, which in many cases has robbed death of its pain and terror.

If the world of Nature is God's plan, Haldane wondered why it is so imperfect. He wrote that Darwin made it reasonable to reject the argument from design and the evil god or gods to which it leads if carried to its logical conclusion. Many of the early Darwinists retained a veneration for Nature, which is justifiable, according to Haldane, if it is God's handiwork. Darwinism was used to justify various forms of human struggle, including war and unrestricted economic competition. T. H. Huxley attempted to reverse this tendency.

Haldane considered it essential that we should study the economic and social origins of religious beliefs and the profound psychological needs that they partially satisfy. But as long as these beliefs are capable of rational proof, it is our duty to study them and see if they really prove anything.

Notes

1. N. Mitchison, "Beginnings," in K. R. Dronamraju, ed., *Haldane and Modern Biology* (Baltimore: Johns Hopkins University Press, 1968), 304.
2. J. B. S. Haldane, *Possible Worlds and Other Papers* (London: Chatto & Windus, 1927), 237–52.
3. J. B. S. Haldane, *The Inequality of Man and Other Essays* (London: Chatto & Windus, 1932), 267–70.
4. Haldane and Religion - Some information provided by Haldane in 1963.
5. *Sankara* or Shankara, also called Shankaracharya (b. 700, India—d. 750?), was a philosopher and theologian, the most renowned exponent of the Advaita school of philosophy, from whose doctrines the main currents of modern Indian thought are derived.

    Sankara was born to the nambudri couple Sivaguru and Aryaamba, in a little village called Kaladi in Kerala, around 8th century CE. The couple had remained childless for a long time, and prayed for children at the temple in nearby Trichur. Siva is said to have appeared to the couple in a dream and promised them a choice of one son who would be short-lived but the most brilliant philosopher of his day, or many sons who would be mediocre at best. The couple opted for a brilliant, but short-lived son, and so Sankara was born.

    Sankara lost his father when quite young, and his mother performed his upanayana ceremonies (which includes sacred thread worn by Hindus) with the help of her relatives. Sankara excelled in all branches of traditional vaidika learning. A few miracles are reported about the young Sankara. Sankara is said to have rerouted the course of the Purna River so that his old mother would not have to walk a long distance to the river for her daily ablutions.

Samnyasa: Sankara was filled with the spirit of renunciation early in his life. Getting married and settling to the life of a householder was never part of his goal in life. To comfort his anxious mother, he promised that he would return at the moment of her death, to conduct her funeral rites, although he would be a *sannyasi*, as one who renounced the world and continued his life as a *yogi*.

Sankara then traveled far and wide, expounding the philosophy of *vedantha* through commentaries on the principal *upanishads*. Sankara traveled to various holy places in India, composing his commentaries in the meantime. At this time he was barely a teenager. He attracted many disciples around him. These commentaries, called bhasyas, stand at the pinnacle of Indian philosophical writing and have triggered a long tradition of subcommentaries. In addition to writing his own commentaries, Sankara sought out leaders of other schools in order to engage them in debate.

6. *Hīnayāna* is a Sanskrit term literally meaning the "smaller vehicle." The term appeared around the 1st or 2nd century. *Hīnayāna* is often contrasted with *Mahayana*, which means the "great vehicle." The word "Hīnayāna" is formed of *hīna*, "little," "poor," "inferior," and *yāna*, "vehicle," where "vehicle" means "a way of going to enlightenment."

There are several ways of interpreting what the term "Hīnayāna" refers to; the "lesser" or "greater" designation does not refer to economic or social status, but concerns the spiritual capacities of the practitioner. Maha Stupa at Thotlakonda Monastic Complex initially flourished as an early Buddhist school of Hinayana, and later developed as the Theravada School of Buddhism, which witnessed peak activity during 2nd century BCE, in Visakhapatnam in Andhra Pradesh, India.

7. N. W. Pirie, "Haldane as Guru: A Review of *Haldane, The Life and Work of JBS Haldane with Special Reference to India*, by K.R. Dronamraju," *Nature* 319 (1986): 630.
8. K. R. Dronamraju, *Haldane: The Life and Work of J.B.S. Haldane with Special Reference to India* (Aberdeen: Aberdeen University Press, 1985).
9. J. B. S. Haldane, *Daedalus, or Science and the Future* (London: Kegan Paul, 1923).
10. J. B. S. Haldane, *The Causes of Evolution* (London: Longmans, Green, 1932).
11. Haldane, *The Causes of Evolution*, 86.
12. Haldane, *The Causes of Evolution*, 83.
13. R. C. Punnett, *Mimicry in Butterflies* (Cambridge: Cambridge University Press, 1915).
14. J. B. S. Haldane, personal communication to the author.
15. C. S. Lewis, *The Screwtape Letters* (London: Macmillan, 1960).

21

Impact of Haldane Today

Haldane made significant contributions to genetics, physiology, biochemistry, biometry, statistics, ethics, and science in developing countries as well as to the popularization of science. How much of his work has survived to this day, and what has been the impact of his contributions on science today?

Major areas of science where his ideas have contributed to progress are: population dynamics of balanced polymorphism[1] and its role in identifying resistance genes to a wide variety of infectious diseases, such as malaria, to develop therapeutic measures; study of mutation rates for a number of genes causing various defects and diseases; linkage and mapping of genes in various species; study of sex-ratio in the offspring of interspecific crosses (Haldane's Rule[2]); physiological effects of diving and inhalation of poisonous gases; kinetics of enzyme action; origin of life on planet Earth; and science in developing countries.

Two major areas related to science that interested Haldane are: (1) what is the impact of science on society and (2) what are the ethical consequences of the applications of science? He was one of the few scientists who examined these subjects in great depth, and he returned to them repeatedly in his popular writings.

Haldane was in the habit of expressing original ideas, even important scientific ideas, in a casual manner, both in speech and in writing, which have led to the development of whole new fields. One example is his 1934 paper on quantum mechanics,[3] an isolated subject that was far removed from the rest of his publications, which was acknowledged by Norbert Wiener to have stimulated his work on *cybernetics*.[4] Haldane's essay on the origin of life was published not in any scientific journal but in the *Rationalist Annual*.[5]

Population Genetics

The most important part of his scientific work was the series of papers he published in the 1920s on the mathematical theory of natural selection, which formed one of the foundations of population genetics. In his "Defense of Beanbag genetics,"[6] Haldane predicted that the existing theories of population genetics would no doubt be simplified and systematized. In the meantime, they would have served a useful role as the steppingstones for the next phase of advancement in genetics.

One example is Haldane's estimate of the cost of natural selection, which was used by Motoo Kimura to justify his "neutral" theory of evolution.

Infectious Disease and Evolution

Haldane's idea that the co-occurrence of malaria and thalassemia[7] in the Mediterranean countries indicated a balanced polymorphism has had an enormous impact on our understanding of the population dynamics of malaria and other infectious diseases. "Haldane's malaria hypothesis" or simply the "Malaria hypothesis" provides an explanation for the co-occurrence of infectious disease and hemoglobin disorders in certain regions of the world. Haldane explained that the unusually high frequencies of the genes for thalassemia (and sickle cell anemia) in the Mediterranean region were due to the higher fitness of the heterozygous[8] carriers. Subsequent investigations in Africa have confirmed the "Haldane's malaria hypothesis."

Potentially lethal manifestations of malaria (cerebral malaria and severe anemia) are found to be rare in sickle-cell heterozygotes. Under conditions of intense *Plasmodium falciparum* transmission, young sickle-cell heterozygotes (AS) survive better than those with normal hemoglobin.

Haldane's ideas and later investigations have led to further research, for developing therapeutic methods by using the genome sequencing data. This method has already proved to be helpful in relation to murine malaria by pinpointing genes for greater resistance to malaria. Extensions of this work to locate resistance genes for a wide variety of bacterial and viral infections that have played an active part in modifying the human genome over centuries are opening new avenues for research.

Mutation Research

Haldane used the equilibrium frequencies of alleles in a population that are maintained by mutation-selection balance as the basis for estimating the mutation rate for hemophilia in humans. Haldane pointed out that the elimination of deleterious genes by natural selection must be balanced by fresh mutations. In 1932, in his book *The Causes of Evolution*, Haldane estimated that the mutation rate for hemophilia in humans is of the order of 10^{-5} per generation. Later, in 1935, he derived a general formula for X-linked recessive genes.

Haldane's estimate was remarkably accurate, even though very little was known about genes and gene action at that time. In a later paper in 1947, Haldane suggested that the mutation rate for hemophilia is higher in males than in females, which has been confirmed by later studies. Haldane's early work on mutation rates developed a method that, along with his other contributions to human genetics, became one of the foundations of human genetics in the early years of its development.

How did Haldane evaluate his own contribution? In his "Defense of Beanbag Genetics," Haldane wrote, "Selection and mutation must balance in the long run, but how long is that? In two rather complicated mathematical papers . . . I showed that while harmful dominants and sex-linked recessives reach equilibrium fairly quickly, the time needed for the frequency of an autosomal[9] recessive to get halfway to equilibrium after a change in the mutation rate, the selective disadvantage, or the mating system, may be several thousand generations. In fact, the verbal argument is liable to be fallacious."[10]

Linkage and Mapping

Devising statistical methods, Haldane led the way to human gene mapping with his first estimate of the map distance between the genes for colorblindness and hemophilia on the X chromosome, at first with Julia Bell in 1937, and improved it, ten years later, in a second estimate with C. A. B. Smith. Haldane laid the foundation for linkage and gene mapping in the human species. In the following decades, improved methods, such as somatic cell genetics, computer facilities, and sequencing methods have led to the rapid sequencing of several species, including humans. His invention of the mapping functions[11] played an important role in mapping studies, and continues to be useful, although modified and extended to complex situations by Newton Morton and others in later years.

Haldane's Rule

One of the most enduring generalizations in biology was proposed by Haldane in 1922, and to this day it defies a satisfactory explanation. It states, "When in the F_1 offspring of two different animal races one sex is absent, rare, or sterile, that sex is the heterozygous [heterogametic] sex." Many explanations have been suggested, but none of them have been widely accepted. Explanations have generally fallen into three categories: (1) the most intuitive explanation was proposed by H. J. Muller in 1942—if some alleles causing sterility or inviability are partially recessive, the heterogametic sex will be more affected in F_1 hybrids, because recessive X-linked alleles are expressed more fully in that sex; (2) a second category involves developmental peculiarities of the X chromosome, such as imprinting or inactivation during spermatogenesis; and (3) the third proposes a higher rate of evolution that can occur for some X-linked genes, causing complications in heterogametic hybrids.

Other theories have been proposed, however, none of them was found to be satisfactory in all situations. Others have proposed that there was no single explanation for Haldane's Rule, a phenomenon that appears to have been manifested by a coincidence of different causes in different taxonomic groups.

Diving and Inhalation of Gases

Haldane's remarkable and courageous experiments in diving physiology and the effects of inhalation of gases have had worldwide influence. He was his own guinea pig in a series of painful and difficult experiments, when he subjected himself and other volunteers, including his wife, Helen Spurway, to extremes of pressure and temperature while breathing a mixture of various gases such as carbon monoxide, carbon dioxide, oxygen, helium, and nitrogen. Several of these experiments resulted in convulsions and sickness, as well as some injuries due to "bends." Early in his life, young Jack assisted his father, physiologist John Scott Haldane, in calculating the diving tables that laid the foundation for diving work.

The fundamental principles enunciated by Haldane continue to have a beneficial impact on those working in these fields, because of the leadership qualities he possessed and also due to the firsthand observations he recorded as the subject-investigator. In a brief review of the field of underwater physiology in 1941, Haldane listed six principal problem areas: mechanical effects, nitrogen intoxication, oxygen intoxication, aftereffects of carbon dioxide, bubble formation during decompression, and cold. These topics have remained the major areas of concern today. In a paper written for the Aero-Medical Society of India,[12] Haldane stated that several of his findings in diving physiology were equally applicable to space medicine.

Enzyme Kinetics

An important contribution of Haldane during his association with the Biochemistry Institute at Cambridge was concerned with enzyme kinetics, especially the Michaelis theory (also known as the Victor Henri theory). This theory assumed that the combination of enzyme and substrate always corresponds to an equilibrium. Haldane considered this to be an unlikely situation. In a paper with G. E. Briggs, published in 1924, Haldane derived the basic law of steady-state kinetics still used for treating enzymatic catalysis. Haldane's book *Enzymes* was first published in 1930 by Longmans, Green and Co., and was reprinted in 1964 by the MIT Press. In his preface to the MIT edition, Haldane listed three advances in enzymology that had occurred since its first publication: first, in purification of numerous enzymes, all turned out to be proteins, many lacking a prosthetic group; second, the structure of an enzyme is exactly specified by that of one or more genes; and finally, metabolic processes, such as oxidations and syntheses, are usually series of enzyme-catalyzed reactions, often involving small molecules, such as the adenosine phosphates. His book has conserved all its didactic value, long after its publication, and still contains all the essential notions of enzymology. It has educated generations of biochemists.

Origin of Life

Haldane's theory of the origin of life, which he proposed independently of A. I. Oparin in 1929, dominated much of the twentieth-century debate on the origin of life. However, in recent years, both the "primordial soup" hypothesis of Haldane as well as the famous "Miller-Urey" experiments have come under attack for several reasons. Among these are the realization that the primitive earth did not contain significant amounts of the reducing gases such as methane, ammonia, and high levels of hydrogen, which were assumed to be present in high concentrations in the Miller-Urey experiments, and the lack of a convincing explanation for the spontaneous evolution of a "genetic code"[13] in nature. Furthermore, fluids released from hydrothermal vents on the seafloor were shown to contain carbon dioxide and large amounts of methane, suggesting that the mixture of sulphur minerals and hydrogen sulfide under high temperatures and pressures results in free hydrogen and free energy, which can reduce carbon dioxide to methane. It has been hypothesized that organic molecules as well as the first living systems appeared not at the earth's surface, as suggested by the Haldane-Oparin theory, but in the ocean depths.

Contribution to Science in India

Upon arriving in India in 1957, Haldane was struck by the need to improve the academic and scientific standards, as well as the need to popularize science among the lay public. His criticisms and recommendations have clearly had a beneficial impact on Indian science in the following categories:

1. *Science administration*: Haldane was often frustrated by the so-called red tape and bureaucratic delays of the administration at the Indian Statistical Institute where we all worked, as well as other centers and agencies outside, which we had to deal with in the course of our daily activities. Because of his social prominence, his complaints were noted at the highest level. One notable example was the delay in obtaining our passports when we were invited to attend the Ceylon Association for the Advancement of Science. After waiting for a couple of months, having received no response to our initial applications, Haldane lost his patience and wrote a sharp letter of complaint to Prime Minister Nehru, who also happened to be the minister for external affairs for the Indian Government. And, as expected, soon we heard from the local passport officer in Calcutta, who came to the institute with some passports, informing us that all we had to sign some papers to receive our passports. Haldane was reluctant to follow that method because it wasted his time and that of others, and he regarded that as an undemocratic procedure that should not have been necessary in the first place. There were other procedural delays at the Institute in ordering reprints of our research publications, obtaining travel funds to attend scientific meetings, and in finding additional office space. However, because his complaints drew the attention of the beurocrats at the highest level, he led the way occasionally to help other colleagues, especially junior workers, who had been accustomed to suffer in silence until then.
2. *Indian Science Congress*: One of his serious complaints was concerned with the inefficiency of the manner in which the annual meetings of the Indian Science Congress were organized. I am happy to note that because of his repeated complaints, these are now much better planned and conducted, approaching international standards. Among Haldane's complaints were the sudden cancellation of scientific sessions, nonadherence to the published program, lack of appropriate audiovisual facilities, and a lack of opportunity for young students to approach and discuss their work with distinguished scientists. In an article in *The Hindu*

newspaper titled "The Scandal of the Indian Science Congress," Haldane listed several problems and some solutions that have been implemented since.
3. *Status of younger scientists*: Haldane urged the Indian Government and various research institutes to provide better facilities for younger workers. But, more important, he deplored the lack of dignity and respect shown to the younger workers by their superiors. He himself set an example by referring to us as his "junior colleagues," and accorded us the respect and dignity that was then lacking in several universities toward their younger employees. To make his point, it was not uncommon for Haldane, while visiting other educational and research institutions, to reject the invitations of vice-chancellors and directors to join them, and instead to join the students in their classrooms and laboratories. He also made his views known in several popular articles in the press, as well as in his conversations with scientific and administrative leaders.
4. *Research projects*: During the years Haldane spent in India (1957–1964), it was not uncommon to see younger scientists remaining idle, although paid by an institution, because they lacked the facilities needed for their research. These were often individuals who were trained in the United States or Europe, where they were able to use expensive and sophisticated equipment, and they wished to continue similar research in India. Haldane suggested several kinds of research projects that they could pursue while waiting for better facilities. They involved the use of local resources, indigenous flora and fauna, and mathematical or statistical methods in which many Indian scientists are trained.

He suggested further that the Indian Government, which is the main source of support, should plan their programs in a way that would facilitate the research of scientists returning from foreign laboratories; for instance, projects that could use local resources rather than those requiring expensive equipment, which Indian laboratories could not afford at that time.

Haldane deplored the method of teaching science by rote, a common practice in India, which did not encourage innovation and originality. He drew attention to his own experience when his father taught him science as a means for solving practical problems in life. He attempted to teach the students how to think for themselves with examples, and not depend on mere transfer of knowledge. It was one of his complaints that the Indian Statistical Institute did not teach courses designed to stimulate originality or innovation in research.

Science in Developing Countries

Haldane's experience in India and the neighboring countries gave him a unique perspective of science teaching and research in developing countries. As a successful and famous scientist from Europe, Haldane saw the deficiencies in their educational system, especially an attitude toward science that he was not familiar with from his previous years in Great Britain. Foremost among the problems he faced were a lack of current textbooks and journals, ill-equipped laboratories, and poorly trained teachers. Lack of funds is the root cause of several problems. However, a more serious problem faced by Haldane was attitude. He observed that many courses were designed to impart knowledge but not lead to first-rate scientific research. Original thinking was discouraged by the teachers. In many instances, learning facts was divorced from practical problems of life. He urged a new way of looking at the world and the world of international science.

Popularization of Science

Haldane was characterized as the "most brilliant scientific popularizer" by Sir Arthur C. Clarke. Haldane's prolific series of popular articles on scientific topics, although first published many years ago, continue to be cited in scientific publications. Several of these have been reprinted in collections of essays in later years, *Science and Life* (1968), *On Being the Right Size* (1985), *Haldane's Daedalus Revisited* (1995),[14] and more recently *What I Require from Life: Writings on Science and Life from J.B.S. Haldane* (2009). Several generations have been educated and have benefited from Haldane's popular essays. Although some are clearly outdated, his approach to science popularization, his method of presenting a scientific subject to the layman in a lucid manner, his ability to simplify complex scientific subjects without distorting their meaning, and his skill in compressing a big topic into a few paragraphs still serve as a model to be emulated by today's science writers. Certain articles of Haldane have retained their fresh perspective and are often quoted today. Two of these, which have had a significant impact on later writers, are "On Being the Right Size," which has been reprinted in several collections, and his classic *Daedalus, or Science and the Future*, which provided the biological ideas of Aldous Huxley's *Brave New World*, and was reprinted in 1995.

Politeness and Unprofessionalism

Haldane explored the reasons for the low level of performance in research and teaching in science in his writings. Here are the opening lines of one of his articles, "What Ails Indian Science?":

> I have already come to the conclusion as to why science in India is developing with disappointing slowness. It is not because Indians are stupid or lazy. It is because they are too polite. They spend hours daily in conversation with others, not on professional matters, but on personal topics. In London, I talked with colleagues for an hour or more, daily, but it was mostly about the details of our work. In the Indian Statistical Institute (Calcutta) the same is true. But it is not true in most academic institutions where I have been in India. Again at scientific meetings and usually in ordinary discussion my Indian colleagues are polite about one another's work. In Europe, we are usually polite about the work of juniors, and highly critical of that of men and women of established reputation. At a recent international meeting on genetics, an American got up after a paper read by my wife and said that he

could not let her highly misleading views pass without criticism. She felt that she had at least reached the status where one is criticized without mercy. She and I at once formed a friendship with the critic. We had something to talk about. In my opinion, only a few branches of Indian science have reached the stage of maturity where this is possible. I may criticize some of my colleagues as I would criticize British colleagues, and hurt their feelings severely. Once again I am up against the choice between politeness and efficiency. I do not know how I shall resolve this dilemma. I hope that as Indian science grows up, it will become less acute.

New Caste System

Haldane observed that a new caste system, based on academic degrees, was evolving in India. One could not teach a number of subjects such as Bengali, chemistry, history, or any other, without a degree in that subject. And a higher degree given for research was necessary if one applied for a professional chair. Referring to his own situation, Haldane wrote that it was only a matter of time before he would be barred from teaching science or statistics, because he possessed no degree of any kind in a scientific subject. But in terms of the new caste system, he was qualified to teach the classics since he had secured a marginal first class in *Literae Humaniores* at Oxford University in 1915, which mostly consisted of a combination of Greco-Roman classics, philosophy, and ancient Greek and Latin. Haldane cited the example of a man in India who was refused a university post to teach his native language because he had no degree in it, although he had published quite a lot in it and possessed a degree in another subject. He mentioned another case where a student was refused admission to a master's degree course because his undergraduate degree was in a different subject. He ridiculed such a practice as calculated to ensure that Indian scientists are too specialized. Successful Indian scientists like Jagadish Bose, Meghnad Saha, and Prasanta Mahalanobis achieved eminence precisely by bridging the gaps between different sciences.

Double Loyalty of a Scientist

Haldane pointed out that every scientist has a double loyalty—to science as well as to his country. He observed that in India junior scientists are not engaged in research but waste their time in filling endless forms about work in contemplation and showing visitors around. Some of those workers could have done some research, even if only mediocre, but were ordered to remain in the laboratory, standing beside an incomplete or inefficient apparatus, and explain the potential project to visitors. These were not isolated cases. Haldane saw them all over India in numerous situations. He wrote that there was widespread jealousy toward younger workers by their superiors. The younger workers, who were at their most productive age, were either discouraged from doing original work or their results were stolen by their supervisor. He mentioned an example, the head of a certain laboratory: "This remarkable man had published over fifty scientific papers in one year. No single human being before him has ever made discoveries at this rate! No doubt junior colleagues had done most of the work, or all of it. But their names were not mentioned. It is not surprising that young men do not care to work under such conditions." They went abroad to continue their research, especially in Europe and North America, in a mass movement that came to be called "brain drain." Another reason why Indian graduate students went abroad was mentioned by Haldane: "They are systematically humiliated by the administrative staff of many institutions, and sometimes by professors, in a manner, which is not tolerated in western Europe or United States. Again I hope to live to see this remedied."

Haldane's outspoken comments and his critical evaluation often offended senior administrators and others in powerful positions in India but endeared him to younger scientists. He championed their cause in a fearless manner when they were unable to speak for themselves. He wrote, "The root cause of all this incompetence and worse is not far to seek. A large number of Indian scientists have no pride in their profession, though they are proud of their salaries and positions. . . . In India today the unworthy successors of Durvasa[15] and Vishvamitra[16] actually invite governors, vice-chancellors, and the like, to address them. This may be a relic of British Rule. If so, it is a regrettable one. Soon, especially after his move to Bhubaneswar, there was a sharp drop in invitations to universities and social functions. This was partly due to the fact that he was now living in a small town, but also because it became common knowledge that he did not want to be disturbed unless it was a matter which was directly related to the research work of the laboratory.

Notes

1. Balanced polymorphism is a situation in which two variants of a gene are maintained in a population of organisms because individuals carrying both kinds are better able to survive than those who have two copies of either kind alone.
2. Haldane's Rule: After a survey of the offspring of crosses between two different species, Haldane formulated the following generalization (later called "Haldane's Rule"): "When in the offspring of two different animal races one sex is absent, rare, or sterile, that sex is the heterozygous [heterogametic] sex." J. B. S. Haldane, "Sex-Ratio and Unisexual Sterility in Hybrid Animals," *Journal of Genetics* 12 (1922): 101–9.
3. J. B. S. Haldane, "Quantum Mechanics as a Basis for Philosophy," *Philosophy of Science* 1 (1934): 78–98.
4. Norbert Wiener defined "cybernetics" as the science of control and communication in the animal and the machine. Norbert Wiener, *Cybernetics* (Cambridge, MA: MIT Press, 1948).
5. The *Rationalist Annual* was the annual publication of the Rationalist Press Association, London.
6. J. B. S. Haldane, "A Defense of Beanbag Genetics," *Perspectives in Biology and Medicine* 7 (1964): 343–59.
7. Thalassemias are a group of inherited disorders characterized by reduced or absent amounts of hemoglobin, the oxygen-carrying protein inside the red blood cells. The two major groups of thalassemias are alpha thalassemia and beta thalassemia. They cause varying degrees of anemia, from insignificant to life threatening.
8. Heterozygous refers to an individual who possesses two different forms of the same gene in his cells, one dominant and the other recessive, each inherited from a different parent, as

opposed to a homozygous condition, where an individual possesses two copies of the same form of the gene.
9. Autosomal refers to any of the chromosomes other than the sex-determining chromosomes or the genes on these chromosomes.
10. Haldane, "A Defense of Beanbag Genetics."
11. When gene mapping started, it became necessary to introduce a mapping function to correct for certain errors. The simplest and the first mapping function was invented by Haldane in 1919; it is a mathematical relationship between the probability of recombination and map units.
12. The Aero-Medical Society of India, now known as The Indian Society of Aerospace Medicine, New Delhi, India. "Physiological Problems at High Pressure," *Aero-Medical Society Journal (New Delhi)* 5 (1960): 1–7.
13. The genetic code provides information encoded within the genetic material (DNA or mRNA sequences), which is translated into proteins by living cells. The code defines how sequences of nucleotide triplets, called codons, specify which amino acid will be added next during protein synthesis. With some exceptions, a three-nucleotide codon in a nucleic acid sequence specifies a single amino acid.
14. K. R. Dronamraju, ed., *Haldane's Daedalus Revisited* (Oxford: Oxford University Press, 1995).
15. Durvasa: In Hindu mythology, Durvasa was an ancient sage, the son of Atri and Anasuya. He is supposed to be an incarnation of Shiva. He is known for his short temper. Hence, wherever he went, he was received with great reverence. According to local tradition in modern Azamgarh, Durvasa's Ashram or hermitage, where many disciples used to go to study under him, was situated in the area, at the confluence of the Tons and Majhuee rivers, 6 kilometers north of the Phulpur Tehsil headquarters.
16. Vishvamitra is one of the most venerated sages of ancient times in India. He is also credited as the author of most of the *Rigveda*. The *Rigveda* contains several mythological and poetical accounts of the origin of the world, including the Hindu prayer *Gayatri mantra*. Vishvamitra was a king in ancient India, also called Kaushika. He was a valiant warrior and the great-grandson of a great king named Kusha, a brainchild of Brahma.

TIMELINE

J. B. S. HALDANE (1892–1964)

November 5, 1892—Born in Oxford, England. Son of John Scot Haldane, physiologist at New College, Oxford, and Louisa Kathleen Haldane.

Attended Oxford Preparatory School (now Dragon School, Oxford). Won a scholarship to go to Eton, where he excelled in Latin, Greek, German, French, History, Chemistry, Physics, and Biology. Julian Huxley was his senior, and the two became friends, establishing a lifelong friendship.

1895—Injured in an accident and questioned the attending physician whether the blood from his wound contained hemoglobin or carboxyhemoglobin.

1990—Started helping father in recording results of experiments. Accompanies father to a lecture by A. D. Darbishire on Mendel's experiments at the Oxford University Junior Scientific Club.

1906—Assisted father by going down mines and diving for physiological experiments.

1911—Went up to New College on a mathematical scholarship; attended E.S. Goodrich's final honors course in zoology, which created a lasting interest in evolutionary biology and genetics. At a seminar organized by Goodrich, Haldane reported the first case of linkage in mice.

1912—First scientific paper in the *Journal of Physiology*, in collaboration with his father and C. G. Douglas on the laws of combination of hemoglobin with carbon monoxide and oxygen.

1915—(with his sister Naomi and A. D. Sprunt) Published a paper in the *Journal of Genetics*, reported the first case of linkage in vertebrates. Double first at Oxford, Classics and Math. Joined the Black Watch battalion, and fought in the trenches of France.

1918—Sent to Simla, India, to recuperate from wounds received in explosions in Mesopotamia (Iraq), fell in love with India, but determined to return when he could associate with Indians "on a footing of equality."

1919—Invented the first mapping function and the unit of map distance, *centimorgan* or *cM*. Appointed fellow in physiology at New College, Oxford.

1920—Described the gene as a self-reproducing nucleoprotein molecule.

1922—Proposed "Haldane's Rule," "When in the F_1 offspring of two different animal races one sex is absent, rare, or sterile, that sex is the heterozygous sex." This rule has been shown to be true for a great number of interspecific crosses across several phyla. The fact that hybrid sterility and inviability can evolve due to Haldane's rule in such a vast array of different organisms is quite impressive. However, the actual explanation of this phenomenon is still undecided. Although "Haldane's Rule" is perhaps the best-known work of Haldane, other contributions such as his mathematical theory of natural selection and impact of mutation are far more important and have had a far greater impact on science and society.

1923—Haldane's first book—*Daedalus, or Science and the Future*, London: Kegan Paul. At his lecture before the Heretics at Cambridge University, Haldane met C. K. Ogden, who was a "scout" for Kegan Paul; *Daedalus* is a futuristic essay, influenced by the books of H. G. Wells, speculating on important future developments in science and technology. Especially noted for predictions in molecular and reproductive biology and human cloning, and for raising ethical and moral issues. Freeman Dyson commented that Haldane and Einstein were the only scientists in early twentieth century who raised ethical and moral questions resulting from scientific progress. Reprinted in *Haldane's Daedalus Revisited* (Ed. K. R. Dronamraju), Oxford University Press 1995. Haldane's ideas from *Daedalus*, especially in vitro fertilization and cloning, were reproduced by Aldous Huxley in his novel *Brave New World* (1932). Appointed "Sir William Dunn Reader in Biochemistry" under Professor F. G. Hopkins (later Nobel laureate and PRESIDENT of the Royal Society of London).

1924—Published the first paper in his series, Mathematical Theory of Natural Selection, "A satisfactory theory of natural selection must be quantitative. In order to establish the view that natural selection is capable of accounting for the known facts of evolution we must show not only that it can cause a species to change, but that it can cause it to change at a rate which will account for present and past transmutations." Haldane's series of ten papers and the independent works of R. A. Fisher (1930) and S. Wright (1931) founded population genetics. Interviewed by the journalist Charlotte Burghes, his future wife.

1925—Cited as a co-respondent in the divorce case involving Charlotte Burghes; *sex viri* trial before a panel of six men chosen from the faculty, dismissed by Cambridge University; Haldane appealed with support from his father and F. G. Hopkins, reinstated as reader in biochemistry. Derived the basic law of steady-state kinetics (with G. E. Briggs), which is still used for treating enzymatic catalysis. Haldane foresaw the value of specific enzymes in synthetic chemistry.

1926—Married Charlotte Burghes (nee Franken); moved to Roebuck House in Cambridge.

1927—*Possible Worlds, and Other Essays*, London: Chatto & Windus; includes Haldane's most famous essay, "On Being the Right Size." Investigated the role of carbon monoxide as a tissue poison, Haldane was the subject-investigator. *Biochemical Journal* 21: 1068. Joined John Innes Horticultural Institution as officer-in-charge of genetical investigations, a part-time position; guided C. D. Darlington's research in cytogenetics.

1929—*Rationalist Annual*, Haldane proposed his theory of the origin of life, independent of A. I. Oparin, who proposed a similar theory earlier in the Soviet Union, suggesting that primitive life arose in an anaerobic prebiotic world, facilitated by the synthesizing action of ultraviolet light on a mixture of gases in the ocean, resulting in the "consistency of hot dilute soup," so that any organism that appeared would have abundant food. In 1954, Stanley Miller's experiments at the University of California in Harold Urey's laboratory confirmed the essential correctness of the Haldane-Oparin hypothesis.

1930—*Enzymes*, London: Longmans, Green.

1932—Elected fellow of the Royal Society of London. *The Causes of Evolution*, London: Longmans, Green. A popular exposition of population genetics, including a summary of Haldane's papers on the mathematical theory of natural selection, first estimation of a human mutation rate, and a quantitative treatment of genes for altruism in human populations as well as the ethical and moral implications of Darwinian evolution. *The Inequality of Man, and Other Essays*, London: Chatto & Windus. An outstanding collection of popular essays, and a science-fiction story ("The Gold Makers") by Haldane. "But until the scientific point of view is generally adopted, our civilization will continue to suffer from a fundamental disharmony, its material basis scientific, its intellectual framework is pre-scientific." "I am a part of nature, and like other natural objects, from a lightning flash to a mountain range, I shall last out my time and then finish. This prospect does not worry me, because some of my work will not die when I do so."

- The time of action of genes, and its bearing on some evolutionary problems.
- Formal genetics of man (A method for investigating recessive characters, using maximum likelihood method)—foundations of human genetics.

1933—Appointed professor of genetics, University College, London.

1934—Quantum mechanics as a basis for philosophy, *Phil. Soc.* 1:78—stimulated Norbert Weiner's idea of cybernetics. *Fact and Faith*, London: Watts.

1935—*Science and the Supernatural* (with Arnold Lunn)—Based on an exchange with Arnold Lunn on science versus religion. *Science and Well-Being*, London: Kegan Paul. Human mutation rate (with Lionel Penrose).

1936—March 14, father John Scott Haldane died. JBS Haldane joins the Spanish civil war, travels to Spain.

1937—*My Friend, Mr. Leakey*, London: The Cresset Press. First human gene map (with Julia Bell)—linkage estimation between the genes for hemophilia and colorblindness.

- Became the Science correspondent for *Daily Worker*.
- The effect of variation on fitness. *American Naturalist* 71: 337. Haldane came up with the surprising finding that the effect of mutation on the fitness of a population depends solely on the mutation rate but not on the deleteriousness of the mutant phenotype. This principle was later used by the US National Academy of Sciences to estimate the genetic effect of radiation at a time when the open air testing of nuclear weapons became a political and social problem.

1938—*A.R.P.*, London: Victor Gollancz. (Haldane campaigned actively to promote a sound policy of air raid protection; subsequent events during the war proved he was correct. His book provided a quantitative treatment of the problem).

Heredity and Politics, London: George Allen & Unwin.
The Marxist Philosophy and the Sciences, London: George Allen & Unwin.
The Chemistry of the Individual. 38th Robert Boyle Lecture, Oxford University Press.

1939—Haldane was invited to investigate the sinking of the submarine HMS *Thetis* off the coast of Liverpool, with the loss of ninety-nine lives, including many civilians. Recruiting some nonscientific volunteers, Haldane tested conditions of escape from a steel chamber that simulated the circumstances in the escape chamber of the *Thetis*.

1940—Taste of oxygen at high pressures; human physiology under high pressure and very cold temperature while breathing nitrogen and carbon dioxide (subject-investigator). Texts published:

Science in Peace and War, London: Lawrence & Wishart.
Science in Everyday Life, London: Macmillan.
Keeping Cool, and Other Essays, London: Chatto & Windus.

1941—*New Paths in Genetics*, London: George Allen & Unwin. Haldane introduced *cis* and *trans* to replace *coupling* and *repulsion* phases in linkage. Physiological work for the Royal Navy and for the Royal Air Force—tested the effects of breathing excess carbon dioxide for a prolonged period. Resulted in convulsions, sickness. War time bombing of London. Moved to Rothamsted Experimental Station.

1942—The selective elimination of silver foxes in Eastern Canada. (*J. Genet.* 44: 296). Proposed civil defense measures against gas attack during the war (A.R.P.)

1944—Mutation and the Rhesus reaction. *Nature, London* 153: 106. Role of helium in deep sea diving. Radioactivity and the origin of life in Milne's cosmology.

1945—Rise of Trofim Lysenko in Soviet science, attack on Mendelian geneticists, and destruction of textbooks in genetics. Haldane was caught in a political crisis between his loyalty to the Communist Party and his loyalty to the science of genetics.

January—Haldane sent a letter to Lysenko, requesting copies of his papers with his experimental results, which led to his "views on genetics" and his conclusions.

1946—Analysis of the interaction of nature and nurture (a classification of the possible types of interaction between two stocks and two environments). Text published: *A Banned Broadcast, and Other Essays*, London: Chatto & Windus.

1947—Genetic effects of atomic bomb explosions. X-chromosome map (with C. A. B. Smith), Beginnings of human gene map. Texts published:

- *Science Advances*, London: George Allen & Unwin.
- *What Is Life?*, New York: Boni & Gaer.

1948—August 26, Meeting of the presidium of the USSR Academy of Sciences abolished existing genetics laboratories, dismissed Mendelian geneticists from all positions—Haldane attempted to obtain translated copies of these resolutions but failed to do so.

Daily Worker—Haldane's article "Where I Disagree with Lysenko."

BBC symposium "Russia Puts the Clock Back," John Langdon-Davies, participants were J. B. S. Haldane, R. A. Fisher, C. D. Darlington, and S. C. Harland. Haldane walked a careful line between defending the scientific integrity of genetics and an outright denunciation of Soviet science.

Royal Society's Croonian Lecture, "The Formal Genetics of Man," *Proc.Roy.Soc.(B)* 135:147.

An important step in the consolidation of "human genetics" as a distinct discipline.

Eighth International Congress of Genetics, Stockholm—The rate of mutation of human genes—greater immunity of carriers for thalassemia against malarial infection.

1949—In defense of genetics (*Modern Quarterly* (NS) 4: 194.

"Disease and Evolution"—Haldane developed a general theory of evolution in which resistance to infectious disease plays a crucial role. Parasitism might have been an important factor in speciation.

1950—*Everything Has a History*. London: George Allen & Unwin.

1951—Invited by Arthur C. Clarke to address the British Interplanetary Society in London; Title of Haldane's lecture: "Life in Space, Space Ships and Other Planets." Beginning of a long friendship with Arthur C. Clarke.

1952—The Haldanes visited India at the invitation of the Indian Science Congress Association and the Indian Statistical Institute in Calcutta.

1953—International Biological Union Symposium (IUBS), "Genetics of Population Structure" at the University of Pavia, Italy. (Haldane's address was noted for its many quotes from the classics, an impressive display of classical knowledge).

1954—"The Statics of Evolution," in *Evolution as a Process*, (Ed. J. Huxley et al). Haldane discussed both the statics and the dynamics of evolution. Measurement of the intensity of natural selection (defined as the logarithm of the ratio of the fitness of the optimal phenotype to that of the whole population).

The Biochemistry of Genetics, London: George Allen & Unwin.

1956—Huxley Memorial lecture and Medal of the Royal Anthropological Institute, London.

1957—"The Cost of Natural Selection," Haldane showed that the cost incurred by the species, during the process of gene substitution, could be as high as thirty times the population number in a single generation. Kimura (1968) used Haldane's estimate as a justification for the "neutral theory of evolution."

May—Karl Pearson Centenary Lecture, University College, London.

July—Haldanes moved to India to work at the Indian Statistical Institute in Calcutta.

December—Haldane delivered the Vallabhbhai Patel Memorial Lectures in memory of India's former deputy prime minister, "Unity and Diversity of Life," All India Radio, New Delhi.

1958—Assam Science Society, Gauhati, Assam.

January—"The Present Position of Darwinism," 45th Indian Science Congress Association, Madras, India.

1959—"Genetics in Relation to Medicine," Yellapragada Subba Rao Memorial lecture, The Academy of Medical Sciences, Hyderabad, India.

University Grants Commission, Committee to review teaching and research.

1960—January—Indian Science Congress meeting, Mumbai.

December—Ceylon Association for the Advancement of Science, Host: governor general of Ceylon Sir Oliver Goonetilleke, meeting with Arthur C. Clarke.

1961—Second International Congress of Human Genetics, Rome, Italy

International Conference on Human Population Genetics, Jerusalem, Israel.

1962—Moved from Calcutta to Bhubaneswar to found the Genetics and Biometry Laboratory with the support of the Orissa state government, India.

1963—International Congress of Genetics, The Hague, "The Implications of Genetics for Human Society." Introduced the term "clone."

Visited Europe and United States, lectured at Rockefeller University, University of Rochester and the University of Wisconsin. Participated with A. I. Oparin in a symposium on the origin of life at the University of Florida.

1964—In response to Ernst Mayr's challenge questioning the significance of the contributions of Fisher, Haldane, and Wright, Haldane wrote a spirited defense of mathematical population genetics (*Perspectives in Biology and Medicine* 7: 343–59).

Underwent surgery for cancer at University College Hospital, London, and returned to Bhubaneswar, India.

Died on December 1, 1964.

INDEX

Adamson, J., 35
Ali, Salim, 256–257
Altruism Equation, 162
altruistic genes, 161
animal behavior, 208
Antheraea mylitta, 210
Apis mellifera, 209
astronomy, 224
asymmetry, 144
Atlantic Monthly, 217
auto-obituary, 307

Bateson, W., 14, 16
BBC, 56
Beadle, G. W., 130
Beanbag genetics, 106–107
BEAR committee, 159
Bernal, J. D., 142
Bethesda, 124
Biochemistry of Genetics, 130
biological, 71–74
biopoiesis, 142
Birkenhead, Lord, 57
Biston betularia, 103
Bose, Nirmal Kumar, 246
Brave New World, 55
Brighter Biochemistry, 42–50

Calcutta, 212
Callan, H. G., 208
cancer, 100
cancer poem, 307–308
Case, E. M., 34, 121
Caspari, E., 129
Causes of Evolution, 98, 99, 101, 104–106
Century Magazine, 217
Chain, B., 39
Chemical Embryology, 248
Cherwell, 12
Clarke, Arthur, C., 213, 217, 252–254
cloning, 72
Conklin, Groff, 281

cost of selection, 108–110
Crick, Francis, 150
Crow, J. F., 80, 84, 159
Cuddly Cactus, 264

Daedalus or Science and the Future, 54–76, 128, 216–217
Daily Worker, 219
Darbishire, A. D., 58, 77
Darlington, C. D., 128, 131–139
Darwin, C., 141
Darwinism, 220
Davis, T. A., 219
death, 232, 304
D'Herelle, F., 143
Disease and Evolution, 159–160
diving, 12, 117–127
Dobzhansky, T., 212, 255–256
dominance, 103
Dragon School, 8
Dronamraju, K. R., 295
Dugatkin, L., 162
Dunn Institute, 23
Dyson, Freeman, 74, 216

Egypt, 270
enzyme kinetics, 23
Enzymes, 23
Eocene, 99
Ephrussi, Boris and Harriet, 207
eugenics, 54–76
evolution, rates of, 107–108
Evolution as a Process, 98
evolutionary biology, 98–112
evolutionary synthesis, 84

family background, 3
farewell dinner, 267
fermentation, 142
Fisher, R. A., 80, 84, 257–258
Fisher, R. B., 26
Fleming, A., 39

Florey, H., 39
Franken, Charlotte, 22–42
Friends and Kindred, 6
friendship, 245

Galton, F., 341
Gamow, G., 212
Gandhi, Mahatma, 230, 246
Garrod, A. E., 131
gene fixation, 91
genetic loads, 159
Gielgud, John, 12
Gielgud, Lewis, 12
Goldmakers, 217
Goodrich, E. S., 14, 77
guppies, 209

Haldane, J. S., 3–5, 8, 12
Haldane, L. K., 3, 56
Haldane, Viscount, 3, 4, 8
Haldane and Modern Biology, 13, 77
Haldane's dilemma, 110
Hall, Daniel, 131
Hardy, G. H., 82–83
Hardy-Weinberg Law, 82
Hill, M. D. (Piggy), 10
Hinduism, 209
HMS Thetis, 13, 117
Hopkins, F. G., 22–24
hot dilute soup, 142
Huxley, Aldous, 40, 55, 62
Huxley, Julian, 10, 56, 61
Huxley, T. H., 213
hydrothermal systems, 149

Inborn Errors of Metabolism, 131
inbreeding, 298
India, 264, 267
Indian Science Congress, 268
Indian Statistical Institute (ISI), 81, 279, 286–313
industrial melanism, 103
International Congress of Human Genetics, 63

James, Ioan, 207
Jayakar, S. D., 98
Jeffries, H., 212

Kalmus, H., 121, 211
Keeping Cool and Other Essays, 113
Kettlewell, H. B. D., 102–103
Kimura, M., 91
King Solomon's Ring, 208

Lacassagne, A., 212
Lady Chatterley's Lover, 61
Lawrence, D. H., 60–61
Lorenz, Konrad, 208
Luck, J. M., 24

Mahalanobis, P. C., 211, 279, 283
Malthus, 323
Man Meets Dog, 208
Manchester Guardian, 217
Martin, Kingsley, 264
Marxism, 240
Mayr, E., 86–87, 212, 246, 255
Medawar, P. B., 256
Mendel, G., 86
meristic variation, 292
Metastable population, 92
microspheres, 142
Miller-Urey experiment, 148
Mimicry in butterflies, 97
Mitchison, N. A., ix
Mitchison, Naomi (Nou), ix, 3, 11, 19, 61, 77–79, 264, 267
Monod, J., 212, 251
Morrell, Ottoline, 60
movies for toads, 228
Muller, H. J., 66–67, 100–101, 143
mutation, 100–101
My Friend Mr. Leakey, 217, 264

natural selection, 93
Needham, J., 24–25, 247
Nehru, J. Prime Minister, 268
new caste system, 311
New Statesman and Nation, 264
New York Times, 70
Neyman, J., 280
Norton, H. T. J. (Harry), 86

Ogden, C. K., 55
On Being Finite, 231
On Being the Right Size, 217
Oparin, A. I., 143
origin of life, 141, 228
Orissa, 246, 301–305
Oxford life, 17–20
Oxford University, 13, 14
oxygen tolerance, 119

panspermia, 150
parthenogenesis, 209–210
Passage to India, 272
Pasteur, L., 142
Patel lectures, 279
path coefficient, 80
Patnaik, Biju, 260
Penrose, L. S., 249
Pirie, N. W., 245
Pleiades and Orion, 227–228
police dog incident, 212
popularizing science, 216–232
population genetics, 79–94
Possible Worlds and Other Essays, 57
Punnett, R. C., 77

Rationalist Annual, 141
Ray-Chaudhuri, S. P., 265
recurrent mutation, 99
Remarkable Biologists, 207
Research and Writing, 221
Royal Society, 77
Russell, Bertrand, 55–57

Sanger, Margaret, 56, 59
science and ethics, 68
scientific predictions, 70
Scott-Moncrieff, Rose, 128
Simla, 264
Smith, C. A. B., 250
Smith, H., 144
Smith, J. M., 251
Snow, C. P., 240
Soviet Academy of Sciences, 309
Spurway, Helen, 122, 207–215
St. Andrews University, 208

starling, 220
Stopes, Marie, 56, 59
Suez crisis, 269
Sutter, J., 90
Syadvada system of predication, 230

Tallahassee, 146

University College London (UCL), 130, 207, 211, 213, 267
University of Wisconsin, 81
urzymes, 146

virgin births, 209
von Frisch, Karl, 209

Weinberg, W., 83–84
Wiener, Norbert, 259–260
Wright, S., 80, 84, 87, 258–259
Wurmser, Rene and Sabine, 207, 245